VW
Owners
Workshop
Manual

Peter G Strasman

Models covered
All VW LT Series vans and light trucks with petrol engine
1984 cc & 2384 cc

Does not cover bodies and conversions produced by specialist manufacturers
Does not cover Diesel engine or 4 x 4 models

ISBN 978 1 85010 323 3

(637-12M1)

Haynes Group Limited
Haynes North America, Inc

www.haynes.com

British Library Cataloguing in Publication Data
Strasman, Peter G.
VW LT series owners workshop manual. –
(Owners Workshop Manual).
1. Volkswagen vans
I. Title   II. Series
629.28'73   TL230.5.V6
ISBN 1-85010-323-2

Acknowledgements

Thanks are due to MAN-VW for the provision of technical information and for the use of certain illustrations. Castrol Limited provided lubrication data, the Champion Sparking Plug Company supplied the illustrations showing the various spark plug conditions, and Sykes-Pickavant Ltd provided some of the workshop tools.

Thanks are also due to the staff at Sparkford who assisted in the production of this manual.

About this manual

Its aim

The aim of this manual is to help you get the best value from your truck. It can do so in several ways. It can help you decide what work must be done (even should you choose to get it done by a garage), provide information on routine maintenance and servicing, and give a logical course of action and diagnosis when random faults occur. However, it is hoped that you will use the manual by tackling the work yourself. On simpler jobs it may even be quicker than booking the vehicle into a garage and going there twice, to leave and collect it. Perhaps most important, a lot of money can be saved by avoiding the costs a garage must charge to cover its labour and overheads.

The manual has drawings and descriptions to show the function of the various components so that their layout can be understood. Then the tasks are described and photographed in a step-by-step sequence so that even a novice can do the work.

Its arrangement

The manual is divided into thirteen Chapters, each covering a logical sub-division of the vehicle. The Chapters are each divided into Sections, numbered with single figures, eg 5; and the Sections into paragraphs (or sub-sections), with decimal numbers following on from the Section they are in, eg 5.1, 5.2, 5.3 etc.

It is freely illustrated, especially in those parts where there is a detailed sequence of operations to be carried out. There are two forms of illustration: figures and photographs. The figures are numbered in sequence with decimal numbers, according to their position in the Chapter – eg Fig. 6.4 is the fourth drawing/illustration in Chapter 6. Photographs carry the same number (either individually or in related groups) as the Section or sub-section to which they relate.

There is an alphabetical index at the back of the manual as well as a contents list at the front. Each Chapter is also preceded by its own individual contents list.

References to the 'left' or 'right' of the vehicle are in the sense of a person in the driver's seat facing forwards.

Unless otherwise stated, nuts and bolts are removed by turning anti-clockwise, and tightened by turning clockwise.

Vehicle manufacturers continually make changes to specifications and recommendations, and these, when notified, are incorporated into our manuals at the earliest opportunity.

Whilst every care is taken to ensure that the information in this manual is correct, no liability can be accepted by the authors or publishers for loss, damage or injury caused by any errors in, or omissions from, the information given.

Introduction to the Volkswagen LT Series

Versatility and ease of maintenance are probably the most important characteristics of this range of vehicles. Strongly constructed and well finished, all models in the range give good performance and roadholding, with acceptable fuel economy.

Driver and passenger comfort is approaching car standards. Overhaul and maintenance operations are simple and can mostly be carried out without the use of special tools.

Contents

VW LT 35 Pick-up. This is the vehicle which was dismantled in our workshop

4

VW LT 40 high-roofed Panel Van

General dimensions, weights and capacities

Dimensions

LT 28, 31, 35 Vans

Overall length	4840 mm (190.6 in)
Overall width (LT 28, 31)	2020 mm (79.5 in)
Overall width (LT 35)	2080 mm (81.9 in)
Overall height (unladen):	
LT 28 Panel Van	2150 mm (84.6 in)
LT 31 Panel Van	2160 mm (85.0 in)
LT 35 Panel Van	2200 mm (86.6 in)
LT 28 high-roofed Van	2560 mm (100.8 in)
LT 31 high-roofed Van	2570 mm (101.2 in)
LT 35 high-roofed Van	2610 mm (102.8 in)
Ground clearance	180 mm (7.1 in)

LT 28, 31, 35 Chassis/cab

Overall length	4810 mm (189.4 in)
Overall width (LT 28, 31)	2040 mm (80.3 in)
Overall width (LT 35)	2080 mm (81.9 in)
Overall height (unladen):	
LT 28	2105 mm (82.9 in)
LT 31	2120 mm (83.5 in)
LT 35	2115 mm (83.3 in)
Ground clearance	180 mm (7.1 in)

LT 35, 45 double cab Pick-up

Overall length	5325 mm (209.6 in)
Overall width	2140 mm (84.3 in)
Height (unladen)	2135 mm (84.1 in)
Ground clearance	150 mm (5.9 in)

LT 40/45 high-roofed Van

Overall length	5290 mm (208.3 in)
Overall width	2055 mm (80.9 in)
Height (unladen)	2635 mm (103.7 in)
Ground clearance	150 mm (5.9 in)

LT 40/45 Chassis/cab

Overall length	5260 mm (207.1 in)
Overall width	2080 mm (81.9 in)
Height (unladen)	2120 mm (83.5 in)
Ground clearance	150 mm (5.9 in)

Unladen weights

Panel Van:	
LT 28	1630 kg (3594 lb)
LT 31	1630 kg (3594 lb)
LT 35	1740 kg (3837 lb)
LT 40/45	2000 kg (2037 lb)
High-roofed Van:	
LT 28	1680 kg (3704 lb)
LT 31	1680 kg (3704 lb)
LT 35	1790 kg (3947 lb)
LT 40/45	2000 kg (4410 lb)
Chassis/cab:	
LT 28	1390 kg (3065 lb)
LT 31	1390 kg (3065 lb)
LT 35 SWB	1450 kg (3197 lb)
LT 35 LWB	1500 kg (3308 lb)
LT 40/45	1665 kg (3671 lb)
Double cab Pick-up:	
LT 35 LWB	1730 to 1800 kg (3815 to 3969 lb)
LT 40/45 LWB	2070 to 2125 kg (4664 to 4686 lb)

Capacities

Engine oil:

  2.0 litre:

      With filter change .... 4.5 litres (7.9 Imp pints)

      Without filter change .... 4.0 litres (7.0 Imp pints)

  2.4 litre:

      With filter change .... 7.0 litres (12.3 Imp pints)

      Without filter change .... 6.0 litres (10.6 Imp pints)

Gearbox .... 3.5 litres (6.2 Imp pints)

Final drive:

  LT 28, 31 .... 1.8 litres (3.2 Imp pints)

  LT 35, 40, 45 .... 2.2 litres (3.9 Imp pints)

Cooling system:

  2.0 litre models .... 9 litres (15.8 Imp pints)

  2.4 litre models .... 13 litres (22.9 Imp pints)

Fuel tank .... 70 litres (15 Imp gallons)

Windscreen washer reservoir .... 2 litres (3.5 Imp pints)

Combined windscreen/headlight washer reservoir .... 8 litres (14 Imp pints)

Steering box (LT 40, 45) .... 0.5 litre (0.9 Imp pint)

Power-assisted steering .... 1.5 litres (2.6 Imp pints)

Buying spare parts and vehicle identification numbers

Buying spare parts

Spare parts are available from many sources, for example: MAN-VW garages, other garages and accessory shops, and motor factors. Our advice regarding spare part sources is as follows:

Officially appointed MAN-VW garages – This is the best source for parts which are peculiar to your vehicle and are otherwise not generally available (eg complete cylinder heads, internal gearbox components, badges, interior trim etc). It is also the only place at which you should buy parts if your car is still under warranty – non-MAN-VW components may invalidate the warranty. To be sure of obtaining the correct parts it will always be necessary to give the storeman your vehicle identification number, and if possible, to take the 'old' part along for positive identification. Remember that many parts are available on a factory exchange scheme – any parts returned should always be clean! It obviously makes good sense to go straight to the specialists on your vehicle for this type of part for they are best equipped to supply you.

Other garages and accessory shops – These are often very good places to buy materials and components needed for the maintenance of your vehicle (eg oil filters, spark plugs, bulbs, drivebelts, oils and greases, touch-up paint, filler paste, etc). They also sell general accessories, usually have convenient opening hours, charge lower prices and can often be found not far from home.

Motor factors – Good factors will stock all the more important components which wear out relatively quickly (eg clutch components, pistons, valves, exhaust system, brake cylinders/pipes/hoses/seals/shoes and pads etc). Motor factors will often provide new or reconditioned components on a part exchange basis – this can save a considerable amount of money.

Vehicle identification numbers

Although many individual parts, and in some cases sub-assemblies, fit a number of different models it is dangerous to assume that just because they look the same, they are the same. Differences are not always easy to detect except by serial numbers. Make sure therefore, that the appropriate identity number for the model or sub-assembly is known and quoted when a spare part is ordered.

The Vehicle Identification Plate is located either on the right-hand rear panel within the driving cab, or on the B pillar of the cab onto which the door locks (photo).

The chassis number is located on the side-member, under the right-hand mudguard and just ahead of the roadwheel.

The engine number is stamped on the flange of the crankcase at its joint with the clutch bellhousing.

Vehicle Identification Plate

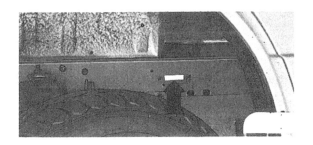

Chassis number (arrowed)

Tools and working facilities

Introduction

A selection of good tools is a fundamental requirement for anyone contemplating the maintenance and repair of a motor vehicle. For the owner who does not possess any, their purchase will prove a considerable expense, offsetting some of the savings made by doing-it-yourself. However, provided that the tools purchased meet the relevant national safety standards and are of good quality, they will last for many years and prove an extremely worthwhile investment.

To help the average owner to decide which tools are needed to carry out the various tasks detailed in this manual, we have compiled three lists of tools under the following headings: *Maintenance and minor repair, Repair and overhaul,* and *Special*. The newcomer to practical mechanics should start off with the *Maintenance and minor repair* tool kit and confine himself to the simpler jobs around the vehicle. Then, as his confidence and experience grow, he can undertake more difficult tasks, buying extra tools as, and when, they are needed. In this way, a *Maintenance and minor repair* tool kit can be built-up into a *Repair and overhaul* tool kit over a considerable period of time without any major cash outlays. The experienced do-it-yourselfer will have a tool kit good enough for most repair and overhaul procedures and will add tools from the *Special* category when he feels the expense is justified by the amount of use to which these tools will be put.

It is obviously not possible to cover the subject of tools fully here. For those who wish to learn more about tools and their use there is a book entitled *How to Choose and Use Car Tools* available from the publishers of this manual.

Maintenance and minor repair tool kit

The tools given in this list should be considered as a minimum requirement if routine maintenance, servicing and minor repair operations are to be undertaken. We recommend the purchase of combination spanners (ring one end, open-ended the other); although more expensive than open-ended ones, they do give the advantages of both types of spanner.

Combination spanners - 10, 11, 12, 13, 14 & 17 mm
Adjustable spanner - 9 inch
Engine sump/gearbox/rear axle drain plug key
Spark plug spanner (with rubber insert)
Spark plug gap adjustment tool
Set of feeler gauges
Brake bleed nipple spanner
Screwdriver - 4 in long x $\frac{1}{4}$ in dia (flat blade)
Screwdriver - 4 in long x $\frac{1}{4}$ in dia (cross blade)
Combination pliers - 6 inch
Hacksaw (junior)
Tyre pump
Tyre pressure gauge
Grease gun (if applicable)
Oil can
Fine emery cloth (1 sheet)
Wire brush (small)
Funnel (medium size)

Repair and overhaul tool kit

These tools are virtually essential for anyone undertaking any major repairs to a motor vehicle, and are additional to those given in the *Maintenance and minor repair* list. Included in this list is a comprehensive set of sockets. Although these are expensive they will be found invaluable as they are so versatile - particularly if various drives are included in the set. We recommend the $\frac{1}{2}$ in square-drive type, as this can be used with most proprietary torque wrenches. If you cannot afford a socket set, even bought piecemeal, then inexpensive tubular box spanners are a useful alternative.

The tools in this list will occasionally need to be supplemented by tools from the *Special* list.

Sockets (or box spanners) to cover range in previous list
Reversible ratchet drive (for use with sockets)
Extension piece, 10 inch (for use with sockets)
Universal joint (for use with sockets)
Torque wrench (for use with sockets)
'Mole' wrench - 8 inch
Ball pein hammer
Soft-faced hammer, plastic or rubber
Screwdriver - 6 in long x $\frac{5}{16}$ in dia (flat blade)
Screwdriver - 2 in long x $\frac{5}{16}$ in square (flat blade)
Screwdriver - 1$\frac{1}{2}$ in long x $\frac{1}{4}$ in dia (cross blade)
Screwdriver - 3 in long x $\frac{1}{8}$ in dia (electricians)
Pliers - electricians side cutters
Pliers - needle nosed
Pliers - circlip (internal and external)
Cold chisel - $\frac{1}{2}$ inch
Scriber
Scraper
Centre punch
Pin punch
Hacksaw
Valve grinding tool
Steel rule/straight-edge
Allen keys
Selection of files
Wire brush (large)
Axle-stands
Jack (strong scissor or hydraulic type)

Special tools

The tools in this list are those which are not used regularly, are expensive to buy, or which need to be used in accordance with their manufacturers' instructions. Unless relatively difficult mechanical jobs are undertaken frequently, it will not be economic to buy many of these tools. Where this is the case, you could consider clubbing together with friends (or joining a motorists' club) to make a joint purchase, or borrowing the tools against a deposit from a local garage or tool hire specialist.

The following list contains only those tools and instruments freely available to the public, and not those special tools produced by the vehicle manufacturer specifically for its dealer network. You will find occasional references to these manufacturers' special tools in the text of this manual. Generally, an alternative method of doing the job without the vehicle manufacturers' special tool is given. However, sometimes, there is no alternative to using them. Where this is the case and the relevant tool cannot be bought or borrowed, you will have to entrust the work to a franchised garage.

Valve spring compressor (where applicable)
Piston ring compressor
Balljoint separator
Universal hub/bearing puller
Impact screwdriver
Micrometer and/or vernier gauge
Dial gauge
Stroboscopic timing light
Dwell angle meter/tachometer
Universal electrical multi-meter
Cylinder compression gauge
Lifting tackle
Trolley jack
Light with extension lead

Buying tools

For practically all tools, a tool factor is the best source since he will have a very comprehensive range compared with the average garage or accessory shop. Having said that, accessory shops often offer excellent quality tools at discount prices, so it pays to shop around.

There are plenty of good tools around at reasonable prices, but always aim to purchase items which meet the relevant national safety standards. If in doubt, ask the proprietor or manager of the shop for advice before making a purchase.

Care and maintenance of tools

Having purchased a reasonable tool kit, it is necessary to keep the tools in a clean serviceable condition. After use, always wipe off any dirt, grease and metal particles using a clean, dry cloth, before putting the tools away. Never leave them lying around after they have been used. A simple tool rack on the garage or workshop wall, for items such as screwdrivers and pliers is a good idea. Store all normal wrenches and sockets in a metal box. Any measuring instruments, gauges, meters, etc, must be carefully stored where they cannot be damaged or become rusty.

Take a little care when tools are used. Hammer heads inevitably become marked and screwdrivers lose the keen edge on their blades from time to time. A little timely attention with emery cloth or a file will soon restore items like this to a good serviceable finish.

Working facilities

Not to be forgotten when discussing tools, is the workshop itself. If anything more than routine maintenance is to be carried out, some form of suitable working area becomes essential.

It is appreciated that many an owner mechanic is forced by circumstances to remove an engine or similar item, without the benefit of a garage or workshop. Having done this, any repairs should always be done under the cover of a roof.

Wherever possible, any dismantling should be done on a clean, flat workbench or table at a suitable working height.

Any workbench needs a vice: one with a jaw opening of 4 in (100 mm) is suitable for most jobs. As mentioned previously, some clean dry storage space is also required for tools, as well as for lubricants, cleaning fluids, touch-up paints and so on, which become necessary.

Another item which may be required, and which has a much more general usage, is an electric drill with a chuck capacity of at least $\frac{5}{16}$ in (8 mm). This, together with a good range of twist drills, is virtually essential for fitting accessories such as mirrors and reversing lights.

Last, but not least, always keep a supply of old newspapers and clean, lint-free rags available, and try to keep any working area as clean as possible.

Spanner jaw gap comparison table

Jaw gap (in)	Spanner size
0.250	$\frac{1}{4}$ in AF
0.276	7 mm
0.313	$\frac{5}{16}$ in AF
0.315	8 mm
0.344	$\frac{11}{32}$ in AF; $\frac{1}{8}$ in Whitworth
0.354	9 mm

Jaw gap (in)	Spanner size
0.375	$\frac{3}{8}$ in AF
0.394	10 mm
0.433	11 mm
0.438	$\frac{7}{16}$ in AF
0.445	$\frac{3}{16}$ in Whitworth; $\frac{1}{4}$ in BSF
0.472	12 mm
0.500	$\frac{1}{2}$ in AF
0.512	13 mm
0.525	$\frac{1}{4}$ in Whitworth; $\frac{5}{16}$ in BSF
0.551	14 mm
0.563	$\frac{9}{16}$ in AF
0.591	15 mm
0.600	$\frac{5}{16}$ in Whitworth; $\frac{3}{8}$ in BSF
0.625	$\frac{5}{8}$ in AF
0.630	16 mm
0.669	17 mm
0.686	$\frac{11}{16}$ in AF
0.709	18 mm
0.710	$\frac{3}{8}$ in Whitworth; $\frac{7}{16}$ in BSF
0.748	19 mm
0.750	$\frac{3}{4}$ in AF
0.813	$\frac{13}{16}$ in AF
0.820	$\frac{7}{16}$ in Whitworth; $\frac{1}{2}$ in BSF
0.866	22 mm
0.875	$\frac{7}{8}$ in AF
0.920	$\frac{1}{2}$ in Whitworth; $\frac{9}{16}$ in BSF
0.938	$\frac{15}{16}$ in AF
0.945	24 mm
1.000	1 in AF
1.010	$\frac{9}{16}$ in Whitworth; $\frac{5}{8}$ in BSF
1.024	26 mm
1.063	$1\frac{1}{16}$ in AF; 27 mm
1.100	$\frac{5}{8}$ in Whitworth; $\frac{11}{16}$ in BSF
1.125	$1\frac{1}{8}$ in AF
1.181	30 mm
1.200	$\frac{11}{16}$ in Whitworth; $\frac{3}{4}$ in BSF
1.250	$1\frac{1}{4}$ in AF
1.260	32 mm
1.300	$\frac{3}{4}$ in Whitworth; $\frac{7}{8}$ in BSF
1.313	$1\frac{5}{16}$ in AF
1.390	$\frac{13}{16}$ in Whitworth; $\frac{15}{16}$ in BSF
1.417	36 mm
1.438	$1\frac{7}{16}$ in AF
1.480	$\frac{7}{8}$ in Whitworth; 1 in BSF
1.500	$1\frac{1}{2}$ in AF
1.575	40 mm; $\frac{15}{16}$ in Whitworth
1.614	41 mm
1.625	$1\frac{5}{8}$ in AF
1.670	1 in Whitworth; $1\frac{1}{8}$ in BSF
1.688	$1\frac{11}{16}$ in AF
1.811	46 mm
1.813	$1\frac{13}{16}$ in AF
1.860	$1\frac{1}{8}$ in Whitworth; $1\frac{1}{4}$ in BSF
1.875	$1\frac{7}{8}$ in AF
1.969	50 mm
2.000	2 in AF
2.050	$1\frac{1}{4}$ in Whitworth; $1\frac{3}{8}$ in BSF
2.165	55 mm
2.362	60 mm

Jacking and towing

Special note

To avoid repetition, the procedure for raising the vehicle in order to carry out work under the chassis/body is not included before each relevant operation where described in this manual.

It is assumed, and certainly recommended, that the vehicle is positioned over an inspection pit or on a lift. Where this equipment is not available, use ramps or jack up the vehicle strictly in accordance with the following guide, and observe the requirement for axle safety stands at all times.

Raising vehicle with ram or hoist

The front lifting point is under the crossmember.

The rear lifting point is at the chassis corner plate on vehicles with single rear wheels, or under the outboard end of the crossmember on vehicles with twin rear wheels.

Raising vehicle with trolley jack

To raise the front of the vehicle, place the jack under the front crossmember or (on LT 40/45 models) under the front axle beam. On LT 35 models, a jacking bracket is located just to the rear of each suspension lower wishbone.

Ram lifting point (front)

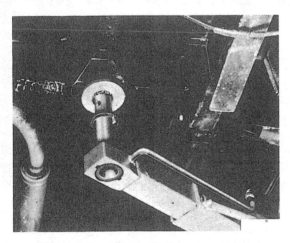

Ram lifting point (rear) (single roadwheels)

Ram lifting point (rear) (twin roadwheels)

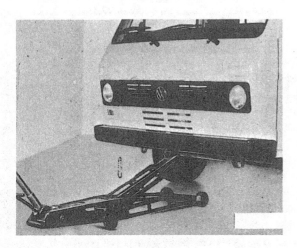

Front jacking point

Front jacking point (LT 35)

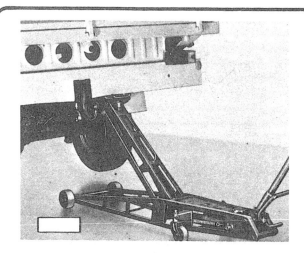

Rear jacking point

Side raising point

Tool kit jack in position

Jacking point (rear) for tool kit jack

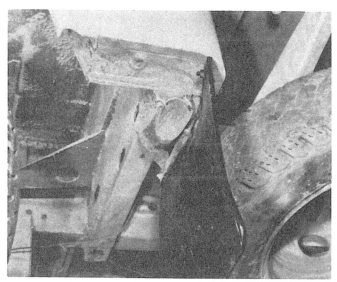

Jacking point (front) for tool kit jack

Spare wheel location (chassis/cab)

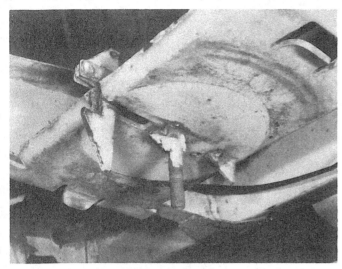

Spare wheel carrier

Front towing eye

To raise the rear of the vehicle, place the jack under the rear crossmember or differential housing.

To raise one side of the vehicle, place the jack under the chassis side-member.

Whenever the vehicle is supported on a jack, supplement it with axle stands.

Raising vehicle with tool kit jack

The jack supplied in the vehicle tool kit should only be used for emergency roadside wheel changing. Slacken the roadwheel nuts slightly and then insert the jack fully into the jacking point closest to the wheel that is being removed. Chock one wheel on the opposite side of the vehicle. Raise the jack, making sure that it is kept fully engaged in its hole (photos).

The spare wheel is located in the right-hand rear corner of the load compartment on van models. On chassis/cab versions, the spare wheel is carried in a cradle attached to the chassis left-hand side-member. To release the wheel unscrew the retaining nut, remove the plate, free the hook and lower the wheel (photos).

Towing

A towing eye is welded to the front and rear of the vehicle (photos). Do not attach a tow-rope to the vehicle bumpers or independent suspension arms.

Use a synthetic fibre rope which will permit a certain amount of stretch.

Unlock the steering with the ignition key before being towed, and remember that the brake servo will be inoperative so greater foot pedal pressures will be required to slow the vehicle.

Rear towing eye

Safety first!

Professional motor mechanics are trained in safe working procedures. However enthusiastic you may be about getting on with the job in hand, do take the time to ensure that your safety is not put at risk. A moment's lack of attention can result in an accident, as can failure to observe certain elementary precautions.

There will always be new ways of having accidents, and the following points do not pretend to be a comprehensive list of all dangers; they are intended rather to make you aware of the risks and to encourage a safety-conscious approach to all work you carry out on your vehicle.

Essential DOs and DON'Ts

DON'T rely on a single jack when working underneath the vehicle. Always use reliable additional means of support, such as axle stands, securely placed under a part of the vehicle that you know will not give way.

DON'T attempt to loosen or tighten high-torque nuts (e.g. wheel hub nuts) while the vehicle is on a jack; it may be pulled off.

DON'T start the engine without first ascertaining that the transmission is in neutral (or 'Park' where applicable) and the parking brake applied.

DON'T suddenly remove the filler cap from a hot cooling system – cover it with a cloth and release the pressure gradually first, or you may get scalded by escaping coolant.

DON'T attempt to drain oil until you are sure it has cooled sufficiently to avoid scalding you.

DON'T grasp any part of the engine, exhaust or catalytic converter without first ascertaining that it is sufficiently cool to avoid burning you.

DON'T allow brake fluid or antifreeze to contact vehicle paintwork.

DON'T syphon toxic liquids such as fuel, brake fluid or antifreeze by mouth, or allow them to remain on your skin.

DON'T inhale dust – it may be injurious to health (see *Asbestos* below).

DON'T allow any spilt oil or grease to remain on the floor – wipe it up straight away, before someone slips on it.

DON'T use ill-fitting spanners or other tools which may slip and cause injury.

DON'T attempt to lift a heavy component which may be beyond your capability – get assistance.

DON'T rush to finish a job, or take unverified short cuts.

DON'T allow children or animals in or around an unattended vehicle.

DO wear eye protection when using power tools such as drill, sander, bench grinder etc, and when working under the vehicle.

DO use a barrier cream on your hands prior to undertaking dirty jobs – it will protect your skin from infection as well as making the dirt easier to remove afterwards; but make sure your hands aren't left slippery. Note that long-term contact with used engine oil can be a health hazard.

DO keep loose clothing (cuffs, tie etc) and long hair well out of the way of moving mechanical parts.

DO remove rings, wristwatch etc, before working on the vehicle – especially the electrical system.

DO ensure that any lifting tackle used has a safe working load rating adequate for the job.

DO keep your work area tidy – it is only too easy to fall over articles left lying around.

DO get someone to check periodically that all is well, when working alone on the vehicle.

DO carry out work in a logical sequence and check that everything is correctly assembled and tightened afterwards.

DO remember that your vehicle's safety affects that of yourself and others. If in doubt on any point, get specialist advice.

IF, in spite of following these precautions, you are unfortunate enough to injure yourself, seek medical attention as soon as possible.

Asbestos

Certain friction, insulating, sealing, and other products – such as brake linings, brake bands, clutch linings, torque converters, gaskets, etc – contain asbestos. *Extreme care must be taken to avoid inhalation of dust from such products since it is hazardous to health.* If in doubt, assume that they *do* contain asbestos.

Fire

Remember at all times that petrol (gasoline) is highly flammable. Never smoke, or have any kind of naked flame around, when working on the vehicle. But the risk does not end there – a spark caused by an electrical short-circuit, by two metal surfaces contacting each other, by careless use of tools, or even by static electricity built up in your body under certain conditions, can ignite petrol vapour, which in a confined space is highly explosive.

Always disconnect the battery earth (ground) terminal before working on any part of the fuel or electrical system, and never risk spilling fuel on to a hot engine or exhaust.

It is recommended that a fire extinguisher of a type suitable for fuel and electrical fires is kept handy in the garage or workplace at all times. Never try to extinguish a fuel or electrical fire with water.

Note: *Any reference to a 'torch' appearing in this manual should always be taken to mean a hand-held battery-operated electric lamp or flashlight. It does NOT mean a welding/gas torch or blowlamp.*

Fumes

Certain fumes are highly toxic and can quickly cause unconsciousness and even death if inhaled to any extent. Petrol (gasoline) vapour comes into this category, as do the vapours from certain solvents such as trichloroethylene. Any draining or pouring of such volatile fluids should be done in a well ventilated area.

When using cleaning fluids and solvents, read the instructions carefully. Never use materials from unmarked containers – they may give off poisonous vapours.

Never run the engine of a motor vehicle in an enclosed space such as a garage. Exhaust fumes contain carbon monoxide which is extremely poisonous; if you need to run the engine, always do so in the open air or at least have the rear of the vehicle outside the workplace.

If you are fortunate enough to have the use of an inspection pit, never drain or pour petrol, and never run the engine, while the vehicle is standing over it; the fumes, being heavier than air, will concentrate in the pit with possibly lethal results.

The battery

Never cause a spark, or allow a naked light, near the vehicle's battery. It will normally be giving off a certain amount of hydrogen gas, which is highly explosive.

Always disconnect the battery earth (ground) terminal before working on the fuel or electrical systems.

If possible, loosen the filler plugs or cover when charging the battery from an external source. Do not charge at an excessive rate or the battery may burst.

Take care when topping up and when carrying the battery. The acid electrolyte, even when diluted, is very corrosive and should not be allowed to contact the eyes or skin.

If you ever need to prepare electrolyte yourself, always add the acid slowly to the water, and never the other way round. Protect against splashes by wearing rubber gloves and goggles.

When jump starting a car using a booster battery, for negative earth (ground) vehicles, connect the jump leads in the following sequence: First connect one jump lead between the positive (+) terminals of the two batteries. Then connect the other jump lead first to the negative (–) terminal of the booster battery, and then to a good earthing (ground) point on the vehicle to be started, at least 18 in (45 cm) from the battery if possible. Ensure that hands and jump leads are clear of any moving parts, and that the two vehicles do not touch. Disconnect the leads in the reverse order.

Mains electricity and electrical equipment

When using an electric power tool, inspection light etc, always ensure that the appliance is correctly connected to its plug and that, where necessary, it is properly earthed (grounded). Do not use such appliances in damp conditions and, again, beware of creating a spark or applying excessive heat in the vicinity of fuel or fuel vapour. Also ensure that the appliances meet the relevant national safety standards.

Ignition HT voltage

A severe electric shock can result from touching certain parts of the ignition system, such as the HT leads, when the engine is running or being cranked, particularly if components are damp or the insulation is defective. Where an electronic ignition system is fitted, the HT voltage is much higher and could prove fatal.

Routine maintenance

For modifications, and information applicable to later models, see Supplement at end of manual

Maintenance is essential for ensuring safety and desirable for the purpose of getting the best in terms of performance and economy from your vehicle. Over the years the need for periodic lubrication – oiling, greasing, and so on – has been drastically reduced if not totally eliminated. This has unfortunately tended to lead some operators to think that because no such action is required, components either no longer exist, or will last for ever. This is a serious delusion. It follows therefore that the largest initial element of maintenance is visual examination. This may lead to repairs or renewals.

Weekly, every 250 miles (400 km) or before a long journey

Check engine oil level and top up if necessary (photos)
Check coolant reservoir level visually
Check brake fluid reservoir level visually, investigate any sudden fall in level

Check and top up washer fluid reservoir (photo)
Check tyre pressures
Check operation of all lights, direction indicators and horn
Check operation of windscreen wipers
Check battery electrolyte level and top up if necessary (photo)

At first 1000 miles (1600 km) – new or rebuilt vehicle

Check cooling system for leaks
Check cylinder head bolt tightness
Check valve clearances
Check alternator drivebelt tension
Adjust idle speed
Check brake pipelines for leaks at unions
Renew engine oil and filter
Check clutch adjustment

Engine oil dipstick

Topping up the engine oil

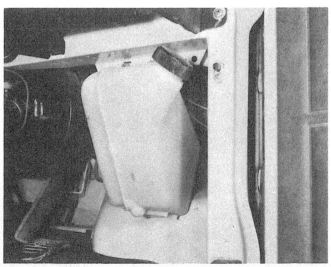

Washer fluid reservoir

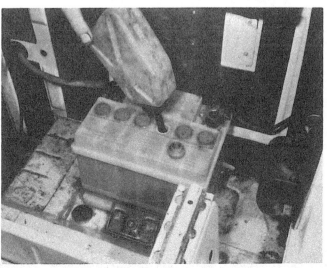

Topping up the battery

Every 5000 miles (8000 km) or 6 months – whichever comes first

Renew engine oil (not filter)
Adjust the clutch
Check brake pad and lining wear
Check tyre wear and condition
Lubricate locks, controls and hinges

Every 10 000 miles (16 000 km) or 12 months – whichever comes first

In addition to, or instead of the work specified in the previous schedules
Check and adjust valve clearances
Renew spark plugs
Check condition and tension of alternator drivebelt
Renew contact breaker points
Check dwell angle and ignition timing
Adjust idle speed and check
Check exhaust afterburning valves (when fitted)
Renew oil filter
Check exhaust system for corrosion
Check for play in balljoints, split gaiters, bushes etc in steering and suspension
Check brake pipelines for damage or corrosion
Check gearbox and final drive oil levels (photo)
Check headlamp beam adjustment
Lubricate kingpins (LT 40/45)

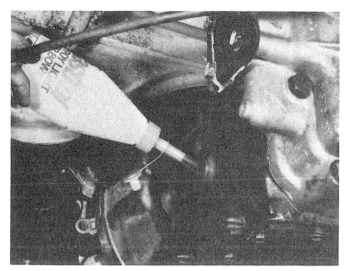

Topping up the gearbox

Every 20 000 miles (32 000 km) or 2 years, whichever comes first

In addition to the work specified in the previous schedules
Renew air filter (more frequently in dusty conditions)
Renew fuel filter

Every 2 years, regardless of mileage

Renew brake fluid
Renew coolant anti-freeze
Renew servo air filter

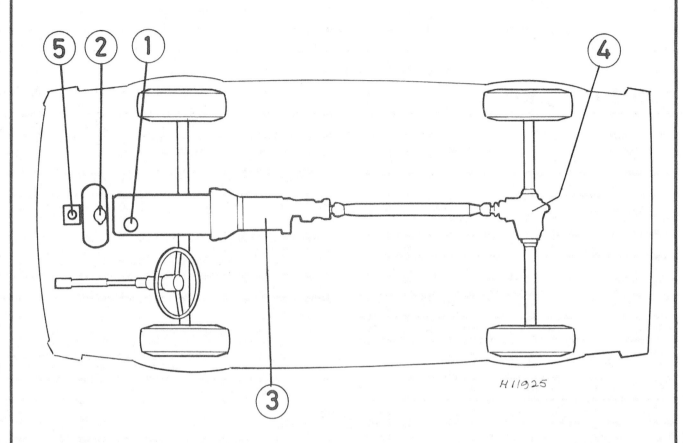

H11925

Recommended lubricants and fluids

Component or system	Lubricant type or specification
1 Engine	Multigrade engine oil SAE 20W/50
2 Cooling system	Antifreeze G11
3 Gearbox	Hypoid gear oil GL4 (MIL-L 2105) SAE 80
4 Rear axle	Hypoid gear oil GL5 (MIL-L 2105B) SAE 90
5 Brake hydraulic system	Hydraulic fluid to DOT 4
Kingpin bushes (LT 40/45)	Multi-purpose lithium based grease
Steering box (LT 40/45)	Hypoid gear oil GL5 (MIL-L 2105B) SAE 90
Power-assisted steering	Dexron® ATF

Note: *The above are general recommendations only. Lubrication requirements may vary from territory to territory and according to operating conditions. If in doubt, consult the operator's handbook supplied with your vehicle*

Fault diagnosis

Introduction

The operator who does his or her own maintenance according to the recommended schedules should not have to use this section of the manual very often. Modern component reliability is such that, provided those items subject to wear or deterioration are inspected or renewed at the specified intervals, sudden failure is comparatively rare. Faults do not usually just happen as a result of sudden failure, but develop over a period of time. Major mechanical failures in particular are usually preceded by characteristic symptoms over hundreds or even thousands of miles. Those components which do occasionally fail without warning are often small and easily carried in the vehicle.

With any fault finding, the first step is to decide where to begin investigations. Sometimes this is obvious, but on other occasions a little detective work will be necessary. The operator who makes half a dozen haphazard adjustments or replacements may be successful in curing a fault (or its symptoms), but he will be none the wiser if the fault recurs and he may well have spent more time and money than was necessary. A calm and logical approach will be found to be more satisfactory in the long run. Always take into account any warning signs or abnormalities that may have been noticed in the period preceding the fault – power loss, high or low gauge readings, unusual noises or smells, etc – and remember that failure of components such as fuses or spark plugs may only be pointers to some underlying fault.

The pages which follow here are intended to help in cases of failure to start or breakdown on the road. There is also a Fault Diagnosis Section at the end of each Chapter which should be consulted if the preliminary checks prove unfruitful. Whatever the fault, certain basic principles apply. These are as follows:

Verify the fault. This is simply a matter of being sure that you know what the symptoms are before starting work. This is particularly important if you are investigating a fault for someone else who may not have described it very accurately.

Don't overlook the obvious. For example, if the vehicle won't start, is there petrol in the tank? (Don't take anyone else's word on this particular point, and don't trust the fuel gauge either!) If an electrical fault is indicated, look for loose or broken wires before digging out the test gear.

Cure the disease, not the symptom. Substituting a flat battery with a fully charged one will get you off the hard shoulder, but if the underlying cause is not attended to, the new battery will go the same way. Similarly, changing oil-fouled spark plugs for a new set will get you moving again, but remember that the reason for the fouling (if it wasn't simply an incorrect grade of plug) will have to be established and corrected.

Don't take anything for granted. Particularly, don't forget that a 'new' component may itself be defective (especially if it's been rattling round in the boot for months), and don't leave components out of a fault diagnosis sequence just because they are new or recently fitted. When you do finally diagnose a difficult fault, you'll probably realise that all the evidence was there from the start.

Electrical faults

Electrical faults can be more puzzling than straightforward mechanical failures, but they are no less susceptible to logical analysis if the basic principles of operation are understood. Vehicle electrical wiring exists in extremely unfavourable conditions – heat, vibration and chemical attack – and the first things to look for are loose or corroded connections and broken or chafed wires, especially where the wires pass through holes in the bodywork or are subject to vibration.

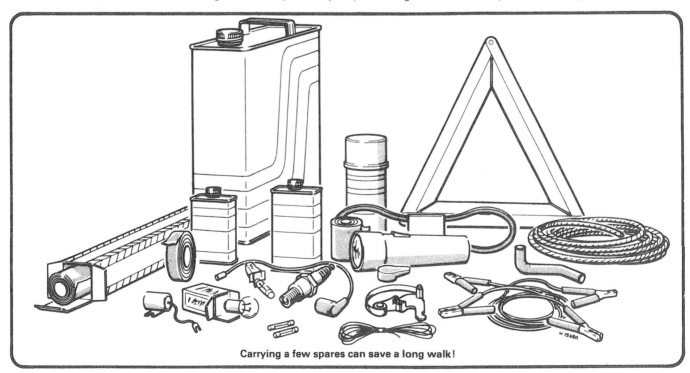

Carrying a few spares can save a long walk!

All vehicles in current production have one pole of the battery 'earthed', ie connected to the bodywork, and in nearly all modern vehicles it is the negative (–) terminal. The various electrical components – motors, bulb holders etc – are also connected to earth, either by means of a lead or directly by their mountings. Electric current flows through the component and then back to the battery via the bodywork. If the component mounting is loose or corroded, or if a good path back to the battery is not available, the circuit will be incomplete and malfunction will result. The engine and/or gearbox are also earthed by means of flexible metal straps to the body or subframe; if these straps are loose or missing, starter motor, generator and ignition trouble may result.

Assuming the earth return to be satisfactory, electrical faults will be due either to component malfunction or to defects in the current supply. Individual components are dealt with in Chapter 10. If supply wires are broken or cracked internally this results in an open-circuit, and the easiest way to check for this is to bypass the suspect wire temporarily with a length of wire having a crocodile clip or suitable connector at each end. Alternatively, a 12V test lamp can be used to verify the presence of supply voltage at various points along the wire and the break can be thus isolated.

If a bare portion of a live wire touches the chassis/bodywork or other earthed metal part, the electricity will take the low-resistance path thus formed back to the battery: this is known as a short-circuit. Hopefully a short-circuit will blow a fuse, but otherwise it may cause burning of the insulation (and possibly further short-circuits) or even a fire. This is why it is inadvisable to bypass persistently blowing fuses with silver foil or wire.

Spares and tool kit

Most vehicles are supplied only with sufficient tools for wheel changing; the *Maintenance and minor repair* tool kit detailed in *Tools and working facilities*, with the addition of a hammer, is probably sufficient for those repairs that most operators would consider attempting at the roadside. In addition a few items which can be fitted without too much trouble in the event of a breakdown should be carried. Experience and available space will modify the list below, but the following may save having to call on professional assistance:

Spark plugs, clean and correctly gapped
HT lead and plug cap – long enough to reach the plug furthest from the distributor
Distributor rotor, condenser and contact breaker points
Drivebelt(s) – emergency type may suffice
Spare fuses
Set of principal light bulbs
Tin of radiator sealer and hose bandage
Exhaust bandage
Roll of insulating tape
Length of soft iron wire
Length of electrical flex
Torch or inspection lamp (can double as test lamp)
Battery jump leads
Tow-rope
Ignition waterproofing aerosol
Litre of engine oil
Sealed can of hydraulic fluid
Emergency windscreen
Tyre valve core

If spare fuel is carried, a can designed for the purpose should be used to minimise risks of leakage and collision damage. A first aid kit and a warning triangle, whilst not at present compulsory in the UK, are obviously sensible items to carry in addition to the above.

When operating abroad it may be advisable to carry additional spares which, even if you cannot fit them yourself, could save having to wait while parts are obtained. The items below may be worth considering:

Clutch and throttle cables
Cylinder head gasket
Alternator brushes

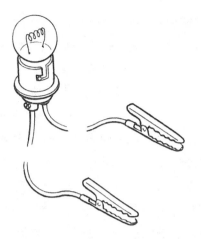

A simple test lamp is useful for tracing electrical faults

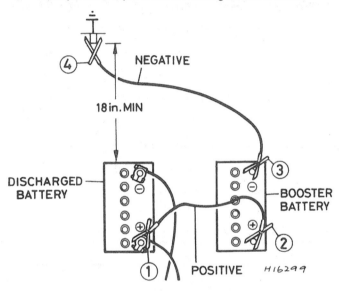

Jump start lead connections for negative earth – connect leads in order shown

---

Engine will not start

Engine fails to turn when starter operated

Flat battery (recharge, use jump leads, or push start)
Battery terminals loose or corroded
Battery earth to body defective
Engine earth strap loose or broken
Starter motor (or solenoid) wiring loose or broken
Ignition/starter switch faulty
Major mechanical failure (seizure) or long disuse (piston rings rusted to bores)
Starter or solenoid internal fault (see Chapter 10)

Starter motor turns engine slowly

Partially discharged battery (recharge, use jump leads, or push start)
Battery terminals loose or corroded
Battery earth to body defective
Engine earth strap loose
Starter motor (or solenoid) wiring loose
Starter motor internal fault (see Chapter 10)

Starter motor spins without turning engine

Flat battery
Flywheel gear teeth damaged or worn
Starter motor mounting bolts loose

Check for fuel delivery to carburettor

A slack or broken drivebelt will cause overheating and battery charging problems

Engine turns normally but fails to start
Damp or dirty HT leads and distributor cap (crank engine and check for spark)
No fuel in tank (check for delivery at carburettor)
Fouled or incorrectly gapped spark plugs (remove, clean and regap)
Other ignition system fault (see Chapter 4)
Other fuel system fault (see Chapter 3)
Poor compression (see Chapter 1)
Major mechanical failure (eg camshaft drive)

Engine fires but will not run
Air leaks at carburettor or inlet manifold
Fuel starvation (see Chapter 3)
Ballast resistor defective, or other ignition fault (see Chapter 4)

Engine cuts out and will not restart

Engine cuts out suddenly – ignition fault
Loose or disconnected LT wires
Wet HT leads or distributor cap (after traversing water splash)
Coil or condenser failure (check for spark)
Other ignition fault (see Chapter 4)

Engine misfires before cutting out – fuel fault
Fuel tank empty
Fuel pump defective or filter blocked (check for delivery)
Fuel tank filler vent blocked (suction will be evident on releasing cap)
Carburettor needle valve sticking
Carburettor jets blocked (fuel contaminated)
Other fuel system fault (see Chapter 3)

Engine cuts out – other causes
Serious overheating
Major mechanical failure (eg camshaft drive)

Engine overheats

Ignition (no-charge) warning light illuminated
Slack or broken drivebelt – retension or renew (Chapter 2) (photo)

Ignition warning light not illuminated
Coolant loss due to internal or external leakage (see Chapter 2)
Thermostat defective
Low oil level
Brakes binding
Radiator clogged externally or internally
Engine waterways clogged
Ignition timing incorrect or automatic advance malfunctioning
Mixture too weak

Note: *Do not add cold water to an overheated engine or damage may result*

Low engine oil pressure

Gauge reads low or warning light illuminated with engine running
Oil level low or incorrect grade
Defective gauge or sender unit
Wire to sender unit earthed
Engine overheating
Oil filter clogged or bypass valve defective
Oil pressure relief valve defective
Oil pick-up strainer clogged
Oil pump worn or mountings loose
Worn main or big-end bearings

Note: *Low oil pressure in a high-mileage engine at tickover is not necessarily a cause for concern. Sudden pressure loss at speed is far more significant. In any event, check the gauge or warning light sender before condemning the engine.*

Engine noises

Pre-ignition (pinking) on acceleration
Incorrect grade of fuel
Ignition timing incorrect
Distributor faulty or worn
Worn or maladjusted carburettor
Excessive carbon build-up in engine

Whistling or wheezing noises
Leaking vacuum hose
Leaking carburettor or manifold gasket
Blowing head gasket

Tapping or rattling
Incorrect valve clearances
Worn valve gear
Worn timing belt
Broken piston ring (ticking noise)

Knocking or thumping
Unintentional mechanical contact (eg fan blades)

Worn fanbelt
Peripheral component fault (generator, water pump etc)
Worn big-end bearings (regular heavy knocking, perhaps less under load)
Worn main bearings (rumbling and knocking, perhaps worsening under load)
Piston slap (most noticeable when cold)

Chapter 1 Engine

For modifications, and information applicable to later models, see Supplement at end of manual

Contents

Specifications

General

Engine type	4-cylinder, in-line, overhead cam, water-cooled
Engine code	CH (August 1975 on), CL (May 1976 on)
Bore	86.5 mm (3.41 in)
Stroke	84.4 mm (3.32 in)
Cubic capacity	1984 cc (121 cu in)
Compression ratio	8.2 : 1
Maximum power output	55 kW (72 bhp SAE) at 4300 rpm
Maximum torque	150 Nm (103 lbf ft SAE) at 2400 rpm

Cylinder compression pressures (engine hot, wide open throttle):

	Engine code CH	Engine code CL
Normal (engine in good condition)	8 to 11 bar (116 to 160 lbf/in^2)	7 to 10 bar (102 to 145 lbf/in^2)
Minimum	6 bar (87 lbf/in^2)	5 bar (73 lbf/in^2)
Maximum variation between cylinders	3 bar (44 lbf/in^2)	3 bar (44 lbf/in^2)

Crankshaft

Number of main bearings	5
Journal diameter (standard):	
Up to engine No CH 010 673	59.958 to 59.978 mm (2.3605 to 2.3613 in)
From engine No CH 010 674	63.958 to 63.978 mm (2.5180 to 2.5188 in)
Undersizes available	0.25, 0.50 and 0.75 mm (0.0098, 0.0197 and 0.0295 in)
Crankpin diameter (standard)	47.958 to 47.978 mm (1.8881 to 1.8889 in)
Main bearing running clearance:	
New	0.02 to 0.08 mm (0.0008 to 0.0032 in)
Service limit	0.16 mm (0.0063 in)
Big-end bearing running clearance:	
New	0.02 to 0.07 mm (0.0008 to 0.0028 in)
Service limit	0.10 mm (0.0039 in)
Endfloat:	
New	0.10 to 0.19 mm (0.0039 to 0.0075 in)
Service limit	0.25 mm (0.0098 in)
Connecting rod side float:	
New	0.05 to 0.30 mm (0.0020 to 0.0118 in)
Service limit	0.40 mm (0.0158 in)

Pistons and piston rings

Standard piston diameter (3 grades)	86.48, 86.49 and 86.50 mm (3.4047, 3.4051 and 3.4055 in)
Corresponding bore diameter	86.51, 86.52 and 86.53 mm (3.4058, 3.4063 and 3.4067 in)
Oversizes available	0.25, 0.50 and 1.00 mm (0.0098, 0.0197 and 0.0394 in)
Piston ring side clearance in groove:	
New	0.04 to 0.07 mm (0.0016 to 0.0028 in)
Wear limit	0.1 mm (0.0039 in)
Piston ring end gap:	
New	0.3 to 0.5 mm (0.0118 to 0.0197 in)
Wear limit	1.0 mm (0.0394 in)
End gap positions	120° apart

Camshaft
Number of bearings 5
Endfloat .... 0.2 mm (0.0079 in) maximum
Run-out at centre bearing .... 0.02 mm (0.0008 in) maximum

Cylinder head
Material .... Light alloy
Block mating surface distortion limit .... 0.1 mm (0.0039 in)

Valves and guides
Valve guide length (circlip to end of guide):
    Inlet .... 17.0 mm (0.669 in)
    Exhaust .... 23.7 mm (0.933 in)
Valve stem to guide clearance (measured as valve head rock, see text):
    Inlet .... 0.9 mm (0.035 in) maximum
    Exhaust .... 1.1 mm (0.043 in) maximum

Valve timing (at 1.0 mm valve lift)
Inlet opens .... 2° BTDC
Inlet closes .... 35° ABDC
Exhaust opens .... 72° BBDC
Exhaust closes .... 11° ATDC

Valve clearances

	Cold	Hot
Inlet	0.10 to 0.15 mm (0.004 to 0.006 in)	0.20 to 0.25 mm (0.008 to 0.010 in)
Exhaust	0.40 to 0.45 mm (0.016 to 0.018 in)	0.45 to 0.50 mm (0.018 to 0.020 in)

Lubrication system
Oil capacity:
    Without filter change .... 4.0 litres (7.0 Imp pints)
    With filter change .... 4.5 litres (7.9 Imp pints)
Oil pressure (oil temperature 80°C (176°F), engine speed 2000 rpm) .... 3.0 to 4.0 bar (43.5 to 58 lbf/in^2)
Oil pump tolerances:
    Gear backlash .... 0 to 0.13 mm (0 to 0.005 in)
    Gear endfloat .... 0.03 to 0.07 mm (0.001 to 0.003 in)

Torque wrench settings

	Nm	lbf ft
Clutch bellhousing bolts:		
M8	25	18
M12	75	55
Engine mounting nuts and bolts	45	33
Crankshaft pulley bolt	300	222
Coolant pump bolts	22	16
Cylinder head bolts (release through 30° before tightening):		
Cold engine	100	74
Warm engine	120	88
Clutch cover bolts	30	22
Sump screws:		
M6	8	6
M8	15	11
Cylinder head rear elbow bolts	10	7
Oil pump pick-up bolts	10	7
Thermostat housing cover bolts	10	7
Main bearing cap bolts:		
Hexagon head	80	59
Socket head	65	48
Connecting rod big-end cap bolts	60	44
Fan blade bolts	25	18
Oil pump mounting bolts	10	7
Camshaft sprocket bolt	80	59
Camshaft bearing cap bolts	10	7
Camshaft bearing cap nuts	20	15
Oil pressure switch	10	7
Coolant temperature sender	8	6
Oil drain plug	40	30
Timing belt tensioner lockbolt	40	30
Timing belt tensioner mounting bolts	20	15
Flywheel bolts	90	66

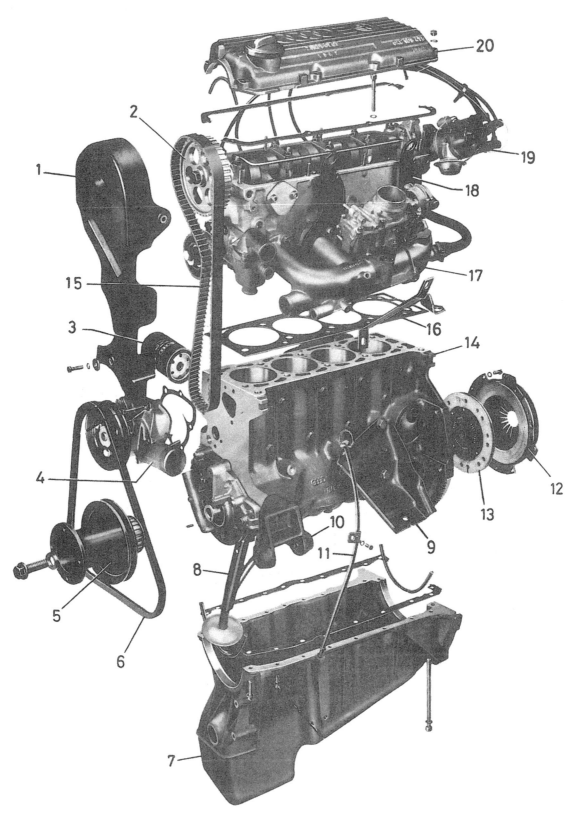

Fig. 1.1 Main components of the engine (Sec 1)

1	Timing belt guard	6	V-belt
2	Camshaft sprocket	7	Sump
3	Oil filter	8	Oil pick-up pipe
4	Coolant pump	9	Engine mounting bracket
5	Crankshaft pulley/belt sprocket	10	Alternator mounting
		11	Dipstick

12	Clutch pressure plate assembly
13	Clutch driven plate
14	Crankcase/cylinder block
15	Timing belt

16	Cylinder head gasket
17	Inlet manifold
18	Cylinder head
19	Distributor
20	Camshaft cover

1 Description

The engine is of four-stroke, four-cylinder in-line water-cooled type.

An overhead camshaft is fitted, which is driven by a toothed belt from a sprocket on the front end of the crankshaft.

The crankshaft is supported on five main bearings with removable shells. The centre shell incorporates thrust flanges to control crankshaft endfloat.

The pistons are fitted with two compression rings and one oil control ring, and are connected to the connecting rod by a gudgeon pin which is located by circlips.

Valve components are conventional, but it should be noted that the exhaust valves are fitted with a roto cap while the inlet valves have a spring seat instead.

Valve clearance adjustment is accomplished by the use of wedge screws inside the tappets (cam followers).

The lubrication system is based upon a gear type oil pump, which is bolted to the front face of the engine cylinder block and is driven by the crankshaft.

A pressure relief valve is incorporated in the oil pump body and a full-flow oil filter is screwed onto the right-hand side of the engine. An oil pressure warning switch is screwed into the rear face of the head (photo).

The materials used for the construction of the engine are light alloy for the cylinder head and sump, and cast iron for the composite crankcase and cylinder block.

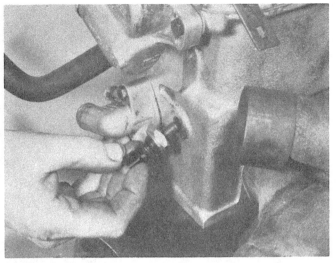

1.0 Oil pressure switch

2 Routine maintenance

1 A small hinged flap is located towards the front of the engine cover within the driving cab. The flap should be opened for routine checking of the engine oil and coolant levels.

2 To open the flap, lift the catch and twist it anti-clockwise.

3 Refer to Routine Maintenance (at the beginning of this manual) for details of regular engine servicing requirements.

3 Major operations possible – engine in vehicle

The following operations may be carried out without the need to remove the engine from the vehicle:

Removal and refitting of toothed timing belt
Removal and refitting of camshaft
Removal and refitting of cylinder head
Removal and refitting of sump
Removal and refitting of pistons/connecting rods
Removal and refitting of oil pump
Renewal of engine mountings

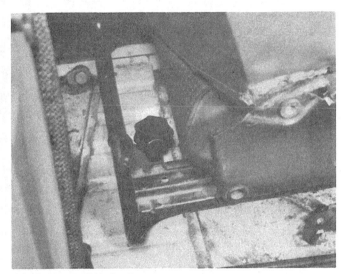

4.1a Front seat fixing knob

4 Engine cover – removal and refitting

1 Before the engine cover can be removed, the driver's seat (RHD) or passenger seat (LHD) should be withdrawn from the cab. Release the driver's seat by unscrewing the plastic knob; release the passenger seat by depressing the stop tab and sliding the seat forward out of its runners (photos).

2 If an additional passenger seat has been installed over the engine cover, remove this too by folding the seat cushion up and releasing the seat from its retainer.

3 If the vehicle is equipped with a radio, open the maintenance flap on the engine cover and release the earthing strap from the camshaft cover.

4 Release the four toggle clips and lift the engine cover from the driving cab (photo).

5 Refitting is a reversal of removal, but engage the rear clips on the engine cover before the front ones.

5 Toothed timing belt – removal and refitting

1 Remove the engine cover as described in the preceding Section.

4.1b Front seat anchor tangs

4.4 Engine cover toggle

5.3 Removing the drivebelt

5.4a Removing top right-hand timing cover bolt

5.4b Removing top left-hand timing cover bolt

5.4c Removing timing cover front bolt

5.4d Removing timing belt cover

2   Unbolt the expansion tank for the cooling system and move it to one side.

3   Release the alternator mounting and adjuster link bolts, push the alternator in towards the engine and slip the belt from the pulleys (photo).

4   Extract the screws and remove the timing belt cover (photos).

5   Unbolt and remove the fan blades from the flange on the front of the crankshaft belt pulley assembly.

6   Now rotate the crankshaft (using a long socket on the pulley bolt, or by inserting two bolts into the holes in the pulley front flange and using a long lever placed between them) until the mark on the rear face of the camshaft sprocket is in alignment with the pointer on the camshaft cover, and the TDC mark on the flywheel is in alignment with the projection on the clutch bellhousing (centre of 'O' aligns with pointer). If this work is being carried out with the engine removed from

the vehicle then the alternative TDC marks will have to be used, the notch on the crankshaft pulley and the pointer on the oil pump.

7   Release the lockbolt which is screwed into a hole offset in the head of the large hexagonal eccentric in the centre of the toothed belt tensioner.

8   Turn the eccentric to release tension on the belt.

9   Slip the belt from the camshaft sprocket, then out of the tensioner pulley groove and remove it from the crankshaft sprocket. If the belt is to be used again, mark it for running direction (photo).

10 **Do not** turn the crankshaft or camshaft from their set positions while the belt is off, or the valve heads and piston crowns may contact each other with consequent damage.

11 Refitting the belt is a reversal of removal, but before doing so, check that the camshaft sprocket and flywheel alignment marks have not moved from their set positions (photo).

12 Once the belt is fitted, tension the belt by turning the eccentric adjuster on the tensioner pulley in an anti-clockwise direction until the belt can only just be twisted through 90° when gripped between thumb and index finger at the centre point of its longest run (photos).

13 Tighten the eccentric lockbolt and refit the cover.

6   Camshaft – removal and refitting

1   Remove the engine cover as described in Section 4.

2   Remove the air cleaner and mechanical fuel pump (early models) after reference to Chapter 3.

3   Remove the distributor (Chapter 4) and the drive gear housing.

4   Release the alternator mounting and adjuster link bolts, push the alternator in towards the engine and slip the belt from the pulleys.

5   Extract the screws and remove the timing belt cover and the camshaft cover. Remove the timing belt as described in Section 5.

6   Loosen the camshaft sprocket securing bolt. In order to prevent the sprocket turning while this is done, either engage top gear and apply the handbrake fully, or pass a rod through one of the holes in the sprocket so that it bears against the top face of the cylinder head.

7   Remove the sprocket.

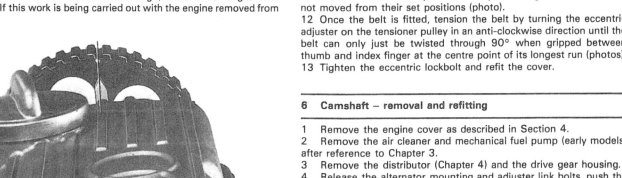

Fig. 1.2 Camshaft alignment marks (Sec 5)

Fig. 1.3 Flywheel/bellhousing and crankshaft pulley/oil pump TDC alignment marks (Sec 5)

5.9 Removing timing toothed belt

5.11 Camshaft sprocket and cover timing marks

5.12a Timing belt correctly fitted

5.12b Checking timing belt tension

6.8 Removing the oil feed pipe

6.12 Camshaft oil seal

6.13a Oiling camshaft bearings

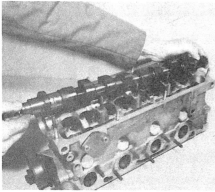
6.13b Installing the camshaft

6.14a Fitting camshaft bearing cap number 2

6.14b Fitting camshaft bearing cap number 1

6.15 Fitting the camshaft sprocket

6.16 Tightening camshaft bearing cap nuts

6.18 Tightening the camshaft sprocket bolt. Note locking bar

6.19 Fitting distributor drive gear housing

6.22a Fitting camshaft cover gaskets

6.22b Fitting camshaft cover sealing strip

8 Remove the oil feed pipe. This will necessitate unscrewing the nuts from bearing caps Nos 2 and 4 (photo).

9 Release the nuts on the other bearing caps by unscrewing them a half a turn at a time in the order 3, 5, 1.

10 Check that the bearing caps are numbered from 1 to 5 from the timing belt end of the engine.

11 Lift the camshaft from the cylinder head.

12 Pull the oil seal from the front end of the camshaft and slide on a new one with lips greased (photo).

13 Refitting commences by lowering the camshaft into liberally oiled bearings (photos).

14 Locate the bearing caps, noting that they have an offset bore to ensure that they go the correct way round. Check that their sequence numbers are in the correct order (photos).

15 At this stage, check that the flywheel/clutch housing TDC marks are in alignment. By offering the sprocket to the camshaft, turn the camshaft to bring the sprocket/camshaft cover marks into alignment (photo).

16 Screw on the bearing cap nuts progressively, a half turn at a time in the following order: 1, 3 and 5, then 2 and 4 with the oil feed pipe. As the cap nuts are screwed down, the resistance of some valve springs will be detected, this is why it is essential to screw the nuts down evenly to eliminate any possibility of distorting the camshaft (photo).

17 It is also essential not to exceed the specified torque wrench figure when tightening the camshaft bearing nuts.

18 Insert the camshaft sprocket bolt and tighten to the specified torque, again holding the sprocket using one of the methods suggested at removal (photo).

19 Install the distributor drive gear housing using a new gasket, then refit the distributor (Chapter 4) (photo).

20 Fit and tension the timing belt as described in Section 5.

21 Refit the belt cover, the alternator drivebelt, the air cleaner and the mechanical fuel pump (where fitted).

22 Fit new camshaft cover gaskets and sealing strips (photos).

23 Fit the camshaft cover (photo).

24 Tighten the camshaft cover nuts, noting the location of the cable clips (photo). Fit the engine cover.

7 Cylinder head – removal and refitting

1 Remove the engine cover as described in Section 4, and the air cleaner (Chapter 3). If the engine is fitted with an exhaust afterburning system (Engine No CH 032 935 onwards), disconnect the hose from the air intake duct.

2 Drain the cooling system as described in Chapter 2.

3 Disconnect the negative lead from the battery, which is located behind the left-hand seat.

4 Remove the camshaft cover.

5 Remove the toothed timing belt as described in Section 5.

6 Disconnect the LT wire which runs between the coil and the distributor (photo).

7 Disconnect the HT wire which runs between the coil and the centre of the distributor cap (photo).

8 Disconnect the lead from the oil pressure switch which is screwed into the rear face of the cylinder head.

9 Disconnect the lead from the coolant temperature switch which is screwed into the coolant elbow on the rear face of the cylinder head.

10 Remove the coolant hose from the elbow.

11 Disconnect the brake vacuum servo hose from the inlet manifold.

12 Disconnect the fuel inlet hose from the carburettor and plug the hose.

13 Disconnect the throttle cable from the carburettor.

14 Disconnect the electrical leads from the automatic choke and the air cut-off solenoid valve on the carburettor. On certain very early models, a manual choke is fitted in conjunction with a 'choke on' warning lamp. Where this is fitted, disconnect the choke operating cable from the lever on the carburettor.

15 Unscrew the nuts and disconnect the exhaust downpipe from the exhaust manifold.

16 Disconnect the coolant hoses from the thermostat housing. Also disconnect the hoses which run to the front end of the inlet manifold (photos).

6.23 Fitting camshaft cover

6.24 Camshaft cover cable clips correctly attached

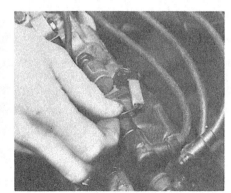

7.6 Disconnecting distributor LT wire

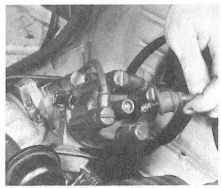

7.7 Disconnecting distributor cap coil lead

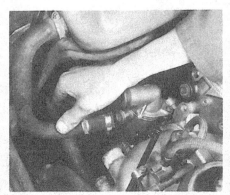

7.16a Disconnecting thermostat housing hoses

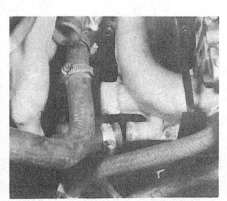

7.16b Disconnecting inlet manifold coolant hoses

17 Disconnect the inlet manifold strut and the crankcase ventilation hose (photo).
18 Unscrew and remove the cylinder head bolts, starting with the centre bolts and working towards each end of the head. Unscrew each bolt one half turn at a time. The bolts are of socket-headed type and will require the use of a special tool.
19 Remove the head from the block, using the inlet manifold as a lever to rock it if it is stuck.

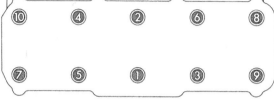

Fig. 1.4 Cylinder head bolt tightening sequence (Sec 7)

20 Peel off the cylinder head gasket and discard it.
21 If the cylinder head is to be decarbonised and the valves ground in, refer to Section 18. If the reason for removing the head was to renew a blown gasket, then carefully clean the mating faces of block and head and brush away any loose material from the combustion chambers in the cylinder head. Remember that the cylinder head is of light alloy and is easily damaged.
22 Place a new gasket on the block, taking care that the holes for coolant passages, oilways, bolt holes and positioning dowels are correctly aligned (photo).
23 Turn the crankshaft to bring the flywheel/clutch bellhousing TDC marks, or the crankshaft pulley notch and pointer, into alignment.
24 Turn the camshaft sprocket to align the mark on the rear face of the sprocket with the pointer on the camshaft cover.
25 Lower the cylinder head into position on the dowels. Screw in the bolts evenly to the specified torque in the sequence shown in Fig. 1.4 (photos).
26 Reconnect all hoses, controls and electrical leads.
27 Refit the timing belt as described in Section 5.
28 Fit the camshaft cover using new side sealing gaskets and end strips.
29 Reconnect the battery.
30 Fit the air cleaner.
31 Fill the cooling system (Chapter 2).
32 Refit the engine cover.

7.17 Inlet manifold strut and (arrowed) crankcase ventilation outlet

7.22 Locating cylinder head gasket

7.25a Installing cylinder head

7.25b Tightening cylinder head bolts

8.4 Disconnecting steering relay rod balljoint

8.5 Unscrewing sump rear flange screws

8 Sump – removal and refitting

1 Drain the engine oil, retaining it in a clean container if required for further use.

2 Drain the cooling system and remove the radiator as described in Chapter 2. This is necessary to provide sufficient clearance to be able to tilt the sump forwards during removal.

3 Remove the alternator adjuster link and remove the drivebelt.

4 Using a suitable balljoint releasing tool, disconnect the balljoint on the left-hand side of the steering relay rod and pull the rod downwards (photo).

5 Unscrew and remove all the sump securing screws, working progressively, a turn at a time in a diagonal sequence. Remove the dipstick (photo).

6 Draw the pan forward and down to clear the oil pump pick-up pipe and strainer (photos).

7 Peel off the side gaskets and end sealing strips and discard them.

8 Refitting is a reversal of removal, but coat the corner sealing joints with gasket cement and tighten the retaining screws to the specified torque.

9 Check that the drain plug is tight and fill the engine with the correct grade and quantity of engine oil. Do not forget to reconnect the steering joint.

10 Refit the radiator and refill the cooling system.

8.6a Removing the sump

9 Pistons/connecting rods – removal and refitting

1 Remove the cylinder head as described in Section 7.

2 Remove the sump as described in Section 8.

3 Examine the top of the cylinder bores for evidence of a wear ridge. If evident, this must be removed by careful scraping in order that the piston rings will not foul on it as the pistons are pushed out of the cylinder block.

4 Working underneath the vehicle, check that the big-end bearing caps are numbered adjacent to the matching numbers on the connecting rod (photo). The bases of the caps are also marked with dots to indicate their fitted sequence, counting from the timing belt end of the engine. The small cast projections will face the timing belt end of the engine.

5 The piston crowns are marked with an arrow which points towards the timing belt end of the engine, but they have no location number. Before removing the pistons, mark them 1 to 4 with quick-drying paint, counting from the timing belt end of the engine (photo).

6 Unbolt and remove the oil pump pick-up pipe after flattening the bolt lockplate tabs.

7 Turn the crankshaft as necessary to bring the big-end of the connecting rod to its lowest point and unscrew the cap bolts. Take off the cap complete with bearing shell.

8.6b Engine oil pick-up pipe

Fig. 1.5 Connecting rod big-end cap alignment (Sec 9)

9.4 Connecting rod big-end matching numbers

Fig. 1.6 Big-end cap sequence marks (Sec 9)

9.5 Piston front directional arrow

Fig. 1.7 Piston crown numbering (Sec 9)

9.11 Removing piston rings

9.14 Piston ring groove oil return hole

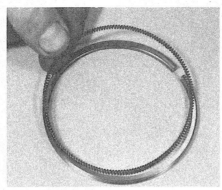

9.16a Oil control ring components

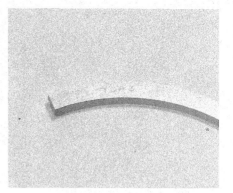

9.16b Compression ring marking

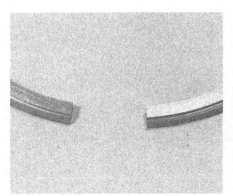

9.16c Step on second compression ring

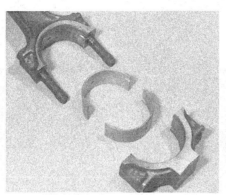

9.19a Connecting rod big-end components

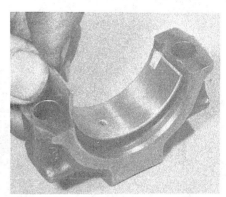

9.16b Fitting a shell bearing to a connecting rod cap

8 Push the piston/rod out of the cylinder bore in an upward direction.

9 If the shells are to be used again, tape them to their respective rod or cap, do not interchange them.

10 Remove the remaining three pistons/rods in a similar way.

11 The piston rings can be removed upwards off the piston using three old feeler blades slid behind each ring at equidistant points. This will prevent breakage, as the rings are very brittle, and also prevent the lower rings dropping into vacant grooves higher up (photo).

12 To separate the piston from the rod, the gudgeon pin must be removed. To do this, first extract the retaining circlips and then immerse the piston in hot water (60°C – 140°F). After a few minutes remove the piston, when the gudgeon pin should push out with finger pressure.

13 Refer to Section 19 for details of piston and cylinder bore examination and renovation.

14 Before fitting the piston rings make sure that the groove in the piston is quite clean. Use a piece of broken piston ring as a scraper to remove carbon from the grooves. Protect your fingers – piston rings are sharp! Clear the groove oil return holes (photo).

15 Push each compression ring down its bore to within 15.0 mm (0.59 in) of the lower edge of the cylinder, and check that its gap conforms to that specified using feeler blades.

16 Fit the lower (oil control) ring, followed by the two compression rings, making sure that the word 'TOP' is on the top face of the compression rings when viewed from the piston crown. No 2 ring has its step facing downwards (photos).

17 Locate the piston on its rod, making sure that with the piston arrow pointing towards the timing belt end of the engine, the matching rod big-end numbers are towards the left-hand (inlet manifold) side of the engine.

18 Fit the gudgeon pin, having again warmed the piston, and fit two new retaining circlips.

19 Wipe the shell bearing seat absolutely clean in both the big-end of the connecting rod and the cap. Fit the shells. If the original shells are being installed, make sure that they are returned to their original positions (photos).

20 Fit a piston ring compressor to the first piston, making sure that the rings and cylinder bores have been well lubricated with engine oil.

21 Insert the connecting rod into the cylinder bore and lower it until the base of the ring compressor is resting squarely on the top face of the block.

22 Position the end of a wooden handle of a hammer against the piston crown and then give the head of the hammer a sharp blow with the hand. This will drive the piston/rod down the bore and release the ring compressor.

23 Draw the rod down. Oil the crankpin, engage the big-end with it, fit the cap, screw on the nuts and tighten to the specified torque wrench setting (photos).

24 Fit the remaining three piston/rod assemblies in a similar way, turning the crankshaft as necessary to bring the crankpins into the most suitable position for connecting the big-ends and caps.

25 Fit the oil pick-up pipe using a new gasket and lockplate. Tighten the bolts and bend up the lockplate tabs.

26 Fit the sump (Section 8).

27 Fit the cylinder head (Section 7).

28 Fill the engine with oil.

10 Oil pump – removal and refitting

1 Remove the timing belt (Section 5).

2 Remove the sump (Section 8).

3 Unbolt and remove the fan blades from the pulley on the front end of the crankshaft.

4 Unscrew and remove the crankshaft pulley securing bolt. In order to prevent the crankshaft rotating while this is being done, either engage a gear and apply the brakes fully, or remove the starter motor and jam the teeth of the flywheel ring gear with a large screwdriver or piece of flat steel.

5 Take off the pulley assembly.

6 Unscrew and remove the oil pump securing screws and withdraw the oil pump from the front face of the engine. Peel off and discard the gasket.

9.23a Fitting a big-end cap

9.23b Tightening big-end cap nuts

10.8a Oil pump oil seal

10.8b Locating oil pump gasket

10.8c Installing the oil pump

12.3 Screwing on the oil filter

7 The oil pump may be dismantled and checked as described in Section 19.

8 Refitting is a reversal of removal, but use a new flange gasket and install a new pump oil seal as a matter of routine, having filled the lips of the seal with grease. As the pump is offered into position, make sure that the driving pins engage in the oil pump driven ring (photos).

11 Engine mountings – renewal

1 The flexible component of the engine mountings can be renewed without removing the engine, provided the weight of the engine can be taken off the mountings.

2 To do this, attach a hoist to the engine lifting hook eyes. Due to the fragile nature of the light alloy sump it is **not** recommended that the engine is jacked up under the sump, even using a block of wood as an insulator.

12 Oil filter – renewal

1 The disposable, cartridge type oil filter should be renewed at alternate oil changes.

2 To remove the old filter, use a strap or chain wrench. If these tools are not available, drive a long screwdriver blade right through the filter at a point furthest from the engine and use it as a lever to unscrew it. Be prepared for some spillage.

3 Smear the rubber sealing ring of the new filter with grease and screw it into position using hand pressure only (photo).

4 Once the engine has been refilled with oil, start the engine and observe that the oil pressure warning lamp goes out after a few seconds. The short delay before the lamp goes out is due to the time required for the new filter to fill with oil.

5 Check for leakage from the filter flange joint.

6 A new filter absorbs approximately 0.6 litre (1.0 Imp pt) of engine oil.

13 Crankcase ventilation system – description and maintenance

1 In order to extract oil fumes and blow-by gases which have passed the piston rings and accumulated in the crankcase, a ventilation system is fitted to all models.

2 The system prevents the vapours being emitted directly to atmosphere. Instead they are drawn into the air cleaner, where they are consumed during the engine combustion process.

3 Basically, the system comprises an oil separator, flame trap and connecting hoses (photo).

4 Maintenance consists of periodically checking the security of the hose connections, and occasionally removing the parts and cleaning them with paraffin to remove any sludge or emulsified condensate ('mayonnaise'). Evidence of long deposits of the latter may be due to cool running of the engine, particularly in winter.

14 Engine – method of removal

1 The engine is removed on its own, leaving the gearbox in the vehicle.

2 As the engine has to be raised and then withdrawn sideways out of the driving cab, a hoist with an extending arm or a jib crane will be required. These can usually be hired locally.

3 It is possible to remove the engine by passing a steel tube through the cab, supporting it at either end. Then using a lightweight hoist attached to a pulley which is free to run on the tube, raise the engine and move it sideways out of the driving cab.

15 Engine – removal

1 Remove the front seats and engine cover as described in Section 4.

2 Disconnect the battery negative lead.

3 Drain the cooling system as described in Chapter 2. Retain the coolant for further use if it is not due for changing.

13.3 Crankcase ventilation system flame trap

Fig. 1.8 Exhaust pipe bracket bolts (4) (Sec 15)

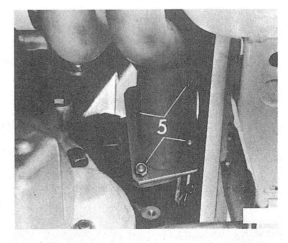

Fig. 1.9 Exhaust downpipe flange nuts (5) (Sec 15)

4 Remove the radiator and air duct (Chapter 2).

5 Unbolt and remove the fan from the crankshaft pulley.

6 Unbolt the exhaust pipe bracket from the gearbox.

7 Unbolt the exhaust downpipe from the exhaust manifold.

Fig. 1.10 Starter motor mounting bolts (6) (Sec 15)

15.10 Alternator wiring connection

Fig. 1.11 Clutch bellhousing bolts (Sec 15)

15.13 Removing expansion tank bracket screw

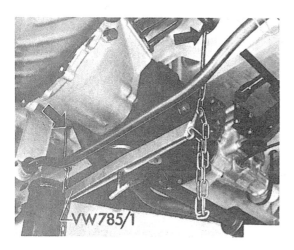

Fig. 1.12 Gearbox supporting cradle (Tool No VW 785/1)
(Sec 15)

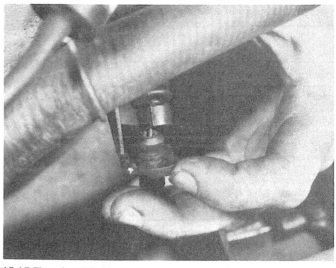

15.15 Throttle cable clip at support bracket

8 Unscrew and remove the oil filter.
9 Unbolt the starter motor. Lift it, with its wiring harness still attached, and rest it in the driving cab.
10 Disconnect the wiring plug from the back of the alternator (photo).
11 Unscrew and remove the bolts which hold the engine to the gearbox/clutch bellhousing, but leave the uppermost bolts in position at this stage. Remove the cover plate from the lower front face of the clutch bellhousing.
12 Support the weight of the gearbox, either on a jack or by using a crossbar hooked up to suitable holes in the body panel as shown (Fig. 1.12).
13 Remove the expansion tank, complete with its mounting bracket, and the coolant hoses which connect with the thermostat housing and the inlet manifold (photo).
14 Remove the air cleaner (Chapter 3) with intake duct and elbow. Disconnect the afterburner hose (Chapter 3), if so equipped.
15 Disconnect the throttle cable from the lever on the carburettor, and release the cable from its support bracket by pulling out the C-clip. Disconnect the cable from the manually-operated choke (if fitted) (photo).
16 Disconnect the electrical leads from the carburettor air cut-off valve and the automatic choke (as applicable).
17 Pull the leads from the temperature sender and the oil pressure switches.
18 Disconnect the HT lead from the coil or distributor centre terminal socket, and the LT lead from the distributor or coil negative terminal.
19 Disconnect the heater hose from the coolant pump.
20 Disconnect the brake servo vacuum hose from the inlet manifold.
21 Disconnect the fuel inlet hose from the carburettor (electric fuel pump) or from the fuel pump (mechanical fuel pump) and plug it with a suitable rod. Where fitted, also disconnect the fuel return hose.
22 Just take the weight of the engine on a hoist and working from inside the driving cab, remove both engine mounting nuts. These are the uppermost nuts visible when looking down on the engine. Now raise the engine until the mounting brackets clear the mounting studs.
23 Remove the top clutch bellhousing bolt not unscrewed previously.
24 Move the engine forwards away from the gearbox until it clears the gearbox input shaft.
25 Raise and turn the engine and remove it from the driving cab.

16 Engine – dismantling (general)

1 With the engine removed from the vehicle, clean away external grease, oil and dirt using either paraffin and a stiff brush or a water-soluble solvent.
2 Position the engine ready for overhaul on a clean bench, or on the floor on a clean sheet of hardboard or plywood.
3 As well as a set of tools, have rags, brushes and a bowl available for cleaning individual parts.
4 Keep associated parts such as sump screws, camshaft bearing nuts, big-end and main bearing nuts, together in tins or boxes. This will save time and confusion during reassembly.
5 Sketch the assembly of complicated components **before** dismantling if there is a likelihood of an error occurring on reassembly.

17 Engine – complete dismantling

1 The need for complete overhaul will probably occur as the result of bearing knock or worn pistons or cylinders, the latter indicated by smoke being emitted from the exhaust pipe.
2 With the engine standing upright, remove the toothed belt (Section 5).
3 Remove the cylinder head (Section 7).
4 Unbolt and remove the alternator and coolant pump.
5 Mark the clutch pressure plate in relation to the engine flywheel, then unbolt and remove it, taking care not to allow the driven plate to drop out as the pressure plate is withdrawn.
6 The flywheel is marked in relation to the mounting flange on the end of the crankshaft. Unscrew and remove the flywheel mounting bolts, preventing the flywheel turning if necessary by locking the starter ring gear teeth with a suitable tool or small metal plate.
7 Unscrew the bolt which secures the crankshaft pulley. In order to hold the crankshaft still, temporarily screw in two flywheel bolts and use a long bar between them as a lever.

Fig. 1.13 Electrical leads and hoses (Sec 15)

12 Automatic choke and temperature sender

13 Heater hose at cylinder head elbow

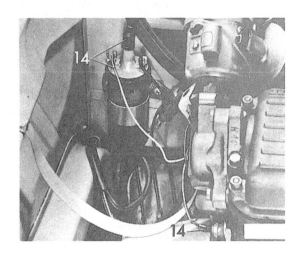

Fig. 1.14 Ignition coil and oil pressure switch leads (14) (Sec 15)

Fig. 1.15 Brake vacuum servo hose (15), fuel inlet hose (16) and throttle cable connection (17) (Sec 15)

Fig. 1.16 Unscrewing flywheel bolts. Note gear locking plate (Tool No. 10-201) (Sec 17)

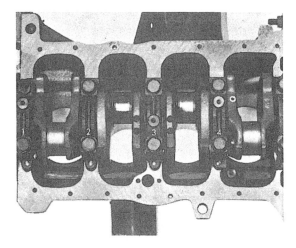

Fig. 1.17 Main bearing cap numbers (Sec 17)

8   Unbolt and remove the oil pump, discard the joint gasket. Remove the dipstick guide tube.
9   Unbolt and remove the engine mounting brackets.
10   Turn the engine on its side and remove the sump (Section 8). Peel off the gaskets and seals.
11   Unbolt and remove the oil pick-up pipe, having first flattened the tabs of the bolt lockplate.
12   Remove the pistons/connecting rods as described in Section 9.
13   Now invert the engine so that it rests on the top face of the cylinder block.
14   The main bearing caps are numbered 1 to 5 from the timing belt end of the engine. The numbers are readable from the oil filter side of the engine.
15   Unscrew and remove the main bearing cap bolts and remove the caps. If they are tight, tap them off carefully using a plastic or copper-faced mallet, but don't tap them sideways as they are located on hollow dowels. Keep the bearing shells with their respective caps if they are to be used again.
16   Carefully lift the crankshaft from the crankcase, taking care that the upper shells which may be sticking to the crankshaft journals do not drop off and get mixed up, as it is again essential to keep them identified as to location if they are to be used again.
17   The cylinder head may be dismantled and serviced as described in the next Section.
18   With the engine completely dismantled, refer to Section 19 concerning examination and renovation.

18 Cylinder head – servicing and decarbonising

1   Unbolt and remove the inlet and exhaust manifolds. Note the hot air collector box on the exhaust manifold.
2   Remove the camshaft as described in Section 6.
3   Remove each of the cam followers complete with wedge adjuster screw. A valve grinding rubber suction cap is useful for this. The followers must be kept in strict order, so use a box subdivided into eight compartments to ensure that they are not mixed up.
4   Using a valve spring compressor, compress the first valve spring until the split collets can be extracted.
5   Take off the spring retainer, the double valve springs and either the spring seat (inlet valve) or the roto cap (exhaust valve).
6   Remove the valve stem oil seals from the ends of the guides and discard them.
7   Repeat the operations on the remaining valves, keeping the valves and associated parts strictly in originally installed order.
8   Remove all carbon deposits from the cylinder head using a blunt tool. Remember that the head is of light alloy construction and is easily damaged.
9   Check for cracks around bolt holes, and for stripped threads. If such damage is found, consult a specialist to see if repair is possible.
10   If there is any doubt about the flatness of the cylinder head, test it for distortion using a feeler blade and a straight-edge or a piece of plate glass. Distortion may occur as the result of overheating, or be due to incorrect tightening of the cylinder head bolts.
11   It may be possible for your dealer or engine reconditioner to skim the head, provided the thickness of the head is maintained in accordance with the dimension given in Fig. 1.19.
12   Wear in the valve guides should now be checked. Do this by inserting the appropriate valve (inlet or exhaust) into its guide and testing for sideways rock. This should not exceed the specified maximum. Ideally the rock should be checked using a dial gauge.
13   If wear exceeds the specified limit, the valve guides should be renewed – a job for your dealer as they will require reaming on completion of installation. Modified valve spring seats will also have to be installed.

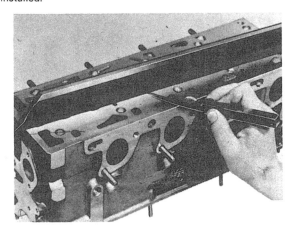

Fig. 1.18 Checking cylinder head for distortion (Sec 18)

Fig. 1.19 Cylinder head refinishing diagram (Sec 18)

a   139.5 mm (5.492 in) minimum

Fig. 1.20 Measuring valve stem 'rock' using a dial gauge (Sec 18)

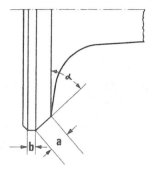

Fig. 1.21 Inlet valve refacing diagram (Sec 18)

a   3.5 mm maximum   b   0.5 mm minimum
                            ⋪   45°

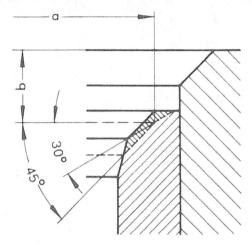

Fig. 1.22 Inlet valve seat recutting diagram (Sec 18)

a   36.5 mm maximum diameter   b   3.5 mm maximum
                                  Seat width 2.0 mm

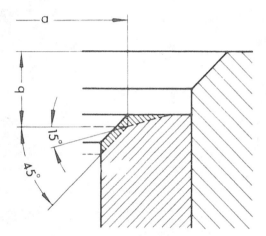

Fig. 1.23 Exhaust valve seat recutting diagram (Sec 18)

a   31.5 mm maximum diameter   b   3.9 mm maximum
                                  Seat width 2.4 mm

Fig. 1.24 Grinding in a valve (Sec 18)

14 Examine the valve seats and heads. If they are in reasonably good condition, with evidence of small black spots on the sealing area, then they can be reconditioned by careful grinding in.

15 If they are severely burned then either renew the valves or reface the inlet valves (the exhaust valves must not be refaced, only renewed). The valve seats may be re-cut in accordance with Figs. 1.22 and 1.23 provided the correct cutters are available.

16 Where the valves and their seats are in good condition, normal grinding in will restore them to a serviceable state.

17 To grind in a valve, stand the cylinder head on the camshaft bearing cap studs.

18 Stick the rubber suction cup of a grinding tool to the first valve to be ground in, oil the valve stem and apply a smear of coarse grinding paste to the chamfered edge of the valve head. Insert the valve into its guide and then using a rotary back-and-forth action, rotate the tool between the palms of the hands. Gradually, the 'gritty' feel of the paste will disappear and the valve should be withdrawn and wiped clean. If the black spots or discoloration are still evident, repeat the operation until they disappear. A single drop of paraffin placed on the grinding paste on the valve seat will speed up the grinding action, but do not overdo this or the paste could run down into the guide, with disastrous results!

19 When the coarse paste has achieved the desired result, finish off with fine paste. A correctly ground-in valve should have a frosty grey band all around its head chamfer, with a corresponding band on the valve seat about 1.5 mm (0.059 in) in width.

20 Repeat the operations on all the remaining valves, then wash away every trace of grinding paste with paraffin and a brush.

21 Commence reassembly by oiling the first valve and inserting it into its guide (photos).

22 Fit the valve stem oil seal. To avoid damage to the seal as it passes over the sharp edge of the valve guide, it is important to use the plastic guide supplied with each set of new seals. Push the seal well down onto the end of the valve guide (photo).

23 Fit the spring seat (inlet) or thicker roto cap (exhaust) (photos).

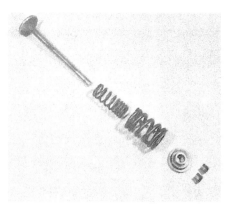

18.21a Valve components – valve, springs, cap and collets. Spring seat or roto cap not shown

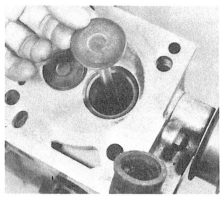

18.21b Inserting a valve into its guide

18.22 Using a guide sleeve to fit a valve stem oil seal

18.23a Inlet valve spring seat (1) and exhaust valve roto cap (2)

18.23b Fitting inlet valve spring seat

18.24 Fitting valve inner coil spring

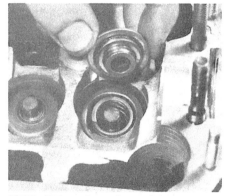

18.25a Fitting valve spring retaining cap

18.25b Valve spring compressed and collets inserted

18.28a Installing cam followers

18.28b Cam follower with edge adjuster screw removed

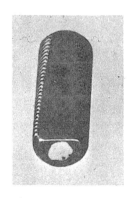

Fig. 1.25 Colour coded valve clearance adjuster screw (Sec 18)

Fig. 1.26 Ground valve clearance adjuster screw (Sec 18)

24 Fit the double coil springs with their closer coils towards the spring retainer. If the springs have been in use for 64 000 km (40 000 miles) or more it is recommended that they are renewed (photo).
25 Fit the spring retainer. Compress the springs and insert the split collets in the valve stem cut-out. A dab of thick grease will hold them in position (photos).
26 Release the spring compressor slowly and remove it.
27 Repeat the operations to install the remaining valves. Once they are all installed, tap the ends of each valve stem using a hammer and a small block of wood to settle the components. When doing this, make sure that there is enough space under the cylinder head for the valve head to project slightly momentarily.
28 Oil and fit the cam followers. It is worthwhile to understand the operation and identification of the wedge type adjuster screws which are used to increase or decrease the valve clearance (photos). Older type screws are colour marked blue, red or yellow which correspond with newer screw markings and matching marks on the cam followers as follows:

Blue	1 notch
Red	2 notches
Yellow	3 notches

29 Later screws are marked with 1, 2, 3 or 4 notches.
30 If during renovation of the valve seats, a seat has to be re-cut, use a plain (unmarked) adjuster screw which is supplied under Part No 046 109 453C.
31 Refit the camshaft as described in Section 6.
32 Adjust the valve clearances as described in Section 20.
33 Using new gaskets, fit the inlet manifold complete with carburettor (photos).
34 Locate a new exhaust gasket and install the exhaust manifold and the hot air collector plate (photos).
35 If the rear coolant elbow has been leaking, remove it and fit a new gasket (photo).

19 Examination and renovation

1 With the engine completely dismantled, clean each component and examine for wear or damage. The points to watch for on the cylinder head are included in the preceding Section.

Crankshaft and bearings
2 Examine the crankpin and main journal surfaces for signs of scoring or scratches. Check the ovality of the crankpins at different positions with a micrometer. If more than 0.001 inch (0.0254 mm) out of round, the crankpins will have to be reground. They will also have to be reground if there are any scores or scratches present. Also check the journals in the same fashion.
3 If it is necessary to regrind the crankshaft and fit new bearings, your local VW garage or engineering works will be able to decide how much metal to grind off and the size of new bearing shells.
4 Big-end bearing failure is accompanied by a knocking from the crankcase, and a slight drop in oil pressure. Main bearing failure is accompanied by vibration which can be quite severe as the engine

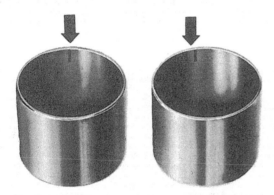

Fig. 1.27 Cam follower markings (Sec 18)

18.33a Locating inlet manifold gasket

18.33b Installing inlet manifold with carburettor

18.34a Locating exhaust manifold gasket

18.34b Installing the exhaust manifold

18.34c Installing hot air collector plate

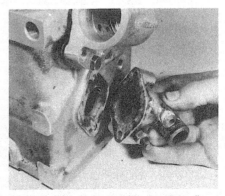

18.35 Fitting cylinder head rear coolant elbow

speed rises. Inspect the big-end, main bearings, and thrust washers for signs of general wear, scoring, pitting and scratches. The bearings should be a matt grey in colour. With lead-indium bearings, should a trace of copper colour be noticed, the bearings are badly worn as the lead bearing material has worn away to expose the indium underlay. Renew the bearings as a matter of course unless they are virtually new.

5   The crankshaft main bearings are available in undersizes to suit journal regrinds of 0.25, 0.50 and 0.75 mm. Your dealer or local engine reconditioner will carry out this work for you and supply the correct bearing shells to give the specified running clearance.

6   The crankpins will also be reground where necessary to the same undersizes as the journals, and suitable bearing shells will be supplied by the reconditioner.

7   Examine the needle roller bearing in the centre of the crankshaft rear flange. If it is noisy when turned or is obviously worn, it should be renewed (photo).

8   To do this, an expanding type extractor is required, although it is possible to remove it by hydraulic pressure if the bearing hole is filled with heavy grease and a close-fitting rod driven in. Replenish the grease and repeat the operation until the bearing is extracted.

9   When fitting the new bearing, make sure that the lettering is visible when installed, and apply pressure to the bearing outer track only until it is 1.0 mm (0.039 in) below the surface of the crankshaft flange.

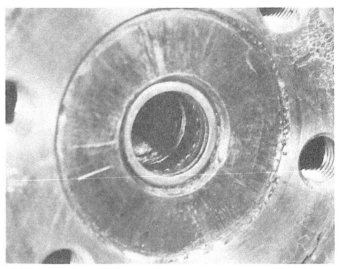

19.7 Crankshaft rear flange needle roller bearing

Cylinder bores

10 The cylinder bores must be examined for taper, ovality, scoring and scratches. Start by carefully examining the top of the cylinder bores. If they are at all worn a slight ridge will be found on the thrust side. This marks the top of the piston ring travel. The owner will have a good indication of the bore wear prior to dismantling the engine, or removing the cylinder head. Excessive oil consumption accompanied by blue smoke from the exhaust is a sure sign of worn cylinder bores and piston rings.

11 Measure the bore diameter just under the ridge with a micrometer and compare it with the diameter at the bottom of the bore which is not subject to wear. If the difference between the two measurements is more than 0.006 in (0.152 mm) it will be necessary to fit special pistons and rings or to have the cylinders rebored and oversize pistons fitted. If a micrometer is not available, remove the rings from each piston in turn (do not mix the rings from piston to piston) and place each piston in its respective bore about ¾ in (19 mm) below the top surface of the cylinder block. If a 0.010 in (0.254 mm) thick feeler gauge can be slid between the piston and the cylinder wall on the thrust side of the bore, then the following action must be taken.

12 Oversize pistons are available as listed in the Specifications. These are accurately machined so as to provide correct running clearances in bores of the exact oversize dimensions.

13 If the bores are slightly worn but not so badly worn as to justify reboring them, then special oil control rings and pistons can be fitted which will restore compression and stop the engine burning oil. Several diferent types are available and the manufacturer's instructions concerning their fitting must be followed closely.

14 If new pistons are being fitted and the bores have not been reground, it is essential to slightly roughen the hard glaze on the sides of the bore with fine glass paper so the new piston rings will have a chance to bed in properly.

15 Newly fitted pistons should be tested for clearance using a feeler gauge and spring balance. Place the feeler gauge between the piston and cylinder wall and having attached the spring balance to it, check the pull required to remove it. A pull of between 7 and 11 lb (3.2 and 4.5 kg) should be needed to withdraw a feeler blade 0.5 in (12.7 mm) wide and 0.0025 in (0.064 mm) thick.

Pistons and piston rings

16 If the old pistons are to be refitted, carefully remove the piston rings and then thoroughly clean them. Take particular care to clean out the piston ring grooves. At the same time do not scratch the aluminium in any way. If new rings are to be fitted to the old pistons then the top ring should be stepped so as to clear the ridge left in the cylinder bore above the previous top ring. If a normal but oversize new ring is fitted, it will hit the ridge and break because the new ring will not have worn in the same way as the old. This will have worn in unison with the ridge.

Fig. 1.28 Installed position of crankshaft rear flange pilot bearing (Sec 19)

a   1.0 mm (0.039 in)

Camshaft and camshaft bearings

17 Carefully examine the camshaft bearings for wear. If the bearings are obviously worn or pitted, then this will mean a new cylinder head as the bearings do not have renewable shells.

18 The camshaft itself should show no signs of wear. If scoring on the cams is noticed, the only permanently satisfactory cure is to fit a new camshaft.

19 Examine the skew gear for wear, chipped teeth or other damage.

Timing belt and sprockets

20 Examine the timing belt for cracks, fraying or tooth deformation. If evident, renew the belt.

21 If the belt has been in operation for 64 000 km (40 000 miles) or more, it is worthwhile renewing it as a matter of routine.

22 Check the camshaft and crankshaft sprocket teeth for wear. If evident renew the crankshaft pulley/sprocket as an assembly, and the camshaft sprocket.

23 Examine the belt tensioner pulley. If it turns noisily or it rocks about its shaft, renew the assembly.

Cam followers

24 Examine the bearing surface of the bucket type cam followers. Any indentation in this surface or any cracks indicate serious wear and the tappets should be renewed. Thoroughly clean them out, removing all traces of sludge. It is most unlikely that the sides of the tappets will prove worn, but if they are a very loose fit in their bores and can readily

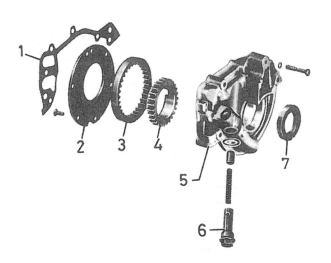

Fig. 1.29 Exploded view of the oil pump (Sec 19)

1	Gasket	5	Body
2	Backing plate	6	Pressure relief valve
3	Outer gear	7	Oil seal
4	Inner gear		

19.27 Oil pump rear plate removed

be rocked, they should be exchanged for new units. It is very unusual to find any wear in the tappets, and any wear is likely to occur only at very high mileages.

Connecting rods

25 When a connecting rod is fitted to its crankpin, the side clearance should not exceed 0.4 mm (0.016 in) when measured with a feeler blade. It is seldom that the side clearance exceeds that specified, but if it does and a new rod has to be purchased, make sure that the weight of the new rod matches the weights of the original ones.

Flywheel and starter ring gear

26 If the flywheel friction plate surfaces have been scored due to damage from the rivets of the plate because it was not renewed in time, or if the teeth of the starter ring gear are worn or chipped, the flywheel should be renewed complete.

Oil pump

27 The oil pump can be dismantled by extracting the countersunk-headed screws and removing the pump backplate (photo).
28 Check the gears for chipped teeth. Measure the gear backlash. If this is outside the specified tolerance, renew the gears or pump complete.
29 Check the gear endfloat using a straight-edge and feeler blades. Again this must conform to the specified tolerance. Remove the pressure relief valve and check for wear or a broken spring (photo).

Oil seals and gaskets

30 These components should always be renewed at time of major overhaul, no matter how good the original ones may appear.

---

20 Valve clearance – adjustment

1 This adjustment is required when rebuilding the engine after overhaul, or at specified routine service intervals. A 3.0 mm Allen key and a special gauge (2038) will be required for this work.
2 The clearances (cold) differ from those when the engine is at normal operating temperature. Apart from initial setting after engine overhaul, the clearances should always be checked with the engine hot. The correct clearances are given in the Specifications.
3 Remove the camshaft cover. Turn the crankshaft pulley, or if the cylinder head is off, turn the camshaft, until both No 1 cylinder camshaft lobes are facing upwards at similar angles.

19.29 Oil pressure relief valve withdrawn

4 Slide a feeler blade between the base circle (larger diameter) of the cam lobe and the cam follower. It should be a stiff sliding fit. If it is not, rotate the cam follower until the screw hole is visible and insert the Allen key. Turn the screws one or more **complete** turns to correct the clearance. Each complete turn of the screw alters the valve clearance by 0.05 mm (0.0020 in). Each turn is verified by a 'click' (photos).
5 Once the adjustment is correct, remove the Allen key and insert the gauge instead. The wall of the cam follower must be between the notched area of the tool (Fig. 1.30). If it is not, remove the screw and fit one of the next largest size. Readjust the clearance.
6 Repeat the operations on the remaining three pairs of valves.
7 Counting from the timing belt end of the engine, the valve sequence is:

1	Inlet (larger diameter head)
2	Exhaust (smaller diameter head)
3	Inlet
4	Exhaust
5	Inlet
6	Exhaust
7	Inlet
8	Exhaust

8 On completion of adjustment at service intervals, fit a new camshaft cover gasket and the cover.

20.4a Checking a valve clearance

20.4b Cam follower adjuster screw and Allen key

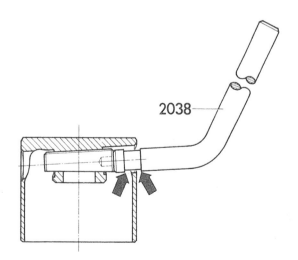

Fig. 1.30 Special tool (2038) for checking position of adjuster
screw in cam follower (Sec 20)

21 Engine – rebuilding

1   With the bottom of the crankcase facing upwards, wipe clean the
shell bearing seats and the backs of the main bearing shells. Fit the
shells, noting that the centre one incorporates the thrust flanges to
control crankshaft endfloat. Oil the shells (photos).
2   Fit the half shells with the oil holes to the crankcase and those
without holes to the bearing caps, the exception being the centre
bearing shell which has holes (photos). Take great care not to leave
any dirt or grit in the cap or on the back of the shell. Lower the
crankshaft into position (photo).
3   Oil the journals and fit the main bearing caps 1 to 4 so that their
numbers can be read from the oil filter side (photo).
4   Tighten the cap bolts to the specified torque.
5   Check the crankshaft endfloat by pushing the crankshaft fully in
one direction and then in the other, and using either a dial gauge or
feeler blades inserted on one side of the centre main bearing (photo).
6   If the endfloat is within the specified tolerance, proceed with
rebuilding. If it is not, then the thrust flanges must be worn at the
centre bearing shells.
7   Grease the lips of a new crankshaft rear oil seal and fit it to the end
of the crankshaft (photo).
8   Fit the rear main bearing cap, having applied sealing compound to
the block mating surface. Screw in and tighten the socket-headed
screw first, and then the hexagon-headed screws, both to the specified
torque (photos).
9   Fit the pistons/connecting rods as described in Section 9.
10 Fit a new oil seal to the oil pump and then locate a new joint
gasket on the front face of the engine. Bolt the oil pump into position,
taking care that the driving pins engage in the cut-outs in the oil pump
driven ring.
11 Bolt on the oil pick-up pipe using a new joint gasket and lockplate.
Bend up the tabs of the lockplate (photo).
12 Fit the sump using new side gaskets and end sealing strips. Apply
jointing compound at the corners where the gaskets and strips overlap
(photos).
13 Fit the engine mounting brackets (if removed).
14 Apply grease to the oil pump oil seal lips and push the crankshaft
pulley/belt sprocket into position (photo).
15 Screw in the pulley retaining bolt, having first applied thread
locking compound to its threads. Hold the crankshaft still using a lever
between two bolts screwed into the crankshaft rear flange (photo).
16 Refit the flywheel, aligning the positioning marks. Jam the starter
ring gear as at removal. Apply thread locking compound to the
flywheel bolts and tighten them to the specified torque (photo).
17 Fit the clutch, centralising the driven plate as described in Chapter
5.
18 Refit the cylinder head as described in Section 7 and the timing
belt (Section 5).
19 Fit the alternator and coolant pump (photo).
20 Fit the drivebelt and tension it as described in Chapter 2. Install
the distributor (Chapter 4).
21 Refit the dipstick guide tube (photo).

22 Engine – installation

1   Hoist the engine into the driving cab and lower it carefully until it
can be pushed to the rear and mated with the clutch bellhousing
(photo). During this operation **do not** allow the weight of the engine to
hang upon the gearbox input shaft. The crankshaft may have to be
turned slightly in order to align the splines on the input shaft with
those in the clutch driven plate hub.
2   Lower the front end of the engine onto the mounting studs and
remove the hoist.
3   Fit the mounting stud nuts and insert the two uppermost retaining
bolts to the clutch bellhousing (photos).
4   Remove the plugs and reconnect the fuel return and inlet hoses to
the carburettor (electric fuel pump) or to the fuel pump (mechanical
pump) (photo).
5   Reconnect the brake servo vacuum hose to the inlet manifold.
6   Reconnect the heater hose to the coolant pump.
7   Reconnect the HT and LT leads to the ignition coil.
8   Reconnect the lead to the oil pressure switch.
9   Reconnect the lead to the coolant temperature sender.

21.1a Fitting a main bearing shell to the crankcase

21.1b Oiling shell bearings

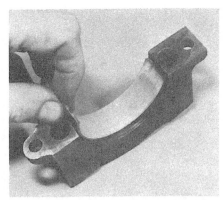

21.2a Fitting a bearing shell to a main bearing cap

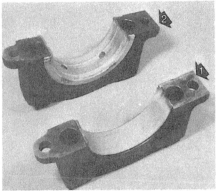

21.2b Main bearing intermediate cap (1) and centre cap (2)

21.2c Installing the crankshaft

21.3 Fitting the centre main bearing cap

21.5 Checking crankshaft endfloat

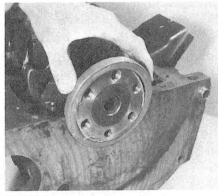

21.7 Fitting crankshaft rear oil seal

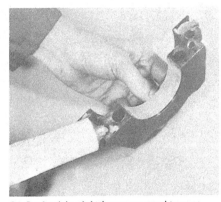

21.8a Applying jointing compound to rear main bearing cap

21.8b Installing crankshaft – rear main bearing cap

21.8c Tightening rear main bearing cap socket screw

21.11 Fitting oil pick-up pipe

21.12a Sump side gaskets

21.12b Sump end seals

21.12c Installing the sump

21.14 Installing crankshaft pulley/sprocket

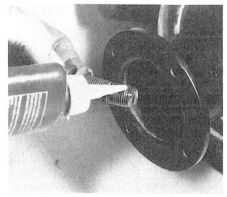
21.15 Apply thread locking fluid to crankshaft pulley bolt

21.16 Tightening flywheel bolts

21.19 Alternator mounting bracket and bolt

21.21 Fitting dipstick guide tube

22.1 Installing the engine

22.3a Engine mounting lower nut (right-hand side)

22.3b Engine mounting lower nut (left-hand side)

22.4 Connecting the carburettor fuel return hose

22.10 Automatic choke lead and terminal

22.13 Air cleaner intake duct

22.19a Exhaust downpipe flange gasket

22.19b Tightening exhaust downpipe flange nut

22.21 Fitting fan blades to crankshaft pulley

10 Reconnect the leads to the air cut-off solenoid valve and the automatic choke on the carburettor (as applicable) (photo).
11 Reconnect the throttle cable to the carburettor and to its retaining bracket.
12 Reconnect the choke cable (manual choke) to the carburettor.
13 Refit the air cleaner, intake duct and elbow. Connect the afterburner and crankcase ventilation hoses (photo).
14 Refit the expansion tank and mounting complete with connecting hoses.
15 Working under the vehicle, remove the gearbox support crossbar or jack and screw in the remaining clutch bellhousing bolts.
16 Connect the wiring plug to the rear of the alternator.
17 Lower the starter motor from its temporary resting place in the driving cab and bolt it into position. Fit the bellhousing lower front cover plate.
18 Fit a new oil filter as described in Section 12.
19 Bolt the exhaust downpipe to the exhaust manifold, using a new flange gasket (photos).
20 Reconnect the exhaust pipe support bracket to the gearbox.
21 Bolt the fan to the crankshaft pulley flange, noting that the ribs are towards the radiator (photo).
22 Install the radiator and air duct.
23 Fill the cooling system (Chapter 2).
24 Fill the engine with oil.
25 Reconnect the battery.
26 Adjust the accelerator cable (Chapter 3).
27 Adjust the clutch cable (Chapter 5).
28 Do not refit the engine cover or front seat at this stage, but wait until the engine has been started and can be visually checked for leaks. Minor adjustments can then be carried out.

23 Engine – initial start-up after major overhaul

1 On later models, there is no reason why the engine should not fire at the first twist of the ignition key. On earlier models with a mechanically-operated fuel pump, it may require a few revolutions of the starter motor until the pump can fill the carburettor bowl.
2 If the engine still does not fire, check the following points:

   (a) *There is fuel in the tank*
   (b) *Ignition and battery leads are correctly and securely connected*
   (c) *The choke is correctly connected*
   (d) *The distributor has been correctly installed*
   (e) *Work systematically through the fault diagnosis chart at the end of this Chapter*

3 Run the engine until normal operating temperature is reached and check the tightness of all nuts and bolts, particularly the cylinder head bolts. This is done by slackening the bolts slightly and retightening to the correct torque.
4 Adjust the idling and carburettor mixture control screws (Chapter 3).
5 Check for any oil or water leaks and when the engine has cooled, check the levels in the radiator and sump, and top up as necessary.
6 Check the ignition timing (Chapter 4).
7 After the first 1000 km (620 miles) check the valve clearances and alternator drivebelt tension.
8 It is important to restrict the engine speeds for the first few hundred miles if major internal components of the engine have been renewed. An oil and filter change may also be of benefit after the new components have bedded-in.

See overleaf for 'Fault diagnosis – engine'

24 Fault diagnosis – engine

Symptom and main cause	Reason(s)
Engine fails to turn when starter operated	
No current at starter motor	Flat or defective battery
	Loose battery leads
	Defective starter solenoid or switch, or broken wiring
	Engine earth strap disconnected
Current at starter motor	Defective starter motor
Engine turns but will not start	
No spark at spark plug	Ignition system damp or wet
	Ignition leads to spark plugs loose
	Shorted or disconnected low tension leads
	Dirty, incorrectly set, or pitted contact breaker points
	Faulty condenser
	Defective ignition switch
	Ignition leads connected wrong way round
	Faulty coil
	Contact breaker point spring earthed or broken
Excess of petrol in cylinder, or carburettor flooding	Too much choke allowing too rich a mixture to wet plugs
	Float damaged or leaking, or needle not seating
	Float level incorrectly adjusted
Engine stalls and will not start	
No spark at spark plug	Ignition failure – sudden
	Ignition failure – in severe rain or after traversing water splash
No fuel at jets	No petrol in petrol tank
	Petrol tank breather choked
	Sudden obstruction in carburettor
	Water in fuel system
Engine misfires or idles unevenly	
Intermittnt spark at spark plug	Ignition leads loose
	Battery leads loose on terminals
	Battery earth strap loose on body attachment point
	Engine earth lead loose
	Low tension leads to (+) and (-) terminals on coil loose
	Low tension lead from (-) coil terminal to distributor loose
	Dirty or incorrectly gapped plugs
	Dirty, incorrectly set, or pitted contact breaker points
	Tracking across inside of distributor cover
	Faulty coil
No fuel at carburettor float chamber or at jets	No petrol in petrol tank
	Vapour-lock in fuel line (in hot conditions or at high altitude)
	Blocked float chamber needle valve
	Fuel pump filter blocked
	Choked or blocked carburettor jets
	Faulty fuel pump
Fuel shortage at engine	Mixture too weak
	Air leak in carburettor
	Air leak at inlet manifold to cylinder head, or inlet manifold to carburettor
Mechanical wear	Incorrect valve clearances
	Burnt-out exhaust valves
	Sticking or leaking valves
	Weak or broken valve springs
	Worn valve guides or stems
	Worn pistons and piston rings
Lack of power and/or poor compression	
Fuel/air mixture leaking from cylinder	Burnt-out exhaust valves
	Sticking or leaking valves
	Worn valve guides and stems
	Weak or broken valve springs
	Blown cylinder head gasket (accompanied by increase in noise)
	Worn pistons and piston rings
	Worn or scored cylinder bores

Symptom and main cause	Reason/s
Incorrect adjustments	Ignition timing wrongly set Contact breaker points incorrectly gapped Incorrect valve clearances Incorrectly set spark plugs Carburation too rich or too weak
Carburation and ignition faults	Dirty contact breaker points Distributor automatic advance and retard mechanisms not functioning correctly Faulty fuel pump giving top end fuel starvation
Excessive oil consumption Oil being burnt by engine	Badly worn, perished or missing valve stem oil seals Excessively worn valve stems and valve guides Worn piston rings Worn pistons and cylinder bores Excessive piston ring gap allowing blow-by Piston oil return holes choked
Oil being lost due to leaks	Leaking oil filter gasket Leaking camshaft cover gasket Leaking sump gasket Loose sump plug
Unusual noises from engine Excessive clearance due to mechanical wear	Worn big-end bearing (regular heavy knocking) Worn main bearings (rumbling and vibration) Worn crankshaft (knocking, rumbling and vibration) Incorrectly tensioned timing belt Worn tensioner pulley shaft or bearing Piston slap (worst when cold)

Chapter 2
Cooling, heating and ventilation systems

For modifications, and information applicable to later models, see Supplement at end of manual

Contents

Specifications

System type .... Pressurised with expansion tank, belt-driven pump and crankshaft mounted cooling fan

Thermostat

Opens	80°C (176°F)
Fully open	93°C (199.4°F)
Valve plate stroke	8.0 mm (0.315 in)

Coolant capacity .... 9.0 l (15.75 Imp pts)

Expansion tank cap pressure .... 0.9 to 1.15 bar (13 to 16.7 lbf/in²)

Torque wrench settings

	Nm	lbf ft
Thermostat housing cover bolts	10	7
Coolant pump bolts	10	7
Coolant pump nuts	22	16
Fan assembly bolts	25	18
Temperature sender switch	8	6
Coolant elbow (rear face of cylinder head) bolts	10	7

1 Description and maintenance

The cooling system is of pressurised type and incorporates a front mounted radiator and a coolant pump driven by a V-belt from the crankshaft pulley.

Cooling fan blades are bolted to the front flange of the crankshaft pulley.

The cooling system is of 'no loss' type, using an expansion tank to accept coolant displaced by expansion. The coolant is stored in the tank until, as the engine cools, it is drawn back into the system.

Topping up is very seldom required, but give a regular visual check to see that the coolant level is up to its mark on the expansion tank. This is visible once the engine cover flap is opened.

A thermostat is located in a housing at the front of the engine to reduce engine warm-up time.

The coolant antifreeze mixture should be renewed at the intervals specified in Routine Maintenance at the beginning of this manual. Refer also to Section 3 regarding antifreeze mixtures.

The interior heater is supplied with coolant from the engine cooling system (see Section 9).

2 Cooling system – draining, flushing and refilling

1 If the vehicle has just come in off the road, allow it to cool down before draining the cooling sytem. *There is a risk of scalding if the expansion tank cap is removed with the system hot.*

2 Move the heater temperature control lever to the maximum heat position.

Fig. 2.1 Expansion tank full mark (Sec 1)

3 Release the cap on the expansion tank. Move the cap to the stop, allow any pressure to escape, then remove it completely.

4 Place a suitable container under the radiator and remove the radiator drain plug.

5 Retain the coolant for further use if it is not due for changing.

6 If the correct antifreeze or anti-corrosion mixture has been maintained in the system, flushing should not be required, and the system may be refilled immediately.

7 Where the coolant is of insufficient mixture strength to have

prevented rust or corrosion occurring, the system should be flushed through until the water runs clear. If necessary, the radiator can be reverse flushed with a cold water hose to remove sediment and blockages caused by loose scale.

8 In severe cases of clogging, use a radiator cleaning compound, but observe the manufacturer's instructions precisely. Such a cleaner must be suitable for mixed metal (iron and aluminium) engines.

9 Now examine the radiator cooling fins. If they are clogged with dirt or flies, use the cold water hose, or an air line in the reverse direction to normal airflow, to clean them.

10 When the system is ready for refilling, screw in the drain plug at the base of the radiator.

11 Pour antifreeze or anti-corrosion mixture into the expansion tank until the coolant level is up to the mark indicated on the tank (photo).

12 Fit the expansion tank cap, start the engine and run it at a fast idle speed for two or three minutes. This will cause any trapped air to escape and the level of coolant in the expansion tank will be seen to fall.

13 Top up the level mark with more coolant mixture and switch off the engine.

3 Coolant mixtures

1 It is recommended that the system is filled with an antifreeze mixture where climatic conditions warrant its use. The cooling system should be drained, flushed and refilled every other autumn. The use of antifreeze solutions for periods of longer than two years is likely to cause damage and encourage the formation of rust and scale, due to the corrosion inhibitors gradually losing their efficiency. If the use of an antifreeze mixture is not necessary because of favourable climatic conditions, never use ordinary water but always fill the system with a corrosion inhibiting mixture of a recommended brand.

2 Before adding antifreeze to the system, check all hose connections and check the tightness of the cylinder head bolts as such solutions are searching. The cooling system should be drained and partly refilled with clean water as previously explained, before adding antifreeze.

3 The quantity of antifreeze which should be used for various levels of protection is given in the table below, expressed as a percentage of the system capacity.

Antifreeze volume	Protection to	Safe pump circulation
25%	–26°C (–15°F)	–12°C (10°F)
30%	–33°C (–28°F)	–16°C (3°F)
35%	–39°C (–38°F)	–20°C (–4°F)

4 Where the cooling system contains an antifreeze or corrosion inhibiting solution, any topping-up should be done with a solution made up in similar proportions to the original in order to avoid dilution.

4 Thermostat – removal, testing and refitting

1 A faulty thermostat can cause overheating or prolong the engine warming-up period. It can also affect the performance of the heater.

2 Drain off enough coolant through the radiator drain plug so that the coolant level is below the thermostat housing joint face. An indication that the correct level has been reached is when 3 litres (5 pts) have been drained.

3 Unscrew and remove the two retaining bolts and withdraw the thermostat cover sufficiently to permit the thermostat to be removed from its seat in the cylinder head (photos).

4 To test whether the unit is serviceable, suspend the thermostat by a piece of string in a pan of water being heated. Using a thermometer, and referring to the opening temperature in the Specifications, its operation may be checked. The thermostat should be renewed if it is stuck open or closed, or if it fails to operate at the specified temperature. The operation of a thermostat is not instantaneous and sufficient time must be allowed for movement during testing. Never refit a faulty unit – leave it out if no replacement is available immediately.

5 Refitting the thermostat is a reversal of the removal procedure. Ensure the mating faces of the housing are clean. Use a new sealing O-ring (photo).

5 Radiator – removal, repair and refitting

1 Drain the cooling system as described in Section 2.

2 Working under the engine compartment, unbolt and remove the protective shield from beneath the radiator.

3 Disconnect the bottom hose from the radiator (photo).

4 Disconnect the top hose from the radiator.

5 Unscrew and remove the two bolts which secure the base of the radiator to brackets or crossmember (photo).

6 Release the self-tapping screws which hold the flexible blanking panels on each side of the radiator and then remove the radiator downwards, complete with foam sealing ring. The radiator is located at its top by a peg which releases automatically (photos).

7 If the radiator is leaking, either have it repaired professionally or exchange it for a rebuilt unit. Unless you are expert at soldering, radiator repair can be a problem due to the need to restrict the heat being applied which might otherwise make any fault worse. Temporary repairs of small leaks may be made using one of the proprietary compounds marketed for the purpose.

8 Refitting is a reversal of removal. Check that the radiator top fixing is correctly engaged in its hole.

9 Refill the cooling system.

6 Drivebelt – tensioning, removal and refitting

1 Periodically, check the tension of the drivebelt which supplies power to the coolant pump and the alternator. Do this by depressing the belt at the midpoint of its longest run using moderate thumb pressure. The belt should deflect by 12.5 mm (0.5 in).

2 If the belt is slack, release the alternator pivot mounting bolt, also the nut and the screw at the ends of the adjuster link. Carefully lever the alternator away from the engine to tension the belt and retighten the link nut and screw.

3 Recheck the tension.

4 When checking the belt tension, examine the belt for fraying, cuts or highly glazed areas on its sides which would indicate it has been slipping. If any of these conditions is evident, renew the belt in the following way.

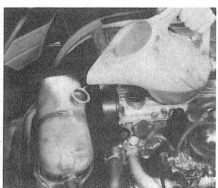
2.11 Filling the coolant system

4.3a Removing the thermostat housing cover

4.3b Removing the thermostat

4.5 Thermostat housing cover O-ring seal

5.3 Disconnecting radiator bottom hose

5.5 Radiator mounting bracket

5.6a Removing radiator side blanking panels

5.6b Removing the radiator

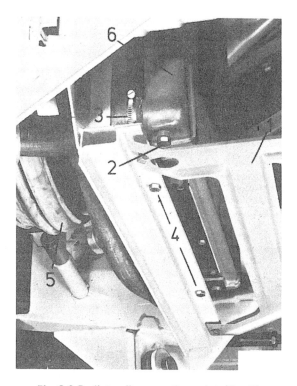

Fig. 2.2 Radiator disconnection points (Sec 5)

1	Protective shield	4	Bracket bolts
2	Drain plug	5	Flexible bellows
3	Bottom hose	6	Radiator

5   Release the alternator mounting and adjuster link bolts and nuts, and push the unit in towards the engine as far as it will go.

6   Slip the drivebelt from the pulleys and then manoeuvre it around and off the fan blades on the crankshaft.

7   Fit the new belt and tension it as previously described. Run the engine for between five and ten minutes, then recheck the tension and further adjust it if necessary.

8   Check the belt tension again after the first 1000 km (620 miles).

7   Coolant temperature sender and gauge – general

1   The temperature sender unit is screwed into the coolant elbow which is bolted to the rear face of the cylinder head (photo).

2   It is rare for the unit to fail. Its operation can only be satisfactorily checked using an ohmmeter. Make sure the connecting lead is securely attached.

3   The temperature gauge used on vehicles built before July 1976 incorporates its own voltage stabliser. Later models have an instrument voltage stabiliser which serves both the fuel and temperature gauges. The stabiliser can be suspected of being faulty if both instruments show incorrect readings at the same time.

8   Coolant pump – removal and refitting

1   Drain the cooling system as described in Section 2.

2   Remove the drivebelt as described in Section 6.

3   Remove the radiator as described in Section 5.

4   Disconnect the expansion tank hoses from the inlet manifold and the thermostat housing, and unbolt the tank mounting bracket.

5   Remove the expansion tank assembly.

6   Disconnect the coolant pump hoses, noting the anti-kink coil spring in the lower hose (photo).

7   Unbolt and remove the pulley from the coolant pump (photo).

8   Unbolt and remove the pump from the front face of the engine cylinder block. Peel off and discard the joint gasket.

7.1 Coolant temperature sender unit

8.6 Disconnecting coolant pump hoses

8.7 Removing coolant pump pulley

8.10 Refitting the coolant pump

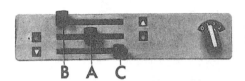

Fig. 2.3 Heater control panel (Sec 9)

A   Temperature control
B   Air distribution
C   Air distribution

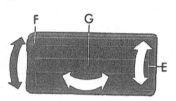

Fig. 2.4 Air outlet grille (Sec 9)

E   Vertical direction
F   Volume control
G   Horizontal direction

Fig. 2.5 Air exhaust control slide (Sec 9)

9.5 Air extractor

9   If the pump is noisy when its shaft is turned, or if it was removed due to a leak from the shaft seal, then it should be renewed. Repair kits are not available.
10   Refitting is a reversal of removal. Use a new joint gasket smeared on both sides with jointing compound (photo).
11   Tension the drivebelt.
12   Refill the cooling system.

9   Heater – description

1   The heater assembly is mounted just below the windscreen inside the driving cab.
2   It is a combination unit to provide both warm air (heated by the engine coolant) and fresh air from outside the vehicle in regulated volume.
3   An electric blower fan is incorporated to supplement airflow when the vehicle is stationary or when it is moving so slowly that the normal ram effect caused by the forward motion of the vehicle is insufficient

to give adequate heating or ventilation.
4   Air distribution is controlled by levers on a control panel, with the necessary air outlets being provided in the facia panel and at the footwells.
5   Stale air is exhausted through the door edges, flow control being provided by slides on the door interior trim panels (photo).

10   Heater – removal and refitting

Control panel and cables

1   Pull the knobs from the levers at the control panel. Remove the blower motor switch knob.
2   Remove the lever slide panel by depressing the retaining tabs behind it and pulling it from the facia panel. Disconnect the leads as it is removed.
3   Release the control cable conduit clips (photo).
4   Extract the two screws which secure the control lever mounting bracket (photo).

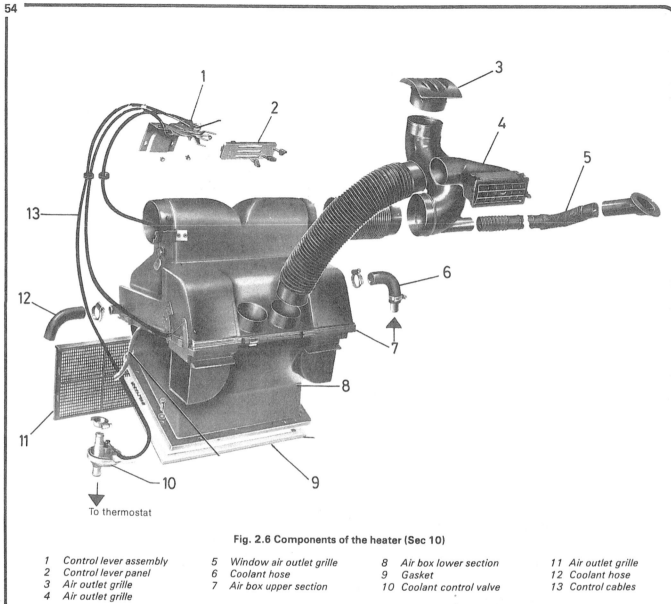

Fig. 2.6 Components of the heater (Sec 10)

To thermostat

1	Control lever assembly	5	Window air outlet grille
2	Control lever panel	6	Coolant hose
3	Air outlet grille	7	Air box upper section
4	Air outlet grille		

8	Air box lower section	11	Air outlet grille
9	Gasket	12	Coolant hose
10	Coolant control valve	13	Control cables

10.3 Heater control cable and clip

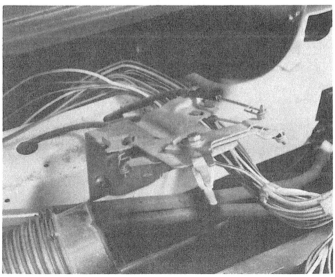

10.4 Heater control levers

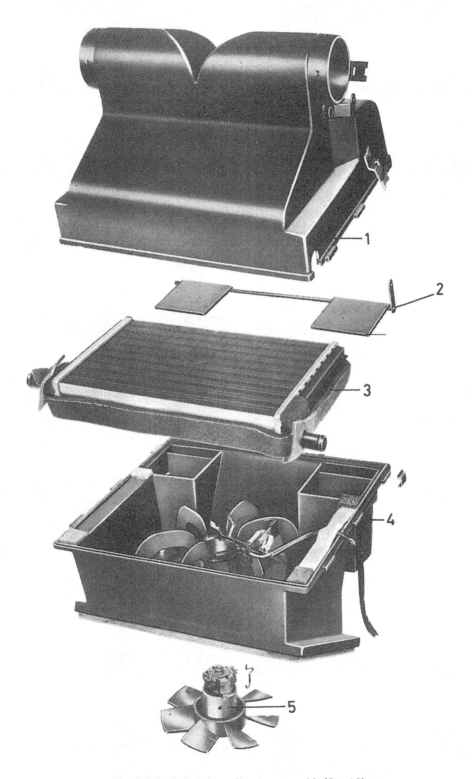

Fig. 2.7 Exploded view of heater assembly (Sec 10)

1 Air box upper section   4 Air box lower section
2 Air control flap   5 Blower motor
3 Matrix

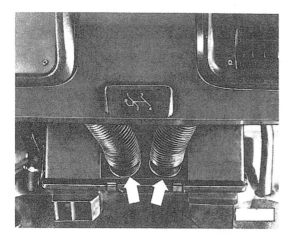

Fig. 2.8 Air ducts attached to upper section of air box (Sec 10)

10.13 Heater coolant control valve

5   Release the ends of the cables from the control levers and the heater flap levers and withdraw the cables from their grommets.
6   Refitting is a reversal of removal, but before clipping the cable conduits in position, set the control levers to off and the heater flap arms and coolant valve to the closed position.

Blower motor
7   Working under the facia panel, prise off the spring clips which hold the lower section of the air box to the upper section.
8   Lower the assembly until the electrical leads can be disconnected from the terminals of the blower motor.
9   The blower motor may now be removed from the air box.
10   Refitting is a reversal of removal.

Matrix
11   The heater matrix can only be removed after the heater has been withdrawn as an assembly.
12   Working under the facia panel, disconnect the convoluted air ducts from the stubs on the heater casing.
13   Drain the cooling system, as described in Section 2 and then disconnect the coolant hoses from the heater (photo).
14   Disconnect the control cables as previously described and disconnect the blower motor electrical harness plug and switch leads.
15   Extract the heater casing mounting screws and withdraw the complete assembly. Avoid spillage of coolant onto the cab floor (photo).
16   Remove the heater, prise off the spring clips which hold the casing sections together, remove the upper casing section and lift out the matrix.
17   If the matrix is clogged, try reverse flushing it with a cold water hose. If it is leaking, have it repaired professionally or exchange it for a new or rebuilt unit.

10.15 Releasing a heater mounting bolt

18   Reassembly and refitting are reversals of removal and dismantling.
19   Reconnect the control cables as previously described and refill the cooling system.

11 Fault diagnosis – cooling system

Symptom	Reason(s)
Overheating	Insufficient coolant in cooling system Drivebelt slipping (accompanied by a shrieking noise on rapid engine acceleration) Radiator core blocked or radiator grille restricted Bottom water hose collapsed, impeding flow Thermostat not opening properly Ignition advance and retard incorrectly set (accompanied by loss of power, and perhaps, misfiring) Carburettor incorrectly adjusted (mixture too weak) Exhaust system partially blocked Oil level in sump too low Blown cylinder head gasket (water/steam being forced down the radiator overflow pipe under pressure) Engine not yet run-in Brakes binding
Engine runs cool	Thermostat jammed open Incorrect grade of thermostat fitted allowing premature opening of valve Thermostat missing
Loss of cooling water	Loose clips on water hoses Water hoses perished and leaking Radiator core leaking Thermostat gasket leaking Expansion tank pressure cap spring worn or seal ineffective Blown cylinder head gasket (pressure in system forcing water/steam down overflow pipe) Cylinder wall or head cracked
Inefficient heater	Incorrect thermostat, or thermostat missing Control cables incorrectly adjusted Faulty or seized heater coolant valve Heater matrix blocked internally

Chapter 3
Fuel, exhaust and emission control systems

For modifications, and information applicable to later models, see Supplement at end of manual

Contents

Specifications

System type ....

Electric fuel pump (mechanical on very early models) with downdraught fixed jet carburettor

Carburettor data

	Solex 35PDSIT			
	060 129 015B	**160 129 015B**	**060 129 015E**	**060 129 015C**
Type				
Model No				
Date of introduction	8-75	4-76	8-77	8-75
From engine No	CH 000 001	CH 013 231	CH 032 935	CL 000 001
Venturi diameter	28 mm	28 mm	28 mm	28 mm
Main jet	162.5	162.5	140	160
Air correction jet/emulsion tube	145	145	85	155
Pilot air jet	200	200	120	200
Auxiliary fuel jet	57.5	57.5	50	57.5
Auxiliary air jet	1.04 mm	1.04 mm	1.04 mm	1.04 mm
Enrichment jet	60	60	60	60
Accelerator pump injection volume per stroke	1.98 to 2.2 cc	1.65 to 2.05 cc	1.65 to 2.05 cc	1.65 to 2.05 cc
Fuel inlet needle valve	1.5 mm	1.5 mm	1.5 mm	1.5 mm
Needle valve washer (thickness)	1.5 mm	1.5 mm	1.5 mm	1.5 mm
Throttle valve gap	0.90 to 1.00 mm	0.90 to 1.00 mm	1.15 to 1.25 mm	0.90 to 1.00 mm
Choke gap	4.5 to 5.5 mm	4.5 to 5.5 mm	3.8 to 4.2 mm	4.5 to 5.5 mm
Choke cover marking	128	128	128	128

Adjustment data

Idle speed	900 to 1000 rpm
CO level at idle	1.0 to 2.0 %

Fuel tank capacity ....

70 litres (15 Imp gallons)

Fuel octane rating ....

85 RON minimum (UK 2-star)

1 Description

The fuel system comprises a fuel tank, an electric fuel pump (some very early models were fitted with a mechanical fuel pump), a downdraught carburettor with automatic choke and an air cleaner with pre-heated air intake.

Emission control is limited to an exhaust afterburning system, which is described in Section 13.

2 Air cleaner – servicing, removal and refitting

1 At the intervals specified in Routine Maintenance, remove the engine cover and unscrew the nut from the air cleaner end cover plate (photo).

2 Remove the cover plate.

3 Unscrew the element retaining nut and withdraw the element (photos).

4 Using compressed air, blow the dirt from the element by applying the air pressure to its inside surfaces.

5 Wipe out the air cleaner casing and refit the element.

6 At the service intervals for element renewal, discard the dirty unit and when fitting the new one, check that the sealing gasket is in good order.

7 To remove the air cleaner complete, disconnect the preheated air duct from the collector plate on the exhaust manifold.

8 Disconnect the air intake ducts from the carburettor and from the

2.1 Unscrewing air cleaner end cover plate nut

2.3a Unscrewing air cleaner element retaining nut

2.3b Withdrawing air cleaner element

2.8a Disconnecting air cleaner cold air duct

2.8b Disconnecting carburettor air intake duct

2.9 Air cleaner fixing clip

3.1 Fuel filter (later models)

cold air spout (photos).

9   Release the air cleaner mounting clip and remove the cleaner from the engine compartment (photo).

10 Refitting is a reversal of removal.

3   Fuel filter – renewal

1   This is a disposable type filter and it is simply a matter of disconnecting the hose clips, removing the old unit and installing the new one. Take care that the directional arrow on the filter matches the direction of the fuel flow (tank to pump) (photo).

2   Start the engine and check for leaks after installing a new filter.

4   Fuel pump (mechanical) – removal and refitting

1   Disconnect and plug the hoses from the pump.

2   Unbolt and remove the pump from the left-hand side of the cylinder head.

3   Withdraw the operating rod which bears against the eccentric cam on the camshaft.

4   Clean away the old joint gasket.

5   Refitting is the reversal of removal. Use a new gasket.

5   Fuel pump (electric) – removal and refitting

Up to Chassis No 286 2506 977

1   Disconnect the battery negative lead.

2   Disconnect the pump electrical connections.

3   Note the hose attachment points and then disconnect and plug them.

4   Remove the pump mounting screws and remove the pump.

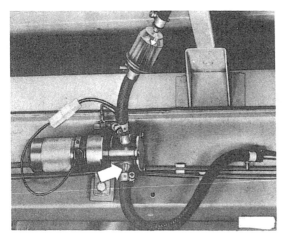

Fig. 3.1 Typical early type electrical fuel pump showing in-line filter (Sec 5)

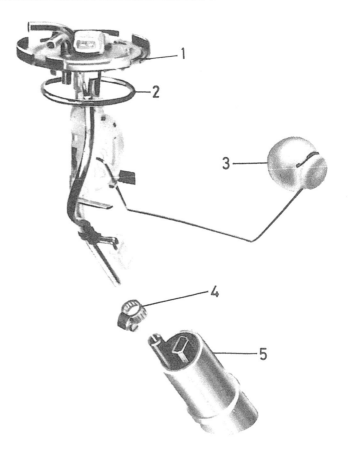

Fig. 3.2 Later type fuel pump combined with tank sender unit (Sec 5)

1	Sender unit	4	Worm drive clip
2	Seal	5	Fuel pump
3	Float		

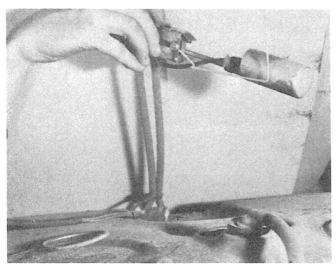

5.6 Removing fuel tank sender unit

From Chassis No 286 2506 978

5   The fuel pump on these models is combined with the fuel tank level sender unit.

6   Disconnect the electrical leads and fuel hoses, and remove the sender unit pump from the fuel tank. To do this, use the special wrench (2012) or a piece of flat steel plate located between two opposite cutouts in the rim of the sender unit (photo).

All models

7   Refitting both types of pump is a reversal of removal. Use a new worm drive hose clip to attach the later type pump to the sender unit, and make sure that the sender unit sealing ring is in good order.

---

6   Fuel level sender unit – removal and refitting

If both the fuel and the temperature gauges give inaccurate readings, their common voltage stabiliser may be at fault. Refer to Chapters 2 and 10 for further details.

1   On sender units without a fuel pump attached, the fuel hose and electrical connections are as shown in Fig. 3.3.

2   On later models combined with a pump, the connections include an electrical plug and three fuel pipe stubs (Fig. 3.4).

3   Removal of both types of sender unit is as described in paragraph 6 of the preceding Section, but on vehicles with a van type body, the

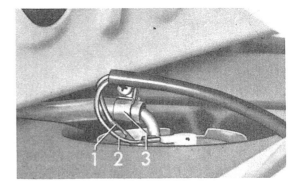

Fig. 3.3 Tank sender unit connections (early type pump) (Sec 6)

1	Earth	3	Fuel hose clip
2	To fuel gauge		

sender unit can only be removed after first having withdrawn the fuel tank (see Section 7).

4   Refitting is a reversal of the removal procedure.

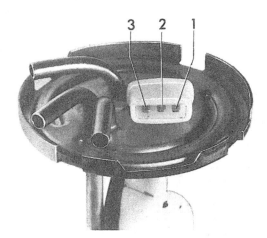

Fig. 3.4 Tank sender unit connections (later type pump) (Sec 6)

1   To fuel gauge   3   Earth
2   Fuel pump (positive)

---

7   Fuel tank – removal, servicing and refitting

1   Disconnecting the lead from the battery negative terminal.
2   If the vehicle is fitted with a platform type body, remove the spare wheel, unbolt the spare wheel holder and swivel it to one side.
3   Drain the fuel tank either by syphoning the fuel into a container or by disconnecting the fuel outlet hose from the pump and switching on the ignition, again catching the ejected fuel in a container which can be sealed.
4   Disconnect the fuel tank filter pipe (photo).
5   Disconnect the tank breather hoses.
6   Disconnect the fuel inlet and (where fitted) return hoses.
7   Disconnect the electrical leads from the tank sender unit.
8   Unbolt the tank and remove it from the vehicle (photo).
9   If the tank is leaking, repair is a job for specialists. Never attempt to weld or solder a fuel tank unless it has been thoroughly purged by steam cleaning first. Temporary repair of small holes may be possible using a proprietary compound.
10 Removal of deposits of sediment, sludge or water can be carried out by pouring in some paraffin and shaking the tank vigorously, but remove the sender unit first to prevent it being damaged. Repeat as necessary with changes of paraffin until the interior of the tank is clean.
11 Refitting is a reversal of removal.

---

8   Carburettor – idle and mixture adjustment

1   The mixture screw on these carburettors is sealed after the mixture setting has been carried out during production. Normally the only adjustment required is to turn the idle speed screw in or out to bring the idle speed within the specified range.
2   Due to changes in operating characteristics of the engine due to wear and carbon build-up, it may be necessary to vary the mixture setting. It will also be necessary to reset it completely after carburettor overhaul. Both operations are covered in the following paragraphs.

Idle speed

3   Have the engine at normal operating temperature.
4   Connect an accurate tachometer to the engine, start the engine and allow it to idle.
5   Turn the idle speed screw (shown in Fig. 3.5) in or out until the engine idle speed is within the specified limits.

Idle mixture

6   The most accurate way to adjust the fuel/air mixture is to use an exhaust gas analyser, with its probe inserted into the exhaust tailpipe and other connections made in accordance with the manufacturer's instructions.
7   Make sure that the engine is still at normal operating temperature, pull off the crankcase breather hose and have all electrical components switched off.
8   On models with an exhaust afterburning system (see Section 13), either pinch the hose at the connecting hose using a pair of self-locking grips, or disconnect the hose and plug it.
9   Break off the tamperproof cap from the idle mixture screw. Satisfy yourself that you are not breaking local or national laws by so doing.
10 With the engine idling at the specified speed, turn the mixture screw in or out until the CO level indicated on the analyser is within the specified tolerance.
11 When the adjustment is correct, remove the test equipment and switch off the engine. Fit a new blue tamperproof cap to the idle mixture screw where this is required by local regulations.
12 Reconnect the crankcase breather hose.

---

9   Carburettor – removal and refitting

1   Remove the engine cover.
2   Remove the air cleaner intake duct from the carburettor.
3   Disconnect the electrical lead from the carburettor automatic choke and the bypass air solenoid cut-off valve. Disconnect the distributor vacuum pipe.
4   Disconnect the fuel hose and plug it.
5   Disconnect the throttle cable from the lever on the carburettor (photo).
6   Unscrew and remove the carburettor mounting nuts.
7   Remove the carburettor from the intake manifold, peel away the flange gaskets and discard them.
8   Refitting is a reversal of removal. Always use new joint gaskets.

---

10   Carburettor – overhaul

1   Complete overhaul of the carburettor should not be required unless the unit has seen considerable service and some components are obviously worn.

7.4 Fuel tank filler pipe

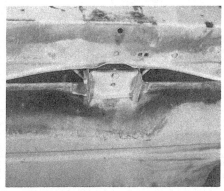

7.8 Fuel tank rear mounting

9.5 Disconnecting throttle control from carburettor

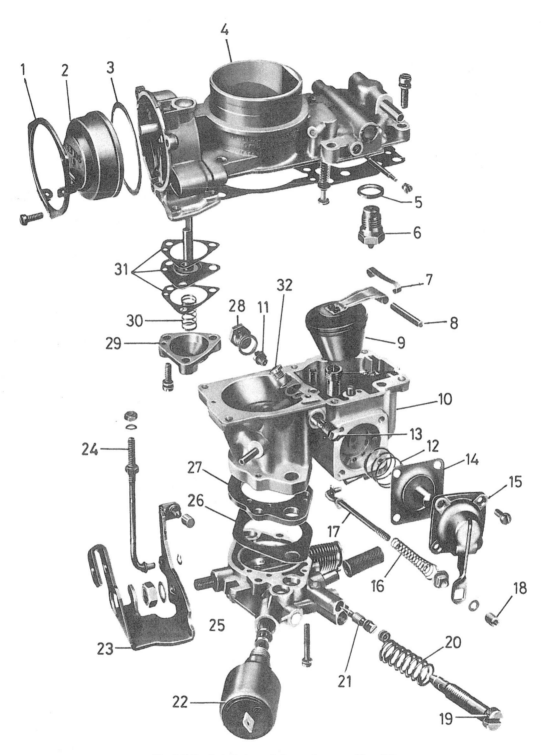

Fig. 3.5 Exploded view of the carburettor (Sec 10)

1	Retainer	10	Main body	18	Pump rod nut	25	Throttle valve block
2	Choke housing cover	11	Main jet	19	Idle speed screw	26	Gasket
3	Gasket	12	Spring	20	Spring	27	Insulator
4	Top cover	13	Pilot jet	21	Mixture screw	28	Main jet plug
5	Washer	14	Accelerator pump diaphragm	22	Air cut-off solenoid valve	29	Choke cover
6	Fuel inlet needle valve	15	Diaphragm cover	23	Throttle lever	30	Spring
7	Pivot spring	16	Spring	24	Link rod	31	Gaskets and shims
8	Pivot pin	17	Accelerator pump rod			32	Air correction jet
9	Float						

2   Generally, removal of the carburettor cover will be sufficient to enable sediment to be mopped out of the fuel bowl and the jets removed and blown through with compressed air.

3   Where complete dismantling is to be carried out, remove the carburettor from the engine as described in the preceding Section.

4   Unscrew the cover securing screws progressively one half turn at a time.

5   Remove the cover and the joint gasket.

6   Prise out the float pivot pin spring clip, remove the float and pivot pin.

7   Tip out the fuel remaining in the bowl and wipe the bowl clean.

8   To remove the throttle valve plate block, invert the carburettor and extract the screws which hold the throttle block to the carburettor body.

9   With the carburettor now separated into its three major sections, additional dismantling can be carried out according to need.

Top cover

10 Extract the screws from the choke housing cover retaining ring and remove the choke components.

11 Unscrew the fuel inlet needle valve and remove it with its sealing washer.

Body

12 Unscrew and remove all the jets, using close-fitting spanners or screwdrivers.

13 If there is any suspicion that the jets may have been changed by a previous owner for ones of incorrect calibration, check the jets against those listed in the Specifications. Clean blocked jets with compressed air or solvent – do not probe them with wire.

14 Unscrew and remove the four screws from the accelerator pump diaphragm cover, remove the cover and take out the diaphragm. If the diaphragm is split or hardened, renew it.

Throttle valve block

15 If the throttle valve plate spindle is worn, dismantling is not advised but rather obtain a new assembly complete. Wear in the throttle valve plate spindle can often cause rough or irregular slow running.

Reassembly

16 Reassembly is a reversal of dismantling, but obtain a repair kit which will contain all the necessary gaskets and other renewable items.

17 As reassembly progresses, carry out the adjustments described in the following Section.

---

11 Carburettor – setting and adjusting

1   The following adjustments will normally only be required after the carburettor has been overhauled and one or more of the original settings disturbed.

Automatic choke

2   The choke housing cover retaining screws should only be tightened after the marks on cover and housing have been aligned. Carry out the choke valve adjustment procedure first if necessary.

Accelerator pump

3   The stroke of the pump and consequent volume of fuel ejected is adjusted by turning the nut on the pump rod.

4   The volume of fuel ejected per stroke can be estimated if a small bore flexible tube is pushed over the pump ejector nozzle to direct the fuel into a suitable measuring glass. A more accurate figure will be obtained if the throttle is opened fully five times and the subsequent volume divided by five to obtain an average figure.

5   On completion, seal the nut to the rod threads with a dab of paint or other sealant.

6   Check that the spout of the fuel injector nozzle directs the fuel into the gap between the edge of the throttle valve plate and the venturi wall. If it does not, bend it carefully.

Choke valve

7   Remove the cover from the automatic choke housing.

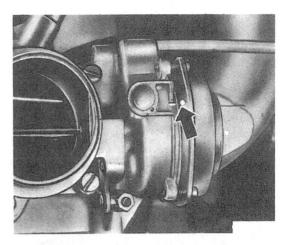

Fig. 3.6 Automatic choke housing and cover alignment marks (Sec 11)

Fig. 3.7 Accelerator pump rod adjuster nut (Sec 11)

a   Decreases volume ejected   *b   Increases volume*

8   Open the throttle to the half-way position, close the choke valve plate with the finger and then release the throttle.

9   Using a screwdriver, press the diapragm pullrod onto its stop and measure the gap between the edge of the choke valve plate and the venturi wall. A twist drill is useful for this.

10 If the gap is outside the specified tolerance, loosen the locknut under the coverplate and turn the adjuster screw as necessary. Retighten the locknut on completion.

Throttle valve

11 Close the choke valve plate with the finger, open the throttle to about the half-way position and then release it.

12 Check the gap between the edge of the throttle valve plate and the venturi wall. A twist drill is useful for this purpose.

13 Where the gap is outside the specified tolerance, adjust the position of the nut on the link rod (Fig. 3.10). Lock the nuts to the threads on the rod on completion.

Adjustments on vehicle

14 The following two adjustments can only be carried out after the carburettor has been refitted to the engine and the engine has reached normal operating temperature.

Throttle valve stop screw

15 This screw is fitted with a tamperproof cap, and should not be altered from its factory-set position unless new components have been fitted.

16 Pull off the distributor vacuum advance hose from the carburettor and substitute a vacuum gauge.

Fig. 3.8 Checking choke valve plate gap (Sec 11)

Fig. 3.10 Checking throttle valve plate gap. Link rod adjuster nut arrowed (Sec 11)

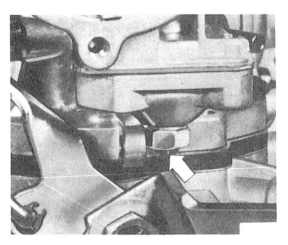

Fig. 3.9 Choke valve gap adjuster screw and locknut (arrowed) (Sec 11)

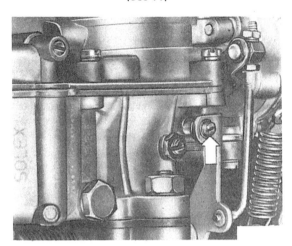

Fig. 3.11 Throttle valve stop screw (Sec 11)

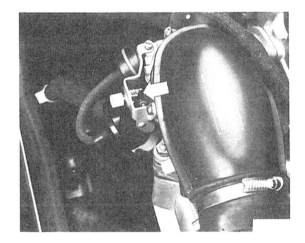

Fig. 3.12 Pressing dashpot control lever (arrowed) against adjusting screw (Sec 11)

17 Start the engine and allow it to idle.

18 Turn the stop screw until a vacuum reading is indicated on the gauge, then unscrew it until the vacuum reading drops to zero. From this setting, unscrew the screw a further quarter of a turn.

19 Adjust the idle speed and mixture as described in Section 8. Fit a new tamperproof cap to the screw.

Dashpot and delay valve (from engine No CH 032 935)

20 With the engine idling, press the control lever against the adjusting screw using light hand pressure.

21 With a tachometer connected, check that the engine speed is between 1350 and 1450 rpm. If not, turn the adjusting screw as necessary.

22 Increase the engine speed to 3000 rpm. The control lever on the dashpot should be pulled into contact with the adjuster screw. As the engine speed is reduced, the control lever should move slowly off the adjusting screw and the throttle close fully.

23 If the delay valve fitted in conjunction with these carburettors is removed, make sure that when refitting it, the connection marked with white is attached to the carburettor.

12 Accelerator cable – adjusting

1 With the accelerator pedal fully depressed, there should be a clearance between the throttle lever and the stop on the carburettor of between 1.0 and 1.5 mm (0.039 and 0.060 in) (Fig. 3.14).

2 Any adjustment required should be made by moving the setting of the cable at its clamp on the carburettor.

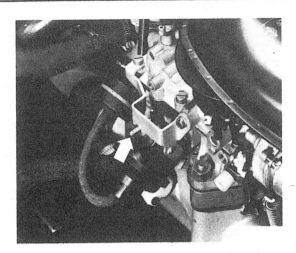

Fig. 3.13 Dashpot adjusting screw (Sec 11)

Fig. 3.14 Throttle lever to stop clearance (9) (Sec 12)

Fig. 3.15 Exhaust afterburning hoses (1) and non-return valves (2) (Sec 13)

distorted, these conditions are due to a fault in the non-return valve. Renew all affected components.

4   Routine maintenance consists of occasionally cleaning out the hoses and checking their connections.

5   Whenever the cylinder head is removed pay particular attention to removal of carbon and other combustion deposits from the channels and passages exposed once the intake manifold has been withdrawn.

13 Exhaust afterburning system – description and maintenance

1   This is really an air injection system in which fresh air is drawn into the intake manifold, then through passages in the cylinder head to mix with gases at the exhaust ports.

2   This causes afterburning of the exhaust gases and reduces the CO-HC content of the exhaust emissions.

3   Inspect the condition of the non-return valves and hoses regularly. If the valves are seen to be blue in colour, or the hoses hard or

14 Manifolds and exhaust system – general

1   The intake manifold is bolted to the left-hand side of the cylinder head.

2   Before it can be removed, the carburettor must be disconnected by uncoupling the throttle cable, electrical leads and fuel hoses.

3   As the intake manifold is coolant-heated, the cooling system must first be drained and the coolant hoses disconnected from it (photo).

4   Disconnect the brake vacuum servo hose and the accelerator cable support bracket (photo).

5   If the intake manifold is removed complete with carburettor, the carburettor can be removed later as described in Section 9.

6   The exhaust manifold is of cast iron construction and located on the right-hand side of the cylinder head. A hot air collector is mounted on the top face of the manifold to provide a source of warmed air for the air cleaner.

7   The exhaust system comprises twin downpipes, a front expansion box, and a main silencer at the rear.

8   Although the system can be separated into four sections, it is recommended that the complete system is withdrawn from the vehicle for renewal of any one section. To release a corroded socket type pipe joint on the vehicle can prove very difficult and may cause damage to adjacent good sections or components.

9   All exhaust system mountings are of flexible type (photo).

14.3 Inlet manifold connecting hose

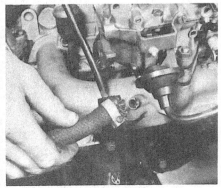

14.4 Brake vacuum servo hose

14.9 Exhaust flexible mounting

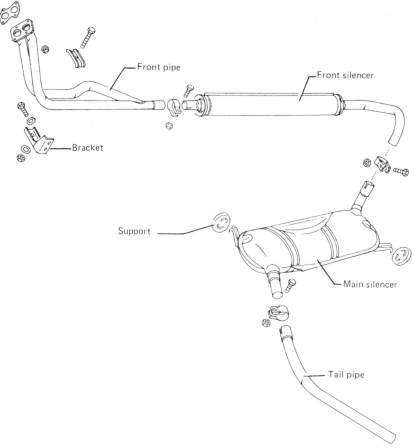

Fig. 3.16 Typical exhaust system (Sec 13)

15 Fault diagnosis – fuel system

Note: Excessive fuel consumption and poor performance are not always due to faults in the fuel system or carburettor. Before attempting to diagnose faults from the table below, make sure that the ignition system is in good condition and properly adjusted, that the tyres are correctly inflated and the brakes not binding, and that the engine is in good mechanical condition.

Symptom	Reason(s)
Excessive fuel consumption	Leakage from tank, pump or lines Float chamber flooding Air cleaner dirty Air cleaner pre-heater malfunctioning Carburettor wrongly adjusted Automatic choke malfunctioning Carburettor generally worn
Insufficient fuel delivery	Insufficient fuel in tank! Fuel filter clogged Fuel pump defective Fuel pipe leaking on suction side of pump
Weak mixture	Air leak at manifold or carburettor Air cleaner element missing Carburettor maladjusted or jets blocked
Backfiring in exhaust system	Air leak in exhaust system Fault in exhaust afterburning system Mixture grossly incorrect Exhaust valve burnt or not seating

Electrode gap check – use a wire type gauge for best results.

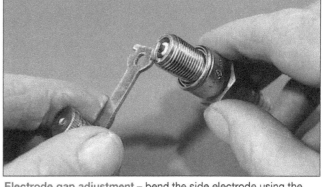

Electrode gap adjustment – bend the side electrode using the correct tool.

Normal condition – A brown, tan or grey firing end indicates that the engine is in good condition and that the plug type is correct.

Ash deposits – Light brown deposits encrusted on the electrodes and insulator, leading to misfire and hesitation. Caused by excessive amounts of oil in the combustion chamber or poor quality fuel/oil.

Carbon fouling – Dry, black sooty deposits leading to misfire and weak spark. Caused by an over-rich fuel/air mixture, faulty choke operation or blocked air filter.

Oil fouling – Wet oily deposits leading to misfire and weak spark. Caused by oil leakage past piston rings or valve guides (4-stroke engine), or excess lubricant (2-stroke engine).

Overheating – A blistered white insulator and glazed electrodes. Caused by ignition system fault, incorrect fuel, or cooling system fault.

Worn plug – Worn electrodes will cause poor starting in damp or cold conditions and will also waste fuel.

Chapter 4 Ignition system

For modifications, and information applicable to later models, see Supplement at end of manual

Contents

Specifications

General

System type .... 12V, negative earth, coil and mechanical contact breaker distributor
Firing order 1–3–4–2
Location of No 1 cylinder .... Timing belt end

Distributor

Direction of rotor rotation .... Clockwise viewed from above
Contact breaker gap .... 0.45 mm (0.018 in)
Dwell angle .... 44° to 50° (50 to 56%)

Ignition timing

Dynamic timing .... 5° BTDC at 850 to 950 engine rpm
Centrifugal advance:
    Commences .... 1100 to 1500 engine rpm
    12° to 16° .... 2400 rpm
    23° to 27° .... 3400 rpm
    Maximum (32° to 36°) .... 4200 rpm
Vacuum advance:
    Commences .... 100 mm Hg
    Maximum (17° to 21°) .... 310 mm Hg

Spark plugs

Type .... Champion N7YC, Bosch W200 T30, Beru 200/14/3A, or equivalent
Electrode gap .... 0.6 to 0.7 mm (0.024 to 0.028 in)

Torque wrench setting

	Nm	lbf ft
Spark plugs	30	22

1 General description

In order that the engine can run correctly it is necessary for an electrical spark to ignite the fuel/air mixture in the combustion chamber at exactly the right moment in relation to engine speed and load.

The ignition system is divided into two circuits, low tension and high tension.

The low tension (LT or primary) circuit, consists of the battery ignition switch, low tension or primary coil windings, and the contact breaker points and condenser, both located at the distributor.

The high tension (HT or secondary) circuit, consists of the high tension or secondary coil winding, the heavy ignition lead from the centre of the coil to the distributor cap, and the rotor arm and the spark plug leads.

The ignition system is based on feeding low tension voltage from the battery to the coil where it is converted to high tension voltage. The high tension voltage is powerful enough to jump the spark plug gap in the cylinders many times a second under high compression pressures, providing that the system is in good condition and that all adjustments are correct.

The wiring harness on most models includes a high resistance wire in the ignition coil feed circuit and it is very important that only a 'ballast resistor' type coil is fitted as a replacement. During starting this ballast resistor wire is by-passed, allowing the full available battery voltage to be fed to the coil. This ensures that during cold starting, when the starter motor current demand is high, sufficient voltage is still available at the coil to produce a powerful spark. Under normal running the 12 volt supply is directed through the ballast resistor before reaching the coil.

The ignition advance is controlled both mechanically and by vacuum, to ensure that the spark occurs at just the right instant for the particular engine load and speed. The mechanical governor comprises two weights, which move out from the distributor shaft as the engine speed rises, due to centrifugal force.

The vacuum control consists of a diaphragm, one side of which is connected via a small bore tube to the carburettor, and the other side to the contact breaker plate. Depression in the inlet manifold and carburettor, which varies with engine speed and throttle opening, causes the diaphragm to move, so moving the contact breaker plate, and advancing or retarding the spark.

2 Contact breaker points – renewal

1 At the intervals specified in Routine Maintenance, the points should be renewed as they will have deteriorated due to erosion and arcing of their contact faces.

2 Remove the engine cover.

3 Release the spring retainers and remove the distributor cap and move it to one side. Pull off the rotor arm.

4 Pull the electrical lead from the LT terminal.

5 Unscrew and remove the screw which holds the fixed contact breaker arm to the baseplate.

6 Lift the contact set from the distributor.

7 Wipe away any oil and dirt from the inside of the distributor and the distributor cap. Check that the contacts in the cap are not severely eroded and that the centre carbon pick-up is in good condition. Examine the cap closely for cracks. If any of these conditions are evident, renew the cap.

8 Fit the new contact breaker set, but leave the fixing screw loose.

9 Turn the crankshaft until the plastic heel of the movable contact arm is central on one of the high points of the distributor cam.

10 Now adjust the points gap to that specified by moving the fixed contact arm. A screwdriver notch is provided in the fixed arm. It is emphasised that this is only a basic setting to enable the engine to be started. A final adjustment must be carried out using a dwell meter as described in Section 3.

11 Apply two drops of engine oil to the felt pad on top of the distributor shaft and one drop to the movable arm pivot. Apply a smear of high melting point grease to the high points of the cam.

12 Fit the rotor arm and the distributor cap.

3 Dwell angle – checking and adjusting

1 On modern engines, setting the contact breaker points gap using feeler gauges must be regarded as a basic adjustment only. For optimum engine performance, the dwell angle must be checked and adjusted.

2 The dwell angle is the number of degrees through which the distributor cam turns during the period between the instants of closure and opening of the contact breaker points. Checking the dwell angle not only gives a more accurate setting of the contact breaker points gap, but this method also evens out any variations in the gap which could be caused by wear in the distributor shaft or its bushes, or differences in height of any of the cam peaks.

3 The angle should be checked with a dwell meter connected in accordance with the maker's instructions. Refer to the Specifications for the correct dwell angle. Some dwell meters operate with the engine idling, others with the engine being cranked on the starter motor.

4 If the dwell angle is found to be too large, increase the points gap; if it is too small, reduce the points gap.

5 The dwell angle should be adjusted *before* adjusting the ignition timing. The ignition timing must always be checked *after* adjusting the dwell angle.

4 Ignition timing

1 This should be carried out using a timing light (stroboscope) connected in accordance with the manufacturer's instructions.

2 Have the engine at normal operating temperature and idling at between 850 and 950 rpm.

3 Pull the vacuum hose from the distributor and plug it.

4 Point the timing light at the aperture close to the clutch bellhousing flange. The line on the flywheel should appear in alignment with the pointer on the bellhousing.

5 If it is not, release the distributor clamp screw and turn the distributor gently until it is in alignment. Retighten the clamp screw.

6 The ignition is now correctly set with the engine timed to its specified BTDC firing point. The O mark on the flywheel is the TDC mark and should be ignored.

7 Switch off the engine, remove the stroboscope and reconnect the vacuum pipe.

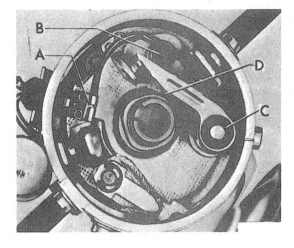

Fig. 4.1 Distributor with cap removed (Sec 2)

A   LT terminal   C   Breaker arm pivot
B   Contact breaker screw   D   Breaker arm plastic head

Fig. 4.2 Adjusting points gap with screwdriver (Sec 2)

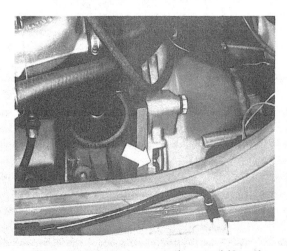

Fig. 4.3 Ignition timing marks (arrowed) (Sec 4)

5 Condenser – removal, testing and refitting

1　The purpose of the condenser (sometimes known as a capacitor) is to ensure that when the contact breaker points open there is no sparking across them which would waste voltage and cause wear.

2　The condenser is fitted in parallel with the contact breaker points. If it develops an internal fault, it will cause ignition failure as the contact breaker points will be prevented from correctly interrupting the low tension circuit.

3　If the engine becomes very difficult to start or begins to miss after several miles of running and the breaker points show signs of excessive burning, then the condition of the condenser must be suspect. One further test can be made by separating the points by hand with the ignition switched on. If this is accompanied by a bright flash, it is indicative that the condenser has failed.

4　Without special test equipment the only safe way to diagnose condenser trouble is to replace a suspected unit with a new one and note if there is any improvement.

5　To remove the condenser from the distributor, take off the distributor cap and rotor arm.

6　Disconnect the condenser lead and release the condenser from the outside of the distributor body.

7　Refitting is a reversal of removal.

6 Distributor – removal and refitting

1　Mark the base of the distributor in relation to the distributor drive housing.

2　Release and remove the distributor clamp screw and the clamp. Release the cap and move it to one side without disconnecting the HT leads.

3　Withdraw the distributor.

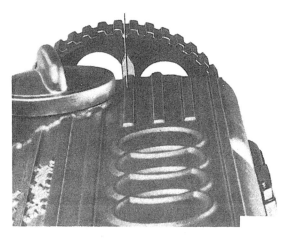

Fig. 4.4 Camshaft sprocket alignment (Sec 6)

4　To refit the distributor, first set the engine so that No 1 piston is at its BTDC firing point. If the engine is in the vehicle this will be achieved when the line on the flywheel is in alignment with the pointer in the clutch bellhousing cut-out. If the engine is out of the vehicle, having been removed for overhaul or exchanged, then establish the firing position by aligning the notch on the crankshaft pulley 5° in advance of the pointer on the front of the oil pump. With either set of alignment marks, the mark on the rear face of the camshaft sprocket must also be in alignment with the pointer on the camshaft cover (Fig. 4.4).

5　Hold the distributor over its drive housing hole so that the alignment marks made before removal are opposite to each other. Now turn the rotor arm so that its contact end is at the No 1 firing position. This is usually marked with a notch in the rim of the distributor body, if not, estimate this by reference to the distributor cap (photo).

6　Due to the fact that the rotor will turn anti-clockwise as the distributor is installed on account of the meshing of the gears, the rotor must now be rotated a few degrees clockwise in anticipation if it is to end up in the correct position (photos).

7　Install the distributor, turn it slightly until the contact points are just about to open and then fit and tighten the clamp.

8　Refit the distributor cap.

9　Check the ignition timing as described in Section 4.

7 Distributor – overhaul

1　The distributor should not be dismantled unnecessarily, in fact a well worn unit is best exchanged for a new or factory rebuilt assembly.

2　Where overhaul is decided upon, it is worthwhile checking the availability of spares beforehand.

3　With the distributor removed from the engine, take off the cap and rotor arm.

4　Remove the externally mounted condenser, and the contact breaker points.

5　Remove the diaphragm unit mounting screws and then take off the E-clip which holds the diaphragm operating rod to the pivot post on the baseplate.

6　Tilt the diaphragm unit and withdraw it, with the operating rod, from the distributor.

7　Extract the screws which hold the baseplate to the distributor body and remove the baseplate.

8　Weak or broken counterweight springs may be renewed, but any more extensive wear in the distributor shaft or bushes should be overcome by renewal of the complete assembly.

9　Apply some engine oil to the counterweight pivots and reassemble by reversing the dismantling operations.

8 Coil – general

1　High tension current should be negative at the spark plug terminals. Check that the LT lead from the distributor connects with the negative (–) terminal on the coil (photo).

2　Some models have a ballast resistor incorporated in the coil circuit. The function of this resistor is described in Section 1.

6.5 Distributor rotor alignment mark

6.6a Rotor position prior to installation

6.6b Rotor position after installation

Fig. 4.5 Distributor installed (No 1 piston at firing point) (Sec 6)

3 The ballast resistor used with some coils may be either externally mounted on the coil or built into the wiring lead.
4 Without special equipment, the best method of testing for a faulty coil is by substitution of a new unit. Before doing this however, check the security of the connecting leads and remove any corrosion which may have built up in the coil HT socket.

9 Spark plugs and HT leads – general

1 The correct functioning of the spark plugs is vital for the correct running and efficiency of the engine.
2 At intervals of 8000km (5000 miles) the plugs should be removed, examined, cleaned, and if worn excessively, renewed. The condition of the spark plugs will also tell much about the overall condition of the engine (see illustrations on page 67).
3 If the insulator nose of the spark plug is clean and white, with no deposits, this is indicative of a weak mixture, or too hot a plug (a hot plug transfers heat away from the electrode slowly – a cold plug transfers it away quickly).
4 If the tip and insulator nose are covered with hard black looking deposits, then this is indicative that the mixture is too rich. Should the plug be black and oily, then it is likely that the engine is fairly worn, as well as the mixture being too rich.
5 If the insulator nose is covered with light tan to greyish-brown deposits, then the mixture is correct and it is likely that the engine is in good condition.

8.1 Ignition coil

6 If there are any traces of long brown tapering stains on the outside of the white portion of the plug, then the plug will have to be renewed, as this shows that there is a faulty joint between the plug body and the insulator, and compression is being allowed to leak away.
7 Plugs should be cleaned by a sand blasting machine which will free them from carbon more thoroughly than cleaning by hand. The machine will also test the condition of the plugs under compression. Any plug that fails to spark at the recommended pressure should be renewed.
8 The spark plug gap is of considerable importance, as, if it is too large or too small, the size of the spark and its efficiency will be seriously impaired. The spark plug gap should be set to the figure given in the Specifications at the beginning of this Chapter.
9 To set it, measure the gap with a feeler gauge, and then bend open, or close, the outer plug electrode until the correct gap is achieved. The centre electrode should never be bent as this may crack the insulation and cause plug failure if nothing worse.
10 When renewing the plugs, remember to refit the leads from the distributor in the correct firing order.
11 The plug leads require no routine attention other than being kept clean and wiped over regularly.
12 At intervals of 8000km (5000 miles) however, pull the leads off the plugs and distributor one at a time and make sure no water has

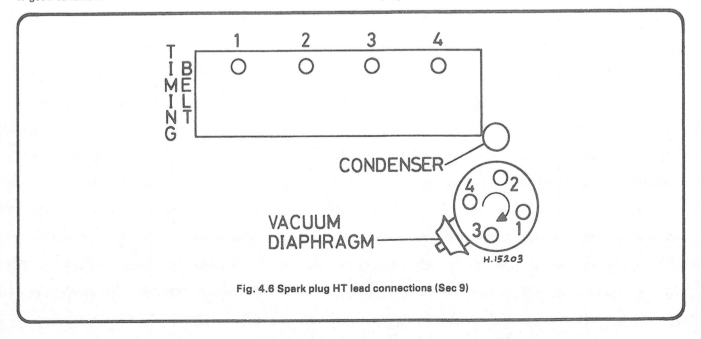

Fig. 4.6 Spark plug HT lead connections (Sec 9)

found its way onto the connections. Remove any corrosion from the end fittings, wipe the collars on top of the distributor, and refit the leads.

13 Every 16 000km (10 000 miles) it is recommended that the spark plugs are renewed to maintain optimum engine performance.

14 All engines are fitted with carbon cored HT leads. These should be removed from the spark plugs by gripping their rubber end covers.

Provided the leads are not bent in a tight loop and compressed there is no reason why this type of lead should fail. A legend has arisen which blames this type of lead for all ignition faults and many owners replace them with the older copper cored type and install separate suppressors. In the majority of cases, it would be more profitable to establish the real cause of the trouble before going to the expense of new leads.

10 Fault diagnosis – ignition system

Symptom	Reason(s)
Engine fails to start	Loose battery connections
	Discharged battery
	Oil on points
	Disconnected leads
	Faulty condenser
	Damp HT leads or interior of distributor cap
	Faulty anti-run-on solenoid
	Fuel system fault
Engine starts and runs but misfires	Faulty spark plug
	Cracked distributor cap
	Cracked rotor arm
	Worn advance mechanism
	Incorrect spark plug gap
	Incorrect points gap
	Faulty condenser
	Faulty coil
	Incorrect timing
	Poor earth connections
Engine overheats, lacks power	Seized centrifugal weights
	Perforated vacuum pipe
	Incorrect ignition timing
Engine 'pinks'	Timing too advanced
	Advance mechanism stuck in advance position
	Broken centrifugal weight spring
	Low fuel octane rating
Engine fires but will not run	Ballast resistor open-circuit (when applicable)
	Fuel system fault

Chapter 5 Clutch

For modifications, and information applicable to later models, see Supplement at end of manual

Contents

Specifications

General
Type	Single dry plate with diaphragm spring. Cable actuation
Release bearing	Grease sealed ball

Driven plate diameter
215 mm (8.5 in)

Pedal free movement
20.0 mm (0.79 in)

Torque wrench settings
	Nm	lbf ft
Clutch cover bolts	25	18

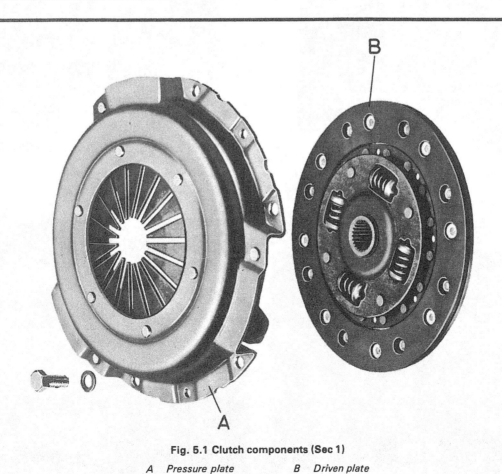

Fig. 5.1 Clutch components (Sec 1)

A   *Pressure plate* B   *Driven plate*

1 Description

The clutch is of single dry plate type, with a diaphragm spring pressure plate.

The release bearing is of sealed ball type.

Actuation is by means of a cable.

2 Clutch – adjustment

1   At the intervals specified in Routine Maintenance at the beginning of this manual, check the clutch pedal free movement. If the pedal is depressed with the fingers it should be felt to travel through the distance specified before resistance is noted to indicate that the clutch is being actuated.

2   Where the free travel is found to be incorrect, release the locknuts at the gearbox lug and adjust their position.

3   Recheck the pedal free movement and when it is correct, retighten the nuts.

3 Clutch cable – renewal

1   Slacken off the tension in the clutch cable by releasing the nuts at the gearbox lug.

2   Once the cable is slackened, unhook its end fitting from the clutch release lever (photos).

3   Pull the coil spring and the dust excluder from the cable.

4   Remove the nut nearest the end of the cable and draw the cable through the gearbox lug.

5   Now disconnect the cable from the clutch pedal arm simply by unhooking it.

6   Remove the radiator air intake grille.

7   Withdraw the cable from below the vehicle, easing it through the cleat on the underside of the body and the rubber grommets (photo).

8   Refit the new cable by reversing the removal operations, and adjust on completion as described in the preceding Section.

4 Clutch pedal – removal and refitting

The clutch pedal pivots on a cross-shaft common to the brake pedal. The removal and refitting operations are similar to those described for the brake pedal in Chapter 9, Section 16, once the clutch cable has been released from the pedal arm as described in Section 3.

5 Clutch – removal

1   Remove the gearbox as described in Chapter 6.

2   Mark the relative position of the clutch pressure plate cover and the flywheel, using a spot of quick-drying paint.

3   Unscrew the clutch cover bolts, progressively, a turn at a time in

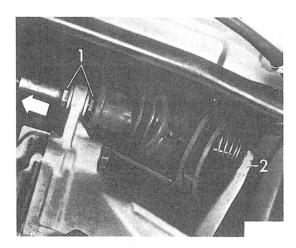

Fig. 5.2 Clutch cable adjusting nuts (1) and release lever (2)
(Sec 2)

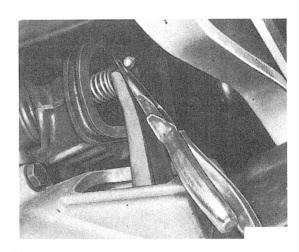

Fig. 5.3 Releasing clutch cable from release lever (Sec 3)

3.2a Releasing clutch cable from lever

3.2b Clutch cable disconnected

Fig. 5.4 Clutch cable body support cleat (Sec 3)

3.7 Withdrawing clutch cable

diagonally opposite sequence.

4   Prise the cover off the flywheel locating dowels, taking care to catch the driven plate which will be released.

6   Clutch – inspection and renovation

1   It is not practical to dismantle the pressure plate assembly and the term 'clutch renewal' is usually used for simply fitting a new clutch driven plate (friction disc).

2   If a new clutch disc is being fitted it is a false economy not to renew the release bearing at the same time. This will preclude having to renew it at a later date when wear on the clutch linings is still very small.

3   If the pressure plate assembly requires renewal, an exchange unit must be purchased. This will have been accurately set up and balanced to very fine limits.

4   Examine the friction linings for wear or loose rivets and the disc for rim distortion, cracks, broken hub springs, and worn splines. The surface of the friction linings may be highly glazed, but as long as the clutch material pattern can be clearly seen this is satisfactory. Compare the amount of lining wear with a new clutch disc at the stores in your local VW garage. If worn, the driven plate must be renewed.

5   It is always best to renew the clutch driven plate as an assembly to preclude further trouble, but, if it is wished to merely renew the linings, the rivets should be drilled out and not knocked out with a

punch. The manufacturers do not advise that only the linings be renewed and personal experience dictates that it is far more satisfactory to renew the driven plate complete rather than to try and economise by only fitting new friction linings.

6   Check the machined faces of the flywheel and the pressure plate. If either is grooved it should be machined until smooth, or renewed.

7   If the pressure plate is cracked or split it is essential that an exchange unit is fitted, also if the pressure of the diaphragm spring is suspect.

8   Check the release bearing for smoothness of operation. There should be no harshness or slackness in it. It should spin reasonably freely bearing in mind it has been pre-packed with grease. **Note:** *When the clutch disc is removed, a certain amount of asbestos dust is likely to be present. This* **should not** *be inhaled: the best method of cleaning is to use a vacuum cleaner.*

9   If the clutch disc is contaminated with oil, it must be renewed. Clean any oil deposits off the flywheel and pressure plate, and rectify the source of contamination (engine or gearbox oil seal) before reassembling.

7   Clutch – refitting

1   Clean away any protective grease from the surface of the pressure plate.

2   Clean the splines of the input shaft with a wire brush and apply a thin smear of molybdenum disulphide grease.

3   If a new pressure plate is being fitted, check it for an alignment (balancing) mark. If one is not evident, match it against the original cover for positioning on the flywheel by aligning the cover dowel holes.

4   Place the driven plate against the flywheel so that the greater projection of the spring hub is away from the flywheel. The plate will normally be marked 'FLYWHEEL SIDE'.

5   Offer the cover to the dowels and screw in the bolts finger tight.

6   The driven plate must now be centralised to accept the input shaft during refitting of the gearbox.

7   Do this using either a clutch alignment tool, available from most motor supply stores, or an old input shaft. An alternative tool can be made up from a stepped mandrel or wooden dowel rod having two diameters, the larger to pass through the splined hub of the driven plate and the smaller to engage in the pilot bearing in the centre of the flywheel. Tape can be used to build up the different diameters on the rod (photo).

8   Due to the fact that the driven plate is only lightly held against the flywheel at this stage, the insertion of the guide tool will move it as necessary to centralise it.

9   Once the driven plate is centralised, tighten the cover bolts to the specified torque in a progressive manner. Remove the guide tool (photo).

10 If the release bearing was removed, smear a little grease on the guide tube, then fit the bearing, retaining springs and clips (photo).

7.7 Clutch centralising tool in position

7.9 Tightening clutch cover bolt

7.10 Connecting release bearing clips

7.11a Reverse side of clutch release lever

7.11b Installing clutch release lever

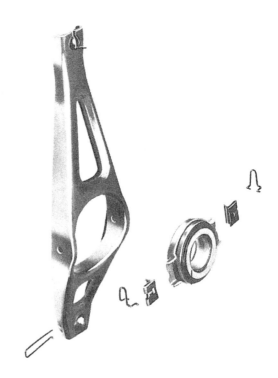

Fig. 5.5 Clutch release components (Sec 7)

11 If the clutch release lever was removed, make sure that the clip at the ball-stud is securely engaged and grease the ball pivot surface (photos).
12 Refit the gearbox (Chapter 6) and adjust the clutch cable (Section 2).

8 Fault diagnosis – clutch

Symptom	Reason(s)
Judder when taking up drive	Loose engine or transmission mountings Badly worn or loose friction linings Oil on friction linings Worn input shaft splines Worn flywheel spigot bush
Clutch spin (failure to disengage) so that gears cannot be engaged	Incorrect cable adjustment Driven plate sticking on input shaft splines due to rust. May occur after standing idle for long periods Damaged or distorted pressure plate assembly
Clutch slip (increase in engine speed does not result in comparable increase in road speed – particularly on gradients)	Incorrect cable adjustment Oil on friction linings Worn friction linings
Noise evident on depressing clutch pedal	Dry, worn or damaged release bearing Broken or weak pedal/cable return spring Wear in driven plate or input shaft hub splines
Noise evident as clutch pedal released	Distorted driven plate Broken driven plate torsion springs Worn input shaft splines

Chapter 6 Gearbox

For modifications, and information applicable to later models, see Supplement at end of manual

Contents

Specifications

General

Gearbox type	4 or 5 forward speeds, all with synchromesh, and one reverse. Floor-mounted gearchange

Unit code:
4-speed	015
5-speed	008

Application:
4-speed	Standard on LT 28, 31 and 35
5-speed	Standard on LT 40 and 45, optional on others

Gear ratios

	4-speed	5-speed
1st	5.01 : 1	6.17 : 1
2nd	2.53 : 1	3.23 : 1
3rd	1.47 : 1	1.84 : 1
4th	1.00 : 1	1.27 : 1
5th	–	1.00 : 1
Reverse	4.96 : 1	5.36 : 1

Oil capacity (all models)

3.5 litres (6.16 Imp pints)

Torque wrench settings

	Nm	lbf ft
Extension housing-to-gearcase bolts	40	30
Clutch bellhousing-to-gearcase bolts	40	30
Bellhousing-to-engine bolts:		
M8	25	18
M12	75	55
Oil drain and filler plugs (4-speed)	35	26
Oil filler plug (5-speed)	20	15
Oil drain plug (5-speed)	25	18
Cover plate bolts	25	18
Reversing lamp switch	30	22
Bearing retainer plate screws	20	15
Extension housing end cap (4-speed)	45	33

1 Description and maintenance

The gearbox may be of 4 or 5-speed type depending upon option and model, except LT40/45 versions which have a 5-speed unit only.

Forward gears are fully synchronised and gear selection is by means of a floor-mounted remote control assembly.

The clutch bellhousing and the extension housing are detachable from the gearcase.

From August 1978, the 5-speed clutch bellhousing and gearcase are used in all models, irrespective of the number of speeds provided.

At the intervals specified in Routine Maintenance, remove the oil level plug from the side of the gearbox. If the oil level is up to the bottom of the plug hole, topping up is not required. If the level is low, bring it up to the correct level using oil of the specified type.

Although the gearbox is 'filled for life' it is recommended that the lubricant is renewed, again at the intervals specified in Routine Maintenance, to offset the normal process of deterioration of the additives in the oil and to eliminate the build-up of metallic particles, which will have contaminated the fluid as the result of friction and tooth contact in spite of the integral magnet.

A socket-headed type plug is used for both the level and drain holes. Always drain gear oil when it is hot and wipe away deposits from the drain plug before refitting it.

Fig. 6.1 Selector linkage adjustment (Sec 2)

a = 15.0 mm (0.59 in)

2 Gearchange linkage (4-speed) – adjusting

1 Select 2nd speed gear.
2 Slacken the pinch-bolt at the clamp on the gearchange rod lever.
3 Move the gearchange hand control lever until there is a clearance between the lever and the edge of the aperture in the floor of 15.0 mm (0.59 in). Do not allow 2nd gear to become disengaged during this operation.
4 With an assistant pushing the hand control lever to the left (looking forwards) against its stop, turn the gearchange rod anti-clockwise using a pair of pipe grips.
5 Holding these set positions, tighten the pinch-bolt at the lever clamp.
6 Check gear selection and check also that the reverse gear safety catch works.

3 Gearchange linkage (4-speed) – removal, dismantling, re-assembly and refitting

1 Working within the driving cab, unscrew and remove the knob from the gear lever.
2 Working under the vehicle, unscrew and remove the pinch-bolt from the clamp on the gearchange rod lever.
3 Pull the gearchange rod lever towards the rear of the vehicle.
4 Unbolt the front bracket from the body and remove the bracket/lever assembly in a downward direction.
5 To remove the gearchange rod, unbolt the rear bracket and pull the rod socket from the ball fitting on the gearbox.
6 The gearchange lever and rod assemblies may now be dismantled by extracting the securing circlips and pins. Renew worn bushes, bearings and rubber gaiters.
7 Reassembly and refitting are reversals of removal and dismantling. Apply multi-purpose grease to all friction surfaces during reassembly.
8 Adjust the linkage as described in Section 2.

4 Gearbox (4-speed) – removal and refitting

1 Position the vehicle over an inspection pit or raise it adequately on a lift, ramps or jacks.
2 Disconnect the battery and remove the engine cover.
3 Mark the alignment of the propeller shaft rear flange relative to the rear axle pinion coupling flange, unscrew the connecting bolts and disconnect the flanges.
4 Pull the front end of the propeller shaft out of the gearbox extension housing. Expect some loss of lubricant as the shaft is pulled out. A heavy quality plastic bag may be slipped over the rear end of the extension housing and secured with a rubber band. If this is done quickly, too much oil leakage will be prevented.

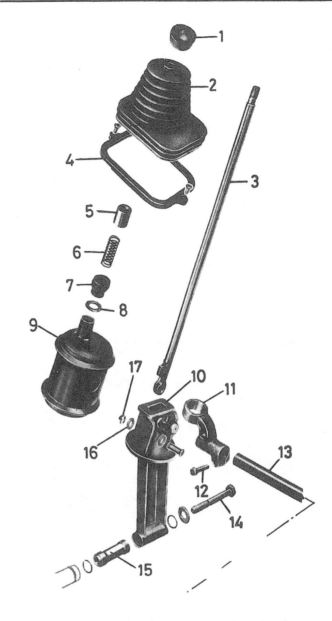

Fig. 6.2 Gear lever components (Sec 3)

1	Knob	10	Front bracket
2	Gaiter	11	Gearchange rod lever
3	Lever	12	Pinch-bolt
4	Retainer	13	Gearchange rod
5	Sleeve	14	Bolt
6	Spring	15	Spacer
7	Bearing	16	Washer
8	Washer	17	Circlip
9	Boot (early models only)		

5 Disconnect the clutch operating cable as described in Chapter 5 and remove the cable guide from the gearbox.
6 Unbolt and remove the cover plate from the lower front face of the clutch bellhousing (photo).
7 Unscrew and remove the bolts which connect the clutch bellhousing flange to the engine.
8 Unbolt the suspension anti-roll bar from the body and move it downwards (only fitted on certain models).
9 Disconnect the exhaust pipe bracket from the gearbox (photo).
10 Unbolt and remove the starter motor and rest it on the floor of the driving cab.

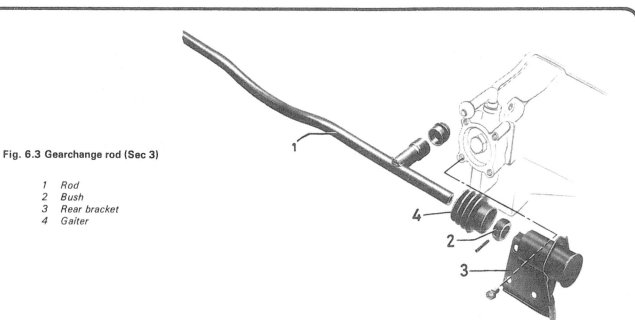

Fig. 6.3 Gearchange rod (Sec 3)

1 Rod
2 Bush
3 Rear bracket
4 Gaiter

Fig. 6.4 Propeller shaft rear flange alignment (Sec 4)

1 Connecting bolt

Fig. 6.5 Gearbox disconnection points (left-hand side) (Sec 4)

| 2 | Clutch cable | 4 | Bellhousing bolts |
| 3 | Cover plate bolts | | |

Fig. 6.6 Gearbox disconnection points (right-hand side) (Sec 4)

| 4 | Bellhousing bolts | 6 | Starter motor |
| 5 | Exhaust bracket | | |

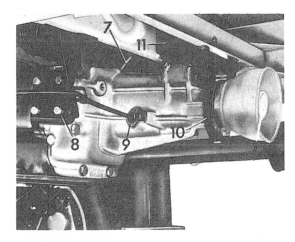

Fig. 6.7 Gearbox disconnection points (rear) (Sec 4)

7	Earth strap	10	U shaped retainer
8	Gearchange rod bracket	11	Rear mounting
9	Speedometer cable		

4.6 Removing bellhousing cover plate

4.9 Removing exhaust bracket

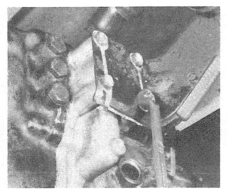
4.12a Unbolting gearchange rod bracket

4.12b Gearchange balljoint disconnected

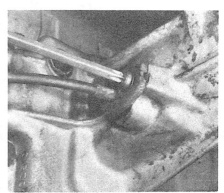
4.13a Disconnecting speedometer cable

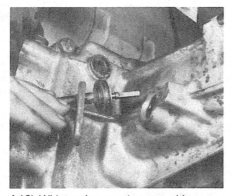
4.13b Withdrawing speedometer cable

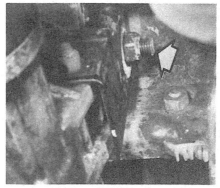
4.17 Gearbox rear mounting bolt (arrowed)

11 Disconnect the gearbox to body earthstrap.

12 Unbolt the gear lever bracket from the gearbox. Pull the balljoint apart after sliding the bellows from it (photos).

13 Disconnect the speedometer cable from the side of the gearcase and plug the hole to prevent loss of oil during removal of the gearbox (photos).

14 Unbolt and remove the U-shaped retainer (where fitted) from the rear of the gearbox extension housing.

15 Support the engine on a stand with a wood block insulator.

16 Support the gearbox on a jack, preferably of trolley type.

17 Unbolt the gearbox from its flexible mounting by withdrawing the retaining crossbolt (photo).

18 With the help of an assistant, pull the gearbox towards the rear of the vehicle off its locating dowels and withdraw it from under the left-hand side of the vehicle.

19 Refitting is a reversal of removal, but if the clutch has been disturbed, centralise the driven plate as described in Chapter 5.

20 Tighten all bolts to the specified torque.

21 Adjust the gearchange linkage (Section 2).

22 Adjust the clutch cable (Chapter 5).

23 Refill the gearbox with the specified lubricant.

5 Gearbox (4-speed) – dismantling into major assemblies

1 With the gearbox removed and the oil drained, clean away external dirt using paraffin and a stiff brush or a water-soluble solvent.

2 Stand the gearbox on a clean workbench, unscrew and remove the flange bolts and withdraw the extension housing from the main gearcase.

3 Peel off the joint gasket and discard it.

4 Remove the clutch release bearing and lever as described in Chapter 5.

5 Tape over the splines on the input shaft to prevent them from damaging the oil seal lips and then unbolt and remove the clutch bellhousing from the gearcase.

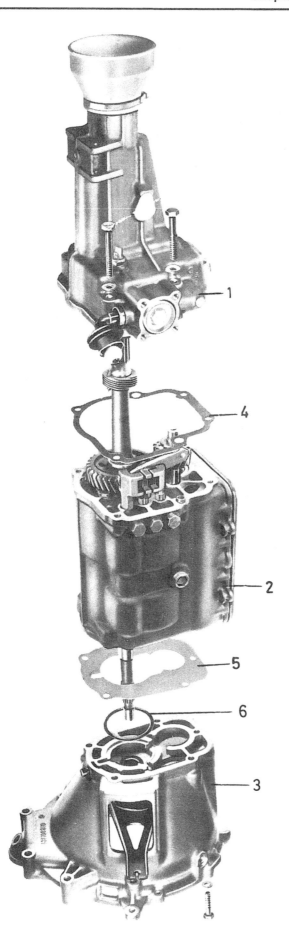

6   Retain the shim which is located between the bellhousing and the bearing.

7   Peel off and discard the joint gasket.

8   Unbolt and remove the oil pan and its gasket from the bottom face of the gearcase.

9   From the rear end of the mainshaft, pull out the spring locking clip and slide the speedometer drive gear from the shaft. Note the chamfered side of the gear is towards the rear end of the shaft.

10 Extract the circlip (b) and remove the relay lever pin (Fig. 8.10).

11 Withdraw the relay lever and reverse sliding gear.

12 Extract the mainshaft circlip.

13 Remove the shim for reverse gear from the shaft.

14 Take reverse gear from the mainshaft.

15 Extract the circlip and remove reverse gear from the countershaft. Note the notches in the gear teeth are nearest the end of the shaft.

16 Unscrew the countersunk screws, these are very tight and will probably require the use of an impact driver to remove them.

17 Remove the bearing retaining plates.

18 Using a suitable bearing extractor, withdraw both countershaft bearings. Without the special tool (6305) this is a very difficult operation and removal is best carried out by destroying the bearing plastic cage so that three small bolts can be passed through the openings left by the removal of the cage and nuts screws onto the bolts. If the bolts are then attached to a puller, the bearing can be withdrawn without damaging the gear teeth or casing. The bearing inner tracks are very tight on the shaft and attempting to drive them from the shaft is pointless (photos).

19 Raise the countershaft with the hand and pull the input shaft from the gearcase.

20 Remove the countershaft. To do this may require lifting the mainshaft geartrain slightly.

21 Unscrew the detent plugs through two to three turns each.

22 Engage 3rd speed gear by moving the 3/4th synchro sleeve.

23 Drive out the roll pin which secures the 3rd/4th fork to the selector rod.

24 Withdraw the 3rd/4th selector rod until the fork can be removed from it.

25 Withdraw the reverse selector rod until the fork-securing roll pin on the 1st/2nd selector rod can be knocked out. Remove the fork.

26 Move all the selector rods to the neutral position. Take care that the interlock plungers have not been displaced due to the selector rods having been pulled out too far. If they have, insert them in their correct positions.

27 Using a press or plastic-faced hammer on the rear end of the mainshaft, displace the shaft from the bearing until the shaft can be pulled rearwards and the bearing removed from the gearcase.

28 Tilt the mainshaft and remove it complete with geartrain from the interior of the gearcase.

29 Wipe out the inside of the gearcase to clean it of swarf and sludge. Clean the magnet.

30 Drive out the oil seal from the clutch bellhousing and fit a new one, having applied grease to its lips (photo).

31 Check that the breather plug is clear.

32 Unscrew and remove the partially released detent plugs, extract the springs and balls.

33 Withdraw each of the selector rods and retrieve the interlock plungers.

34 If necessary, the reverse idler shaft can be removed from the gearcase if a bolt is screwed into the tapped hole in its end and a piece of tubing used as a pressure sleeve while the bolt is tightened against a thrust washer.

35 If necessary, the relay lever support can be unscrewed and removed from the gearcase.

36 Now turn your attention to the extension housing. Any wear in the gearchange components will necessitate removal of the lever or finger and renewal of the bushes, which can be screwed out of their seats or prised from them. Access to the gearchange finger is obtained after the end cap is unscrewed (photos).

37 Renew the extension housing rear oil seal and apply grease to the

Fig. 6.8 Gearcase sections (Sec 5)

1	Extension housing	4	Gasket
2	Gearcase	5	Gasket
3	Clutch bellhousing	6	Shim

Fig. 6.9 Exploded view of gearbox (Sec 5)

1	Oil drain plug	6	3rd/4th selector fork	11	Mainshaft bearing
2	Pressed steel bottom cover	7	1st/2nd selector fork	12	Bearing retainer plate
3	Gasket	8	Mainshaft	13	Countershaft bearing
4	Countershaft	9	Countershaft bearing	14	Relay lever
5	Input shaft	10	Gearcase	15	Countershaft reverse gear

16	Reverse sliding gear
17	Mainshaft reverse gear
18	Shim
19	Circlip
20	Speedometer drive gear

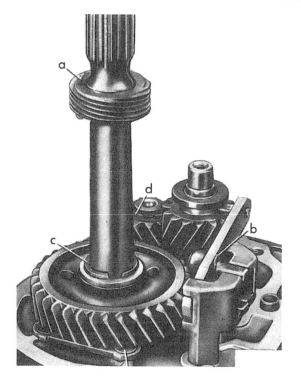

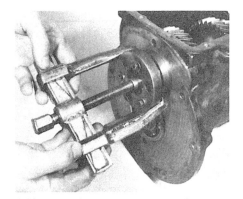

5.18a Removing a countershaft bearing

Fig. 6.10 Rear of mainshaft (Sec 5)

a   Locking clip tab (speedo drive gear)
b   Circlip (relay lever pin)
c   Circlip (mainshaft)
d   Circlip (countershaft reverse gear)

5.18b Method of connecting puller to bearing

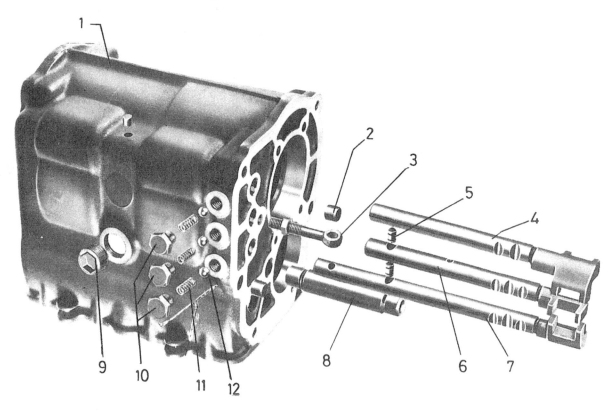

Fig. 6.11 Selector details (Sec 5)

1 Gearcase	4 Reverse selector rod	7 3rd/4th selector rod	10 Detent plugs
2 Hollow dowel	5 Interlock pin	8 Reverse idler shaft	11 Detent springs
3 Relay lever support	6 1st/2nd selector rod	9 Oil filler/level plug	12 Detent balls

5.30 Bellhousing oil seal

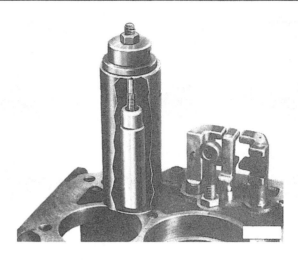

Fig. 6.12 Method of removing reverse idler shaft (Sec 5)

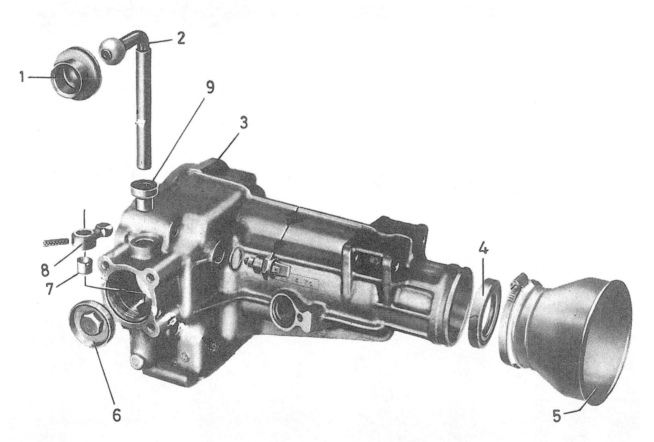

Fig. 6.13 Gearchange housing components (Sec 5)

1	Gaiter	4	Oil seal	6	End cap	8	Selector lever
2	Inner lever	5	Dust excluder	7	Inner lever bush	9	Bush/seal assembly
3	Extension housing						

lips of the new seal.

38 Withdraw the speedometer drive gear and pinion from the extension housing (photo).

39 With the gearbox now completely dismantled, examine and take apart the geartrains as necessary and as described in the following Sections. The countershaft cannot be dismantled and if it is damaged or worn, it will require renewal complete. Renew the countershaft bearings if they were damaged during removal.

5.36a Unscrewing gearchange mechanism end cap

5.36b Gearchange finger exposed

5.38 Removing speedometer drive gear

6 Mainshaft (4-speed) – dismantling and reassembly

1   From the front end of the mainshaft, extract the circlip from the groove in the shaft and then remove the 3rd/4th synchro and 3rd speed gear. These components simply slide off the shaft.
2   Remove 3rd speed gear needle bearing from the mainshaft.
3   From the rear end of the mainshaft, remove 1st speed gear thrust washer, 1st speed gear, 1st/2nd speed synchro unit and 2nd speed gear. Do this either by supporting the rear face of the 2nd speed gear and pressing the shaft from the components, or by using a puller located behind the gear. Take care not to damage the teeth of the gear.
4   Remove the needle bearings for 1st and 2nd gears and the 2nd gear thrust washer.
5   With the mainshaft dismantled, check for chipped or worn gears, slack or noisy bearings and faulty synchro units as described in the following paragraphs.
6   The need for partial or complete renewal of the synchro units will usually be known in advance, due to evidence of noisy gearchanging or the fact that the synchro could be easily 'beaten' during gearchanging.
7   Press the synchro baulk ring onto the cone of the gear and twist it. It should stick to the cone and the gap between ring and tooth band should not exceed 2.5 mm (0.098 in). If it does, renew the baulk ring.
8   Check the remaining components for wear in hub and sleeve splines and teeth, also the fit of the sliding keys.
9   When reassembling the synchro note the following points (photos):c

Fig. 6.14 Synchro baulk ring checking diagram (Sec 6)

a   2.5 mm (0.098 in) max

(a)   1st speed gear baulk ring has no notches on its outer edge, while 2nd speed gear baulk ring has three notches
(b)   When engaging the synchro hub with the sleeve, try several different positions until a setting is found which provides the minimum 'rock' or backlash. Note also that with 1st/2nd synchro, the groove on the sleeve is next to 2nd speed gear when installed on the mainshaft
(c)   The sliding keys are offset at 120° to each other, and the circular springs must be fitted to run in opposing directions in relation to each other with the spring angled ends engaged in a key

6.9a 1st/2nd synchro unit

6.9b 3rd/4th synchro unit

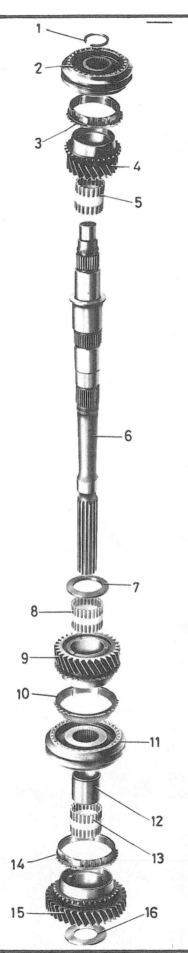

Fig. 6.16 1st gear baulk ring (Sec 6)

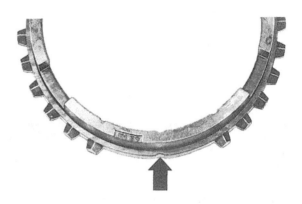

Fig. 6.17 2nd gear baulk ring. Note notch (arrowed) (Sec 6)

Fig. 6.15 Mainshaft components (Sec 6)

1 Circlip
2 3rd/4th synchro
3 Baulk ring
4 3rd speed gear
5 Needle bearing
6 Mainshaft
7 Thrust washer
8 Needle bearing
9 2nd speed gear
10 Baulk ring
11 1st/2nd synchro
12 Bush
13 Needle bearing
14 Baulk ring
15 1st speed gear
16 Thrust washer

Fig. 6.18 1st/2nd synchro unit. Sleeve groove arrowed (Sec 6)

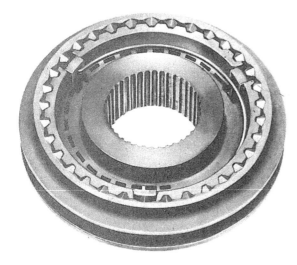

Fig. 6.19 Synchro spring arrangement (Sec 6)

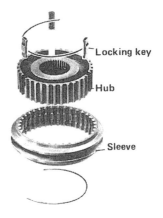

Fig. 6.20 Components of 1st/2nd synchro (Sec 6)

15 Fit 1st speed gear needle bearing complete with inner track, both well lubricated with gear oil (photo).
16 Position 1st speed gear baulk ring (photo).
17 Fit 1st speed gear and its thrust washer (photos).
18 To the front end of the mainshaft fit 3rd speed gear needle roller bearing (well oiled) (photo).
19 Fit 3rd speed gear (photo).
20 Fit 3rd speed baulk ring (photo).
21 Fit 3rd/4th synchro unit so that the wider shoulder on the hub is towards 3rd speed gear (photo).
22 Fit a new circlip (of the thickest possible width from the sizes available) into the shaft groove (photo).

10 Commence reassembly by fitting 2nd speed gear thrust washer to the rear end of the shaft. Make sure that the non-chamfered face of the washer will be against 2nd speed gear (photo).
11 Oil and fit the 2nd speed gear needle roller bearing (photo).
12 Fit 2nd speed gear (photo).
13 Fit 2nd speed bcaulk ring (photo).
14 Fict 1st/2nd synchro, making sure that the groove on the sleeve is towards 2nd speed gear (photo).

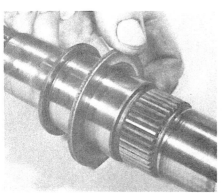

6.10 Fitting 2nd speed thrust washer to mainshaft

6.11 Fitting 2nd speed gear needle bearing to mainshaft

6.12 Fitting 2nd speed gear to mainshaft

6.13 Fitting 2nd speed gear baulk ring to mainshaft

6.14 Fitting 1st/2nd synchro to mainshaft

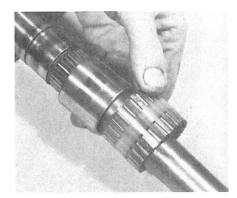

6.15 Fitting 1st speed gear needle roller bearing to mainshaft

6.16 Fitting 1st speed gear baulk ring to mainshaft

6.17a Fitting 1st speed gear to mainshaft

6.17b Fitting 1st speed gear thrust washer

6.18 Fitting 3rd speed gear needle bearing to mainshaft

6.19 Fitting 3rd speed gear to mainshaft

6.20 Fitting 3rd speed gear baulk ring

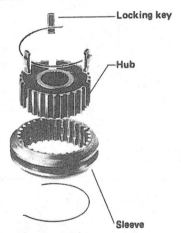

Locking key

Hub

Sleeve

Fig. 6.21 Components of 3rd/4th synchro
(Sec 6)

6.21 Fitting 3rd/4th synchro unit

6.22 Fitting 3rd/4th synchro circlip

7 Input shaft (4-speed) – dismantling and reassembly

1 Remove the shaft circlip and the shims.
2 Press the shaft out of the bearing, but make sure that it is the centre bearing track that is supported. If a bearing puller is used on the outer track, then the bearing must be discarded and a new one fitted.
3 Extract the needle bearing and take off the 4th gear synchro baulk ring, which is probably stuck to the cone of the input shaft gear.
4 Examine the gear teeth for chipping and the bearings for wear or slackness when turned.
5 Support the bearing inner track and press the input shaft into it. Make sure that the bearing outer circlip groove is towards the front end of the shaft.
6 The thickest possible circlip must now be fitted into the shaft groove, but before this is done select a shim from the three thicknesses available which will give a bearing-to-shim clearance of between 0.05 and 0.15 mm (0.002 and 0.006 in). This is a trial and error operation to find the correct shim.
7 Grease the needle roller bearing and locate it in the input shaft recess.
8 Fit the baulk ring to the gear cone on the input shaft.

8 Gearbox (4-speed) – reassembly

1 If the extension housing was dismantled, reassemble by reversing operations and apply grease to the gearchange mechanism bushes.
2 Screw in the end cap until its rim is 1.0 mm (0.039 in) below the gearcase.
3 If the reverse idler gear shaft was removed, refit it now by driving it into the gearcase using a copper-headed hammer (photo).
4 Fit reverse selector rod.
5 Lower the assembled mainshaft into the gearcase (photo).

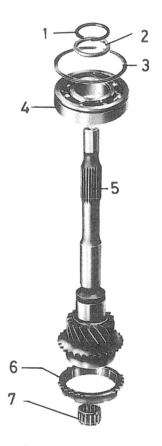

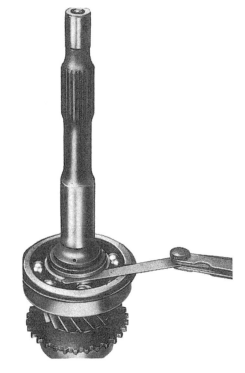

Fig. 6.23 Checking input shaft bearing to shim clearance (Sec 7)

Fig. 6.22 Input shaft detail (Sec 7)

1	Circlip	5	Input shaft
2	Shim	6	4th gear baulk ring
3	Bearing outer circlip	7	Needle bearing
4	Bearing		

6   Push the mainshaft bearing over the shaft and into the gearcase. Make sure that the closed side of the bearing faces inwards (photo).
7   Locate the smaller of the two bearing retainer plates and insert its screws but do not tighten at this stage.
8   The mainshaft will now require drawing through its bearing inner track into its final position. Do this by screwing a bolt into the end of the shaft and using a piece of tubing as a distance piece with a thrust washer in a similar manner to that described for removal of the reverse idler shaft (photo).
9   Using a pencil magnet or a pair of long-nosed pliers, insert the first

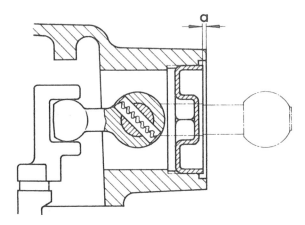

Fig. 6.24 Extension housing cap installing diagram (Sec 8)

a   1.0 mm (0.039 in) maximum

8.3 Installing reverse idler shaft

8.5 Installing mainshaft assembly

8.6 Locating mainshaft bearing

8.8 Drawing mainshaft through bearing. Bearing retainer plate should be fitted first

8.10a Installing 1st/2nd selector rod

8.10b Pinning 1st/2nd fork to selector rod

interlock plunger next to the reverse selector rod.

10 Insert 1st/2nd selector rod and push it in far enough to be able to slide the fork onto it and to pin it. The 1st/2nd fork is the one with the arms furthest apart (photos).

11 Fit the remaining interlock plunger.

12 Install 3rd/4th selector rod and fork and pin the fork in position. To facilitate pinning the fork, engage 3rd speed gear by moving the synchro sleeve (photo).

13 Install the input shaft complete with needle roller bearing and baulk ring. Turn the baulk ring as necessary to align the grooves with the synchro sliding keys. Fit the shim and circlip (see Section 7) (photos).

14 Pass the countershaft into the gearcase and fit the front and rear bearings using a piece of tubing as a drift applied to the bearing centre track. When fitting the bearings, make sure that the numbers are visible from outside the gearcase (photos).

15 Fit the bearing retaining plate and the screws.

16 Tighten the bearing retainer plate screws to the specified torque and stake them with a punch (photos).

17 Slide the countershaft reverse gear into position, noting that the grooved area on the teeth must be nearer the circlip. If the gear is tight to push on, heat it in boiling water for a few minutes before fitting. Fit the circlip (photos).

18 Heat the mainshaft reverse gear in a similar way and slide it into position, noting that the chamfer on the teeth must be towards the circlip when fitted. A suitable shim must now be selected from the six thicknesses available to give a clearance between gear and shim after the circlip is fitted. This is a trial and error operation to establish the shim thickness required. Fit the circlip (photos).

19 If the relay lever support was removed, refit it now and adjust it to comply with the dimension shown in Fig. 6.28. Tighten the locknut on completion. This will probably necessitate the use of an open-ended spanner which has been ground away to reach it (photos).

20 Fit the relay lever and the sliding gear (photo).

21 Secure the relay lever pin with a circlip (photo).

22 Fit the detent balls, the coil springs and their retaining plugs (photos).

23 Check for correct and positive gear selection. It should not be

Fig. 6.25 Pinning 3rd/4th selector fork to its rod (Sec 8)

possible for two adjacent selector rods to be moved simultaneously. Should this occur then the interlock plungers have been displaced.

24 Install the speedometer drive gear (bevelled edge nearer the end of the shaft), retaining it in position with the special locking clip (photos).

25 Fit the cover plate to the gearcase using a new gasket and making sure that the plate is the correct way round, with the magnet towards the rear end of the gearbox (photos).

26 Before the clutch bellhousing can be bolted to the gearcase, the thickness of the shim which is to be fitted between the bellhousing and the input shaft bearing must be established. Measure the projection of the bearing outer track from the gearcase and the depth of the recess in the bellhousing. Use a dial gauge, a depth gauge, or feeler blades and a straight-edge. Add 0.1 mm (0.0039 in) to the recess depth for the thickness of a new gasket. Subtract one measurement from the

8.12 Installing 3rd/4th selector rod and fork

8.13a Fitting input shaft needle bearing to mainshaft

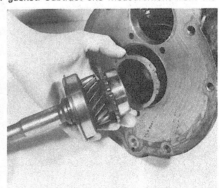

8.13b Installing input shaft

8.13c Fitting input shaft shim

8.13d Fitting input shaft circlip

8.14a Installing countershaft assembly

8.14b Installing countershaft bearings

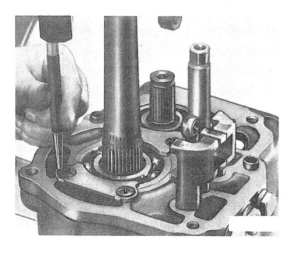

Fig. 6.26 Staking bearing retainer plate screws (Sec 8)

8.16a Tightening bearing retainer plate screws

8.16b Staking bearing retainer plate screws

8.17a Sliding reverse gear onto countershaft

8.17b Fitting countershaft reverse gear circlip

8.18a Sliding reverse gear onto mainshaft

Fig. 6.27 Checking clearance between mainshaft reverse gear and shim (Sec 8)

8.18b Fitting reverse gear shim

8.18c Fitting reverse gear circlip

8.19a Fitting relay lever support

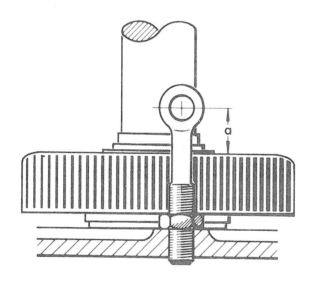

Fig. 6.28 Relay support lever adjusting diagram (Sec 8)

a   15.7 to 16.5 mm (0.618 to 0.650 in)

8.19b Measuring relay lever support projection

8.20 Fitting relay lever and pin

8.21 Securing relay lever pin with circlip

8.22a Fitting detent ball and spring

8.22b Tightening detent plug

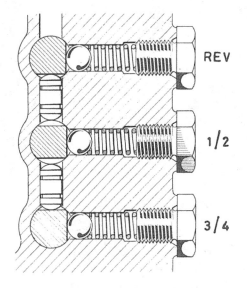

Fig. 6.29 Cross-sectional view of detents (Sec 8)

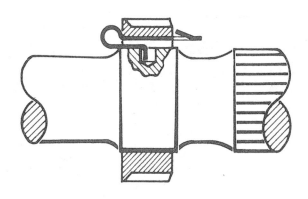

Fig. 6.30 Speedometer drive gear retaining clip (Sec 8)

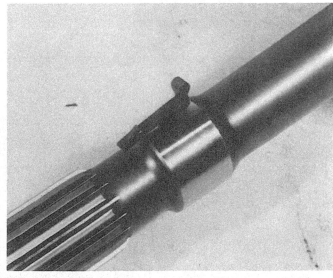

8.24a Speedo drive gear clip

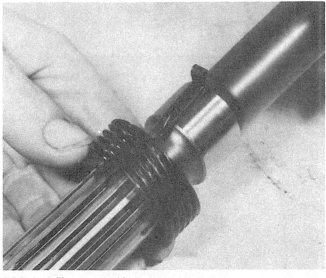

8.24b Installing speedo drive gear

8.24c Speedo drive gear locked on shaft

8.25a Gearcase cover and gasket

8.25b Installing gearcase cover plate

8.26 Fitting bellhousing shim

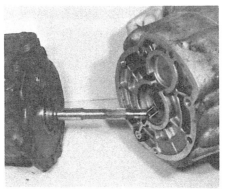

8.27 Connecting bellhousing to gearcase. Shaft splines should be covered with tape to protect oil seal

8.28 Tightening bellhousing bolts

8.30 Connecting extension housing to gearcase

8.31 Tightening extension housing bolts

other and the result is the thickness of the shim required. Four different shim thicknesses are available (photo).

27 Before bolting the clutch bellhousing into position, tape the input shaft splines to prevent damage to the oil seal during installation. Apply sealant to the threads of those bolts which will enter the gearcase in order to prevent possible oil seepage (photo).

28 Using a new gasket, bolt on the bellhousing and tighten the bolts to the specified torque (photo).

29 Fit the clutch release lever and bearing as described in Chapter 5.

30 Using a new gasket, lower the extension housing into position, taking care not to damage the oil seal lips with the mainshaft splines. As the extension housing moves into place, engage the selector finger in the selector rod dog cut-outs. The gears must be in the neutral mode (photo).

31 Apply sealant to the bolt threads to prevent any oil seepage and screw them in. Tighten the bolts to the specified torque (photo).

9 Gearchange linkage (5-speed) – adjusting

1 Place the gear control lever in the neutral mode.

2 Using a piece of metal of suitable thickness, check that dimension (a) is as shown in Fig. 6.31. If not, release the finger pinch-bolt and slide the finger up or down as necessary.

3 Working under the vehicle, slacken the pinch-bolt in the gearchange selector rod clamp and adjust the relative position of control lever and rod to provide a gap (b) between the tip of the finger and the stop plate as shown in Fig. 6.31.

4 On completion of adjustment, press the gear lever to the right and tighten the clamp pinch-bolt.

5 Engage 3rd speed gear by pulling the gearchange lever upwards approximately 10.0 mm (0.39 in) and then moving it rearwards.

6 Release the selector rod clamp pinch-bolt for the second time and

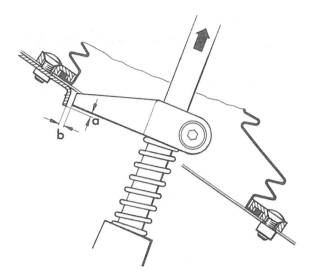

Fig. 6.31 Selector finger adjustment (Sec 9)

a   2.0 mm (0.079 in)   b   3.0 mm (0.118 in)

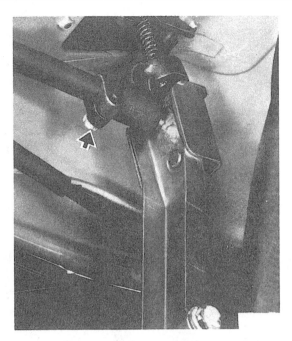

Fig. 6.32 Gearchange rod pinch-bolt (Sec 9)

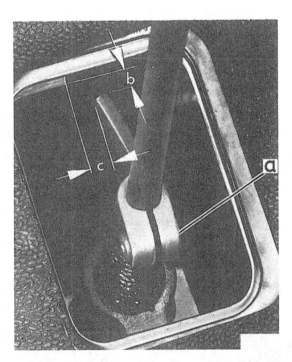

Fig. 6.33 Finger adjustment at base of gear control lever (Sec 9)

a   Pinch-bolt   c   3.0 mm (0.118 in)
b   3.0 mm (0.118 in)

press the gear control lever to the left against its stop. Retighten the clamp pinch-bolt.

7   With the gearchange lever again in neutral press the lever to the left and check the gap (c) between the corner of the finger and the edge of the stop plate as shown in Fig. 6.33. If adjustment is required, release the finger pinch-bolt and twist the finger, but do not alter the finger height setting. On completion, tighten the pinch-bolt.

8   With all the adjustments carried out, select each gear. It must not be possible to move the gearchange lever directly from 1st speed to reverse gear. This action is prevented by the finger striking the stop plate when correctly adjusted.

Fig. 6.34 Remove interlock finger against stop plate (Sec 9)

10 Gearchange linkage (5-speed) – removal, dismantling, re-assembly and refitting

The operations are very similar to those described in Section 3. Reference should be made to Fig. 6.35 for details of design differences in components.

11 Gearbox (5-speed) – removal and refitting

1   The operations are virtually identical to those described in Section 4 for the 4-speed unit.
2   The help of at least one assistant will be required to take the weight of the gearbox during removal and refitting.

12 Gearbox (5-speed) – dismantling into major assemblies

1   With the gearbox removed and the oil drained, clean away external dirt using paraffin and a stiff brush or a water-soluble solvent.
2   Stand the gearbox on a clean workbench, and unscrew and

Fig. 6.35 Gearchange linkage (5-speed) (Sec 10)

1　Knob
2　Gaiter
3　Retainer
4　Finger (1st/reverse
　　interlock)
5　Pinch-bolt
6　Spring
7　Bearing
8　Washer
9　Gearchange lever
10　Selector rod lever
11　Selector rod (part of 16)
12　Pivot pin
13　Front bracket
14　Spacer
15　Pivot bolt
16　Selector rod (part of 11)
17　Gaiter
18　Gaiter
19　Rear bracket
20　Bush

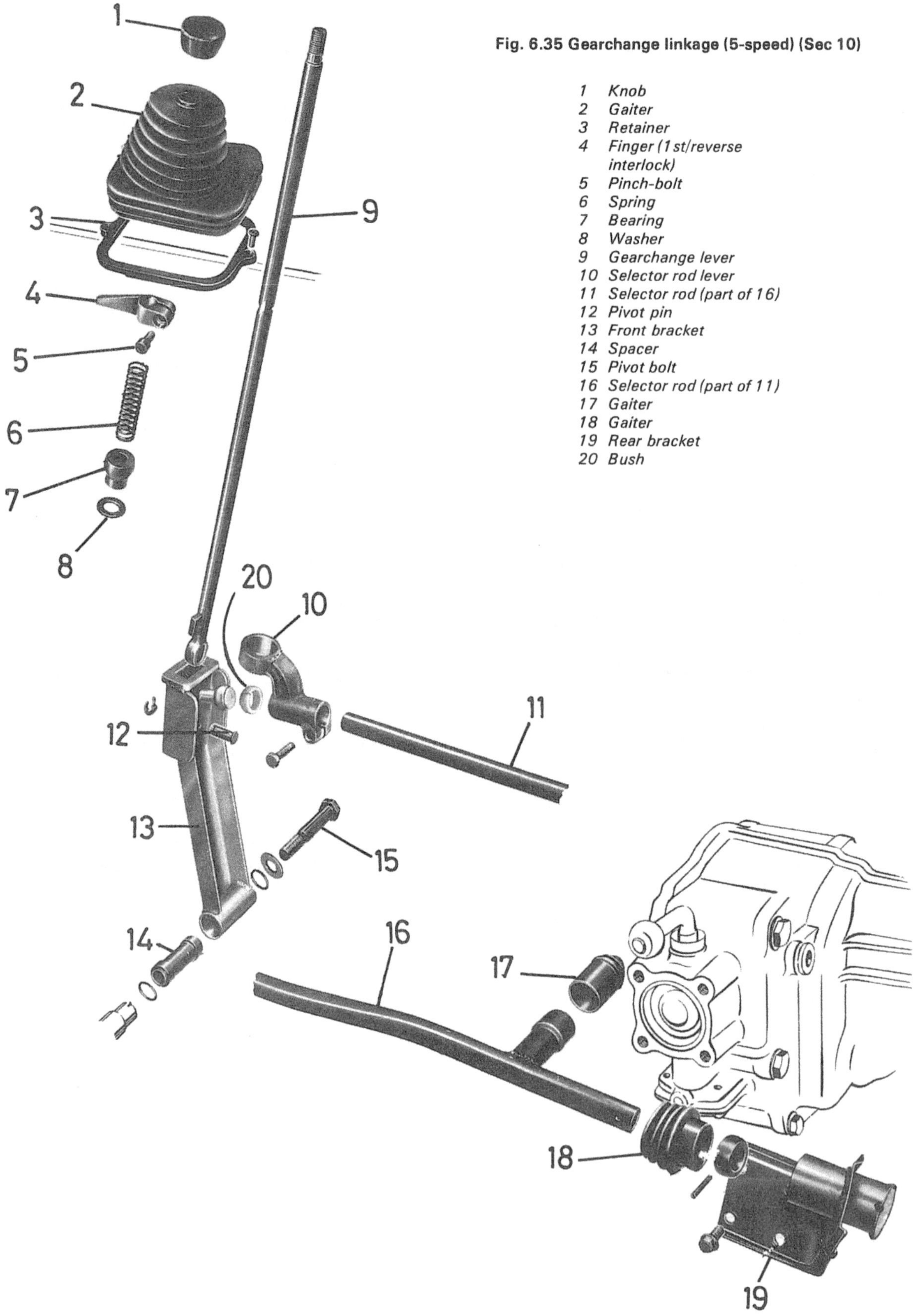

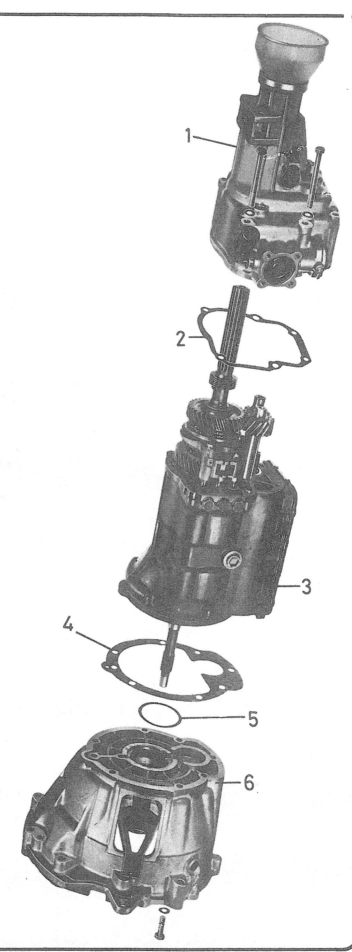

Fig. 6.36 Casing sections of 5-speed gearbox (Sec 12)

1   Rear extension housing   4   Gasket
2   Gasket   5   Shim
3   Main gearcase   6   Clutch bellhousing

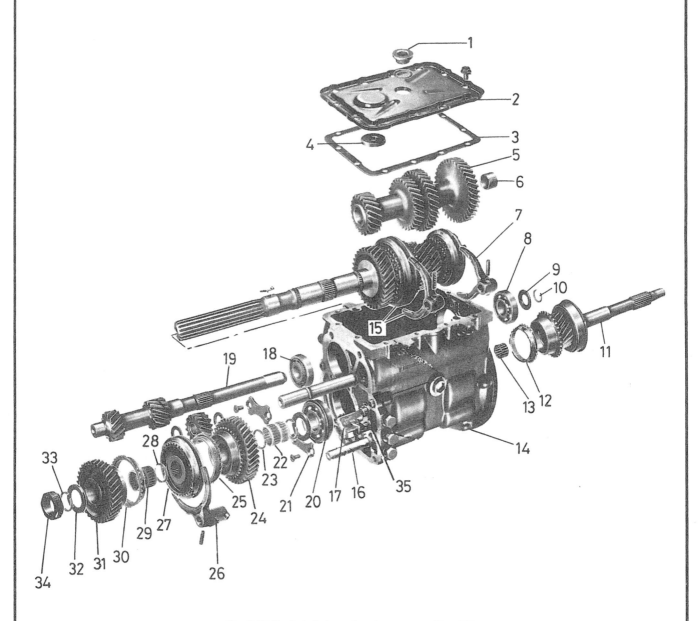

Fig. 6.37 Exploded view of main gearcase (Sec 12)

1	Drain plug
2	Oil pan cover
3	Gasket
4	Magnet
5	Countergear
6	Bush
7	4th/5th selector fork
8	Countershaft front bearing
9	Dished washer
10	Circlip
11	Input shaft
12	Baulk ring
13	Needle bearing
14	Gearcase
15	2nd/3rd selector fork
16	1st/reverse selector shaft
17	4th/5th selector shaft
18	Countershaft rear bearing
19	Countershaft
20	Mainshaft bearing
21	Bearing retainer plate
22	1st gear needle bearing
23	Thrust washer
24	1st speed gear
25	Baulk ring
26	1st/reverse selector fork
27	1st/reverse synchro
28	Circlip
29	Needle bearing
30	Baulk ring
31	Reverse gear
32	Thrust washer
33	Circlip
34	Speedometer drive gear
35	2nd/3rd selector shaft

remove the extension housing flange bolts.

3 Engage 2nd speed gear by pulling the inner gearchange lever slightly out and twisting it.

4 Tap the extension housing off its locating dowels using a plastic-faced mallet.

5 Peel off the joint gasket and discard it.

6 Remove the clutch release bearing and lever as described in Chapter 5.

7 Take over the splines on the input shaft to prevent them from damaging the oil seal lips and then unbolt and remove the clutch bellhousing from the gearcase.

8 Retain the shim which is located between the bellhousing and the bearing.

9 Peel off and discard the joint gasket.

10 Take off reverse gear and its thrust washer.

11 Unbolt and remove the oil pan and remove the joint gasket.

12 Unscrew each of the three detent plugs through three complete turns.

13 From the rear end of the mainshaft pull out the locking spring clip and slide the speedometer drive gear from the shaft.

14 Extract the shaft circlip, remove the thrust washer, reverse gear with synchro baulk ring and the needle bearing, all from the mainshaft.

15 Engage 1st speed gear by moving the synchro sleeve.

16 Knock out the roll pin from 1st/reverse selector fork and slide the fork from its selector shaft.

17 Remove the circlip which secures 1st/reverse synchro hub.

18 Remove the circlip and the dished washer from the end of the countershaft.

19 Remove the countershaft front bearing. Without the special tool (6305) this is a very difficult job, and removal is best carried out by destroying the bearing plastic ball cage so that three small bolts can be passed through the openings left by the removal of the cage, and nuts screwed onto the bolts. If the bolts are then attached to a puller, the bearing can be withdrawn without damaging the gear teeth or casing. The bearing inner track is very tight on the shaft, and attempting to drive it from the shaft is pointless.

20 Using a plastic-faced mallet, tap the countershaft out of the casing, at the same time pulling out 1st/reverse synchro unit.

21 Remove the countergear from the gearcase.

22 Remove the input shaft from the gearcase.

23 From the rear end of the mainshaft, take off the circlip, 1st speed gear with baulk ring and the needle bearing.

24 Remove the next circlip and thrust washer.

25 Extract the countersunk head screws from the bearing retaining plates. These are likely to be very tight and may require the use of an impact driver.

26 Remove the bearing retainer plate.

27 Remove the countershaft rear bearing from the gearcase.

28 Engage 4th speed gear by moving the synchro sleeve.

29 Drive out the roll pin from 4th/5th selector fork.

30 Pull the 4th/5th selector shaft back and take off the selector fork.

31 Withdraw 1st/reverse selector shaft until the roll pin for 2nd/3rd selector fork can be knocked out. Remove the fork.

32 Move all the selector shafts to the neutral position.

33 Unscrew the detent plugs completely and extract the springs and balls. Withdraw the selector shafts.

34 Remove the interlock plungers from their holes between the selector shafts using a pencil magnet.

35 Now press the rear end of the mainshaft out of its bearing. The use of a large plastic-faced hammer may substitute for a press.

36 Tap the mainshaft bearing from the casing and withdraw the mainshaft geartrain from the gearcase.

37 Wipe out the inside of the gearcase to remove swarf and sludge. Clean the magnet.

38 Drive out the oil seal from the clutch bellhousing and fit a new one, having applied grease to its lips.

39 If necessary, the reverse idler shaft can be removed from the gearcase if a bolt is screwed into the tapped hole in its end and a piece of tubing used as a pressure sleeve while the bolt is tightened against a thrust washer.

40 Now examine the extension housing. Any wear in the gearchange components will necessitate removal of the lever or finger and renewal of the bushes and seals. Access to the gearchange finger is obtained after unscrewing the end cap.

41 Renew the extension housing rear oil seal and apply grease to the lips of the new seal.

Fig. 6.38 Rear end of mainshaft (Sec 12)

1 Speedometer drive gear locking clip
2 Thrust washer

Fig. 6.39 Removing 4th/5th selector fork roll pin (Sec 12)

42 Withdraw the speedometer driven gear and pinion from the extension housing.

43 With the gearbox now completely dismantled, examine and take apart the geartrains as necessary and as described in the following Sections. The ball-bearings on the countershaft can be renewed, also the shaft or gear assembly as separate components.

13 Mainshaft (5-speed) – dismantling and reassembly

1 From the front end of the mainshaft, take off the shaft circlip.

2 Remove 4th/5th synchro unit, 4th gear baulk ring and 4th speed gear. If the components are tight on the shaft, use a puller with its claws located behind the rear face of 4th speed gear, or support the gear and press the mainshaft out of the components.

3 Take the 4th gear needle bearing from the shaft.

4 From the rear end of the mainshaft, take off 2nd gear thrust washer, 2nd speed gear and baulk ring.

5 Remove 2nd gear needle roller bearing.

6 Remove the needle bearing inner track. If this is tight on the shaft, use a puller to draw off 2nd/3rd synchro and 3rd speed gear. Position the claws of the puller behind 3rd speed gear.

7 Remove 3rd speed gear needle bearing.

8 With the mainshaft dismantled, check for chipped or worn gears, slack or noisy bearings and faulty synchro units.

9 Refer to Section 6 for general details of synchro overhaul, but note

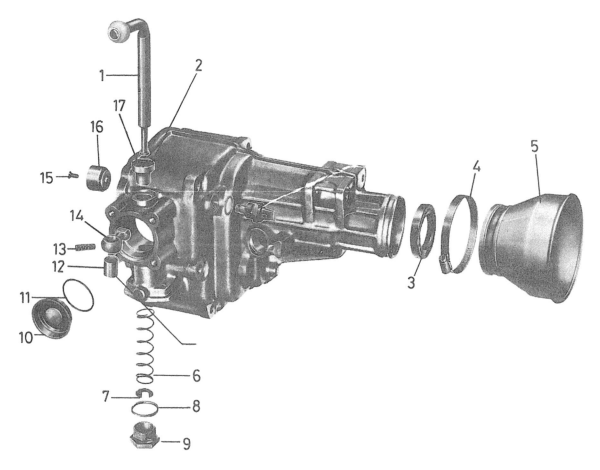

Fig. 6.40 Extension housing components (Sec 12)

1	Inner gearchange lever	6	Spring
2	Extension housing	7	Dished washer
3	Oil seal	8	Sealing ring
4	Clip	9	Plug
5	Cap	10	Cap (renew if removed)
		11	Sealing ring
		12	Bush
		13	Spring
		14	Gearchange finger
		15	Countersunk screw
		16	Needle bearing for countershaft
		17	Bush/seal

Fig. 6.41 1st/reverse baulk ring (Sec 13)

Fig. 6.42 2nd/3rd baulk ring. Note groove (arrowed) (Sec 13)

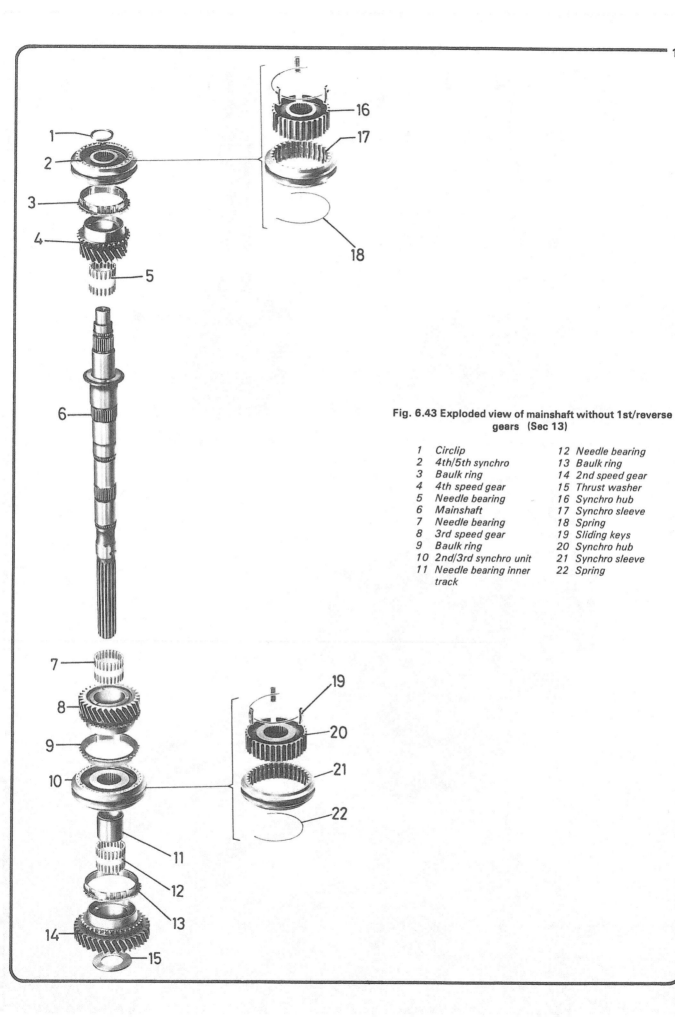

Fig. 6.43 Exploded view of mainshaft without 1st/reverse gears (Sec 13)

1	Circlip	12	Needle bearing
2	4th/5th synchro	13	Baulk ring
3	Baulk ring	14	2nd speed gear
4	4th speed gear	15	Thrust washer
5	Needle bearing	16	Synchro hub
6	Mainshaft	17	Synchro sleeve
7	Needle bearing	18	Spring
8	3rd speed gear	19	Sliding keys
9	Baulk ring	20	Synchro hub
10	2nd/3rd synchro unit	21	Synchro sleeve
11	Needle bearing inner track	22	Spring

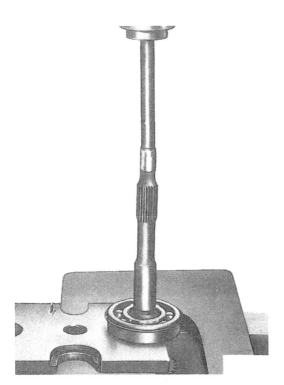

Fig. 6.44 Pressing input shaft out of bearing (Sec 14)

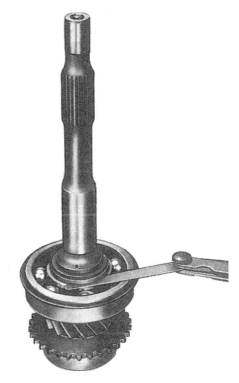

Fig. 6.45 Measuring input shaft bearing clearance (Sec 14)

Fig. 6.46 Selector components (Sec 15)

1 Gearcase
2 Dowel
3 1st/reverse selector
　shaft

4 2nd/3rd selector shaft
5 4th/5th selector shaft
6 Circlip

7 Reverse sliding gear
　shaft
8 Interlock plungers

9 Detent plugs, springs
　and balls

that with a five speed gearbox, the identification of the baulk rings is different.

10 Commence reassembly by oiling and installing 4th gear needle bearing to the front end of the mainshaft.

11 Fit 4th speed gear, 4th gear baulk ring and 4th/5th synchro unit, making sure that the wider shoulder on the hub is towards 4th speed gear. Use a press or puller as necessary.

12 Oil 3rd gear needle bearing and push it onto the rear end of the mainshaft.

13 Fit 3rd speed gear, 3rd gear baulk ring and 2nd/3rd synchro. Note that the groove in the synchro sleeve is towards 2nd speed gear. Use a puller to draw these components into position on the shaft, or press the mainshaft into the synchro hub, having first well supported its rear face.

14 Fit the needle bearing inner track for 2nd speed gear.

15 Fit 2nd speed gear needle bearing, well oiled.

16 Fit 2nd gear baulk ring.

17 Fit 2nd speed gear and thrust washer.

14 Input shaft (5-speed) – dismantling and reassembly

1 Only the bearings can be renewed on this shaft.

2 Extract the shaft circlip and press or draw off the bearing.

3 When installing the new bearing, make sure that the large circlip groove on the outer track is towards the front of the shaft.

4 Once the bearing is fitted, the thickest possible circlip must be used which will engage in the shaft groove.

5 Fit the original shim and the circlip and then using feeler blades, check the clearance between the bearing inner track and the shim. This should be between 0.05 and 0.15 mm (0.002 and 0.006 in). If the clearance is incorrect, change the shim for one of alternative thickness from the three thicknesses available (3.0, 3.1 and 3.2 mm).

6 Renew the needle bearing if necessary by extracting it from its recess with the finger.

15 Gearbox (5—speed) – reassembly

1 Pass the mainshaft geartrain into the gearcase.

2 Fit the mainshaft rear bearing so that the sealed side of the bearing faces inwards.

3 Loosely fit the bearing lockplate.

4 Unless a press is available, it is recommended that the mainshaft is drawn into position through its bearing, using a piece of tubing as a distance piece and then screwing a bolt into the tapped hole in the end of the shaft so that it will apply pressure to a thrust washer located on the end of the tubing.

5 Insert the selector shaft interlock plungers into their casing holes. Insert the selector shafts and position in the neutral mode.

6 Fit the balls and springs and partially screw in the detent plugs.

7 Slide the fork onto the 2nd/3rd selector shaft and tap in the roll pin. This will necessitate slightly withdrawing 1st/reverse shaft to gain access to the roll pin hole.

8 Slide 4th/5th fork onto its selector shaft and install the fixing roll pin. To do this, engage 4th gear by moving the synchro sleeve.

9 Install the countershaft rear bearing to the gearcase. Before this can be done, the loosely fitted mainshaft bearing retainer plate must be removed. Make sure that the sealed side of the bearing plastic cage faces inward.

10 Fit the bearing retainer plates, tighten the screws to the specified torque and stake the screws using a centre punch.

11 To the rear end of the mainshaft fit the thrust washer and the circlip.

12 Fit the needle bearing, 1st speed gear with baulk ring and the securing circlip.

13 Lower the input shaft into the gearcase and making sure its needle bearing and baulk ring are in position (keys aligned), locate it on the end of the mainshaft and tap the bearing into the casing.

14 Lower the countergear with its bush into the gearcase.

15 Slide the countershaft through the countergear assembly and at the same time, push 1st/reverse synchro onto the mainshaft. This synchro is located between two of the countershaft gears and this is the only way it can be installed.

16 Using a plastic-faced hammer, tap the countershaft and 1st/reverse synchro up against their stop. Fit the shaft circlip to secure

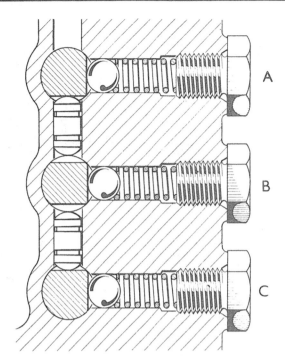

Fig. 6.47 Sectional view of detents (Sec 15)

A 1st/reverse   C 4th/5th
B 2nd/3rd

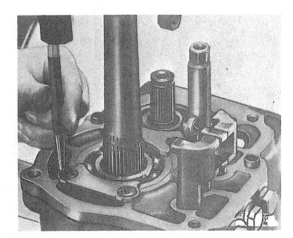

Fig. 6.48 Staking bearing retainer plate screws (Sec 15)

the synchro hub.

17 Fit the countershaft front bearing, applying pressure only to the bearing inner tracks and making sure that the sealed side of the plastic cage is inwards.

18 Fit the bearing circlip, making sure that it is fully engaged in its groove.

19 To the mainshaft fit reverse gear needle bearing, reverse gear and baulk ring, the thrust washer and circlip.

20 Fit 1st/reverse selector fork and shaft and secure with the roll pin.

21 To the rear end of the mainshaft fit the speedometer drive gear and lock it in position with the spring clip.

22 Fit the oil pan using a new gasket. The magnet must be located towards the rear of the gearbox.

23 If removed, install reverse sliding gear shaft, then slide on the gear with the thrust washers.

24 If the input shaft bearing has not been changed, the original shim

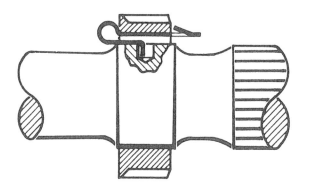

Fig. 6.49 Position of speedometer drive gear locking clip (Sec 15)

may be fitted between the bearing and the clutch bellhousing. If the bearing has been changed the thickness of a new shim must be determined.

25 Determine shim thickness by measuring the projection of the bearing outer track from the surface of the gearcase and the depth of the recess in the bellhousing. Use either a dial gauge, a depth gauge, or feeler blades and a straight-edge for this operation. Add 0.1 mm (0.0039 in) to the recess depth to compensate for the thickness of a new gasket. Subtract one measurement from the other and the result is the thickness of the shim required. Four different shim thicknesses are available.

26 Using the new gasket, bolt on the bellhousing and tighten to the specified torque. Use sealing compound on the bolt threads to prevent oil leakage.

27 Fit the clutch release lever and bearing as described in Chapter 5.

28 Using a new gasket, prepare the extension housing for connecting to the gearcase. This must follow a certain sequence. First move the selector linkage to take up 2nd gear position. Do this by pulling the inner lever on the extension housing slightly out and then twisting the ball end towards the housing.

29 Offer the extension housing into position, checking that the selector finger is on the 4th/5th selector shaft. Now turn the extension housing anti-clockwise and engage the finger with the 2nd/3rd selector rod.

30 When the extension housing is finally bolted into position, check the selection of all gears.

31 Tighten the extension housing bolts to the specified torque.

16 Fault diagnosis – gearbox

Symptom	Reason(s)
Weak or ineffective synchromesh	Worn baulk rings General wear in synchro teeth or splines
Jumps out of gear	Weak detent springs or wear at shaft detent grooves Worn selector forks Worn synchro sleeve grooves Worn gear shaft thrust washers Loose selector fork
Excessive noise	Low oil level Incorrect type or grade of lubricant Worn shaft bearings Worn gearteeth
Difficulty in engaging gears	Fault in clutch mechanism Wear in gearchange linkage

Chapter 7 Propeller shaft

For modifications, and information applicable to later models, see Supplement at end of manual

Contents

Specifications

Type	Single section, tubular, with sealed universal joints

Torque wrench setting	Nm	lbf ft
Rear flange bolts	25	18

1 Description and maintenance

The propeller shaft is of single piece tubular type, with a sealed universal joint at each end.

Wear in the joints can only be rectified by complete renewal of the shaft which must therefore be regarded as disposable.

Maintenance consists of occasionally checking the security of the rear flange bolts.

2 Propeller shaft – removal and refitting

1 Before removing the propeller shaft, check that the relative position of the shaft rear flange and the rear axle companion flange has been marked. This has usually been done during production, but if not, mark them with quick-drying paint.

2 Unscrew and remove the flange bolts, pull the shaft slightly forwards and down and then withdraw it from the rear end of the gearbox extension housing.

3 Before refitting the shaft, apply multi-purpose grease to the front sliding section of the shaft.

4 Align the rear flange marks and tighten the bolts to the specified torque.

3 Propeller shaft universal joints – examination

1 Wear in the needle roller bearings is characterised by vibration in the transmission, 'clonks' on taking up the drive, and in extreme cases of lack of lubrication, metallic squeaking, and ultimately grating and shrieking sounds as the bearings break up.

2 It is easy to check if the needle roller bearings are worn with the propeller shaft in position, by trying to turn the shaft with one hand, the other hand holding the rear axle flange when the rear joint is being checked, and the front half coupling when the front joint is being checked. Any movement between the propeller shaft and the front and the rear half couplings is indicative of considerable wear.

3 Examine the propeller shaft splines for wear. If worn, it will be necessary to purchase a new propeller shaft.

4 If the propeller shaft splines are worn, check also the condition of the splines on the end of the gearbox output shaft. If these are also worn, a new mainshaft will be required.

5 Worn universal joints will necessitate a new shaft complete as the old joints cannot be reconditioned, unless advantage is taken of the service offered by some companies to remove staked type joints and substitute circlip type joints as replacements (photo).

6 Proprietary kits for the overhaul of staked type joints may be available. If contemplating the purchase of such a kit, make sure that you have the necessary workshop equipment to use it – typically, a large metal-working vice and a substantial hammer will be required. Remember also that if the universal joint housings are worn or damaged, renewal of the joints is useless.

3.5 Staked type universal joints

4 Fault diagnosis – propeller shaft

Symptom	Reason(s)
Vibration when car running on road	Out-of-balance shaft Wear in splined sleeve Loose flange bolts Worn shaft joints
'Clonk' on taking up drive or on overrun	Loose flange bolts Worn shaft joints Worn splined coupling Axle or gearbox fault

Chapter 8 Rear axle

For modifications, and information applicable to later models, see Supplement at end of manual

Contents

Specifications

General

Type	Hypoid, differential integral with axle casing
LT 28, 31	Semi-floating
LT 35, 40, 45	Fully floating

Final drive ratios

	Standard		Optional	
Vans:				
LT 28	4.44 : 1	4.875 : 1	5.375 : 1	
LT 31, 35	4.875 : 1	5.375 : 1	4.44 : 1	
LT 40, 45	5.375 : 1	5.857 : 1	4.875 : 1	
Chassis cab:				
LT 31, 35	5.375 : 1	4.875 : 1	4.44 : 1	
LT 45	5.857 : 1	5.375 : 1	4.875 : 1	

Lubricant capacity

LT 28, 31	1.8 l (3.2 Imp pints)
LT 30, 40, 45	2.2 l (3.9 Imp pints)

Torque wrench settings

	Nm	lbf ft
LT 28, 31		
Bearing retainer nuts	40	30
Roadspring U-bolts	120	89
Panhard rod fixing bolts	60	44
Shock absorber lower mountings	45	33
Roadwheel nuts	200	148
LT 35		
Axleshaft flange bolts	55	41
Roadspring U-bolts	120	89
Anti-roll bar to rear axle	60	44
Shock absorber lower mounting	60	44
Roadwheel nuts	320	236
LT 40, 45		
Axleshaft flange bolts	70	52
Roadspring U-bolts	170	125
Shock absorber lower mounting	170	125
Anti-roll bar to axle	80	59
Roadwheel nuts	320	236

1 Description

The rear axle incorporates the final drive and differential, which is integral with the axle casing and cannot be removed as a separate assembly.

All operations can be carried out without the need to remove the axle from the vehicle, but in view of the essential requirement of special gauges and tools for overhaul, the work described in this Chapter should be regarded as the limit to which the home mechanic should go. Leave all other operations to your VW dealer.

The rear axle is of rigid type with hypoid gears and it is supported on leaf springs.

Depending upon model, the axle may be of semi-floating or fully floating design.

Fig. 8.1 Typical rear axle components (Sec 1)

1	Differential cover	4	Tapered roller bearing	7	Drain plug	10 Gasket
2	Gasket	5	Bearing outer track	8	Axle casing	11 Pinion assembly
3	Bearing cap	6	Shim	9	Filler/level plug	12 Differential

2 Maintenance

1   At the intervals specified in Routine Maintenance at the beginning
of this manual, remove the oil filler/level plug from the rear axle
housing and check the oil level. Top up if necessary with oil of the
specified type until it is level with the bottom of the hole and just
begins to run out. (The vehicle must be on level ground).

2   Although the rear axle is 'filled for life' it is recommended that the
lubricant is renewed at the intervals specified in Routine Maintenace,
to offset the normal process of deterioration of the additives in the oil
and to eliminate the build-up of metallic particles, which will have
contaminated the fluid as the result of friction and tooth contact in
spite of the integral magnet.

3   A socket-headed type plug is used for both the filler/level and drain
holes. Drain the lubricant when it is hot and wipe away the deposits

from the drain plug magnet before refitting it.

4   Keep the casing breather clear by probing with wire after the
rubber cap has been pulled off (photo).

3 Pinion oil seal – renewal

1   The need for renewal of the pinion oil seal will be obvious from the
amount of oil which is thrown out of the seal just to the rear of the
pinion coupling flange.

2   Remove the propeller shaft as described in Chapter 7.

3   Carefully mark the fitted position of the pinion nut in relation to the
end of the pinion.

4   The pinion flange must now be held stationary while the pinion nut
is unscrewed and removed. In the absence of the special retaining tool
(2032), use a length of flat steel with two holes drilled at one end

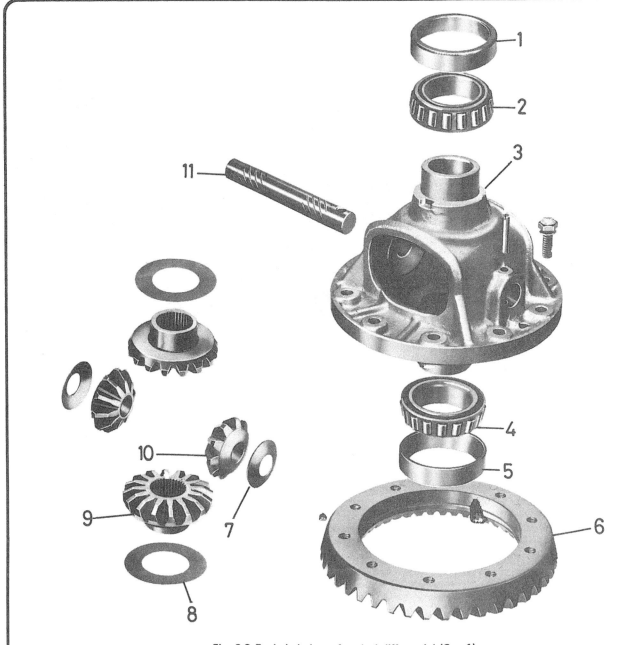

Fig. 8.2 Exploded view of typical differential (Sec 1)

1	Bearing outer track	4	Roller cage	7	Dished washer	10 Pinion gear
2	Roller cage	5	Bearing outer track	8	Thrust washer	11 Pin
3	Differential case	6	Crownwheel	9	Side gear	

2.4 Removing axle casing breather cover

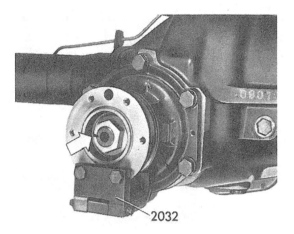

Fig. 8.3 Pinion nut alignment mark (arrowed) (Sec 3)

2032   Pinion holding tool

which can be bolted to the pinion flange. This tool will prove quite satisfactory provided the steel flat is of good length to give adequate leverage, as the nut is very tight. Count the number of turns required to remove the nut and record the number for future reference.

5   With the nut removed, use a puller to remove the pinion flange. Be prepared for some oil spillage.

6   Lever out the oil seal. If this proves difficult, drill two holes in the oil seal casing front face at opposite points, screw in self-tapping screws and use them as leverage points. Take care not to damage the seal housing.

7   Fill the lips of the new oil seal with multi-purpose grease and tap it squarely into position using a piece of tubing as a drift.

8   Fit the pinion flange.

9   Clean the threads of the nut and pinion shaft and apply thread locking fluid.

10   Screw on the nut to its position exactly as marked before removal. *This is very important in order to maintain the pinion bearing preload as originally set.*

11   Refit the propeller shaft (Chapter 7).

12   Check the rear axle oil level and top up if necessary.

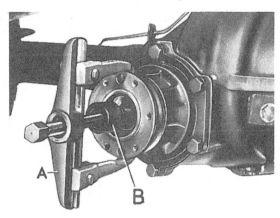

Fig. 8.4 Removing the pinion companion flange (Sec 3)

A   Puller   B   Thrust cup

---

4   Axleshaft and bearings (LT 28, 31) – removal, overhaul and refitting

---

1   Remove the roadwheel.

2   Take out the retaining screw and remove the brake drum. If it is

tight, screw three M10 x 40 screws into the tapped holes in the drum to force it off. If the drum tends to jam on the brake linings, slacken the adjuster by passing a screwdriver through the hole in the brake backplate. Refer to Chapter 9 if necessary.

3   Unscrew the four nuts which secure the bearing retainer and pull the halfshaft out of the axle casing. Be prepared for some oil spillage.

4   If the bearing is to be renewed, use a sharp cold chisel to cut a

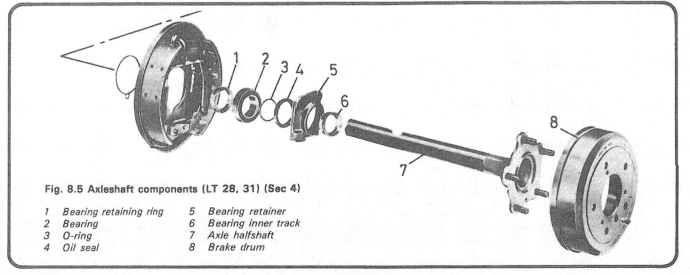

Fig. 8.5 Axleshaft components (LT 28, 31) (Sec 4)

1	*Bearing retaining ring*	*5*	*Bearing retainer*
2	*Bearing*	*6*	*Bearing inner track*
3	*O-ring*	*7*	*Axle halfshaft*
4	*Oil seal*	*8*	*Brake drum*

Fig. 8.6 Withdrawing the brake drum (LT 28, 31) (Sec 4)

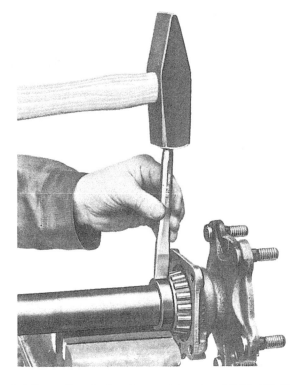

Fig. 8.7 Expanding the bearing retaining ring (LT 28, 31) (Sec 4)

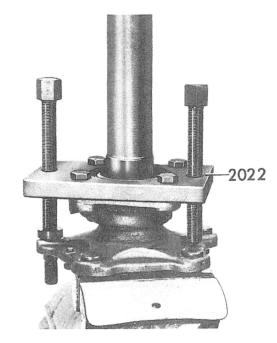

Fig. 8.8 Method of removing bearing and retaining ring from LT 28, 31 axleshaft with puller 2022 (Sec 4)

groove in the bearing retaining ring. This will have the effect of slightly expanding the retaining ring to ease removal. Remove the roller cage and O-ring.

5   In the absence of a press, an extractor can be made up similar to the one shown and bolted to the bearing retainer (Fig. 8.8). Screwing in the bolts will then exert sufficient pressure to draw off the bearing and the retainer together.

6   Remove the bearing outer track from the axle casing using a suitable clawed extractor.

7   Fit the new bearing outer track into the axle casing by driving it into position using a piece of tubing applied to its outer rim.

8   The new bearing inner track and the bearing retaining ring should be pressed on together, but heat them before installation as follows:

Bearing inner track – 100°C (212°F)
Retaining ring – 250° to 280°C (482° to 536°F) – cherry red to violet colour

9   Before pressing the components onto the axleshaft, fit a new oil seal to the bearing retainer and install it on the shaft together with the bearing roller cage and a new O-ring seal.

10 Press on the heated components together.

11 Apply grease to the bearing rollers and then insert the axleshaft into the axle casing. Keep the shaft horizontal and turn it slightly in both directions until the inboard end of the shaft can be felt to engage with the splines in the differential bevel gears.

12 Push the shaft fully home. Fit and tighten the bearing retainer nuts to the specified torque.

13 Fit the brake drum and the roadwheel.

14 If the brake adjustment was slackened off to remove the drum, readjust the brakes after reference to Chapter 9.

15 Check the axle oil level and top up if necessary.

---

5   Axleshaft and bearings (LT 35, 40, 45) – removal, overhaul and refitting

1   If only the axleshaft is to be removed, this can be done without having to take off the roadwheel or brake drum (paragraphs 3 and 4).

2   If the axleshaft bearings require renewal, then the brake drum and the roadwheel must first be removed.

3   Remove the roadwheel.

4   Extract the securing screw and take off the brake drum. If it is tight, screw four M10 x 40 bolts into the tapped holes provided to force it off. If it still jams, slacken off the brake adjuster by passing a screwdriver through the hole in the brake backplate.

5   Unscrew and remove the axle flange bolts and withdraw the shaft from the axle casing. Be prepared for some oil spillage.

6   Using a large socket spanner, unscrew and remove the bearing adjusting nut. Remove the thrust washer.

7   Pull the hub from the axle casing complete with outer bearing, inner bearing and oil seal.

8   Examine the spacer sleeve on the axle casing for scoring. If

evident, drive it off and tap a new one into position using a piece of tubing.

9   Prise out the oil seal from the hub and remove the bearing roller cages.

10 Drive out both bearing outer tracks using a brass or copper drift.

11 Install the bearing tracks, the inner roller cage and a new oil seal. Tap the oil seal into position until it is flush with the end of the hub.

12 Pack multi-purpose grease within the hub to half fill it.

13 Fit the hub to the axle casing and push on the outer bearing.

14 Fit the thrust washer and screw on the nut. If the nut has already

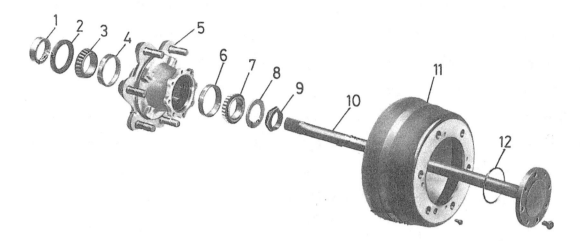

Fig. 8.9 Axleshaft components (LT 35, 40, 45) (Sec 5)

1 Spacer sleeve	4 Bearing outer track	7 Outer bearing	10 Axle halfshaft
2 Oil seal	5 Hub	8 Thrust washer	11 Brake drum
3 Inner bearing	6 Bearing outer track	9 Adjuster nut	12 O-ring

Fig. 8.10 Bolts (arrowed) for withdrawing brake drum (LT 35, 40, 45) (Sec 5)

Fig. 8.11 Bearing adjuster nut staked in position (Sec 5)

been staked twice before, fit a new one.

15 Spin the hub and tighten the nut finger tight until all endfloat just disappears.

16 Stake the nut into the casing groove without altering the adjustment.

17 Check that the axleshaft O-ring seal is in good condition, if not, renew it.

18 Insert the axleshaft into the casing. Hold it horizontally, turning it slightly until it engages in the splines of the bevel gear, and then push it fully home.

19 Screw in and tighten the flange bolts to the specified torque.

20 Fit the brake drum and roadwheel. Check the axle oil level and top up if necessary.

21 Adjust the brake if the adjuster was slackened off at dismantling.

6 Rear axle – removal and refitting

1 Raise the rear of the vehicle, support it securely and remove the roadwheels.

2 Remove the propeller shaft (Chapter 7).

3 Unhook the spring for the brake pressure regulator, but do not disturb the setting of the adjusting screw (see Chapter 9).

4 Disconnect the handbrake cables from the rear brake drums and tap the cable sleeves out of the brake backplates.

5 Disconnect the brake hydraulic pipelines from the wheel cylinders.

6 Unbolt the brake hydraulic 3-way union from the rear axle casing and release the pipelines from their clips. Plug the open lines.

7 On models LT 28 and 31, disconnect the Panhard rod (transverse link) from both the rear axle casing and the body.

8 On models LT 35, 40 and 45, disconnect the anti-roll bar from the rear axle.

9 Lower the vehicle slightly and then raise the rear axle on a trolley jack.

10 Disconnect the shock absorber lower mountings.

11 Unscrew the nuts from the leaf spring U-bolts.

12 Remove the complete rear axle assembly from the vehicle.

13 Installation is a reversal of removal, but observe the following points:

(a) Tighten all nuts and bolts to the specified torque
(b) Remember to reconnect the pressure regulator spring
(c) Bleed the brake hydraulic system (Chapter 9)
(d) Check the axle oil level and top up if necessary.

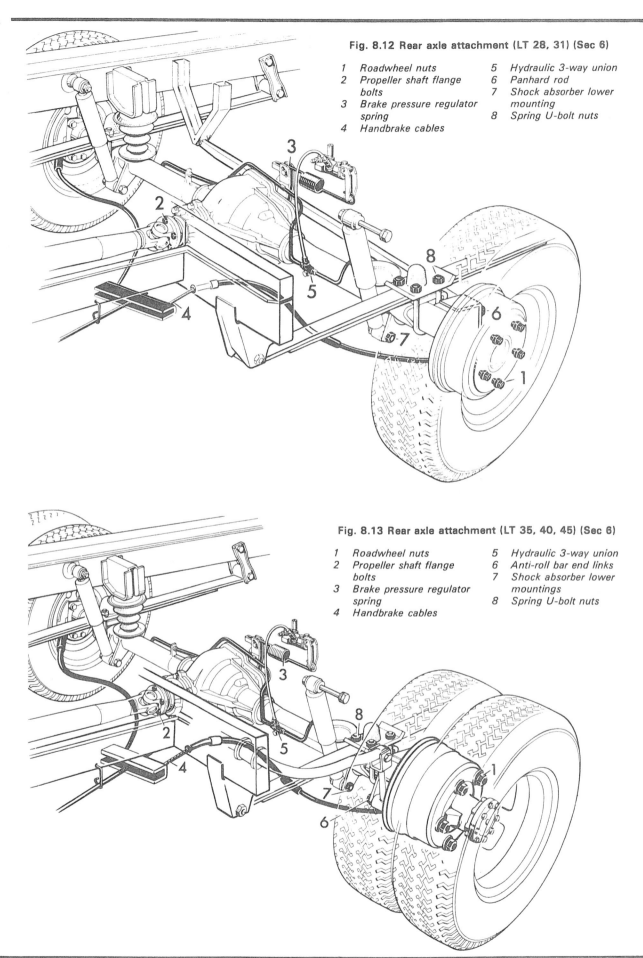

Fig. 8.12 Rear axle attachment (LT 28, 31) (Sec 6)

1   Roadwheel nuts
2   Propeller shaft flange
    bolts
3   Brake pressure regulator
    spring
4   Handbrake cables

5   Hydraulic 3-way union
6   Panhard rod
7   Shock absorber lower
    mounting
8   Spring U-bolt nuts

Fig. 8.13 Rear axle attachment (LT 35, 40, 45) (Sec 6)

1   Roadwheel nuts
2   Propeller shaft flange
    bolts
3   Brake pressure regulator
    spring
4   Handbrake cables

5   Hydraulic 3-way union
6   Anti-roll bar end links
7   Shock absorber lower
    mountings
8   Spring U-bolt nuts

7   Fault diagnosis – rear axle

Symptom	Reason(s)
Oil leakage	Faulty pinion oil seals Faulty axleshaft oil seals Defective cover gasket Blocked axle casing breather
Noise	Lack of oil Worn bearings General wear
'Clonk' on taking up drive, and excessive backlash	Incorrectly tightened pinion nut Worn components Worn axleshaft splines Elongated roadwheel bolt holes

Chapter 9 Braking system

For modifications, and information applicable to later models, see Supplement at end of manual

Contents

Specifications

System type .... 4-wheel hydraulic, dual circuit with servo assistance and pressure regulating valve. Discs front, drums rear, with mechanical handbrake to rear wheels

Disc brakes
Disc diameter	281.0 mm (11.1 in)
Disc thickness:	
New	20.0 mm (0.79 in)
Minimum after refinishing	18.0 mm (0.71 in)
Pad friction material minimum thickness	2.0 mm (0.08 in)

Drum brakes
Drum internal diameter (new):	
LT 28, 31	252.0 mm (9.9 in)
LT 35, 40, 45	270.0 mm (10.6 in)
Refinishing limit	1.5 mm (0.06 in) increase on new diameter
Shoe friction material minimum thickness	2.5 mm (0.10 in)

Hydraulic fluid type .... To SAE J 1703c (DOT 3)

Torque wrench settings
	Nm	lbf ft
Brake disc to hub (M10)	50	37
Brake disc to hub (M14)	165	122
Caliper to steering knuckle (M14)	160	118
Caliper to steering knuckle (M16)	220	163
Backplate to axle (LT 28, 31)	30	22
Backplate to axle (LT 35)	80	59
Backplate to axle (LT 40, 45)	160	118
Rear wheel cylinder bolt (LT 28, 31)	30	22
Rear wheel cylinder bolt (LT 35, 40, 45)	20	15
Roadwheel nuts (LT 28, 31)	200	148
Roadwheel nuts (LT 35, 40, 45)	320	237
Servo unit mounting nuts	15	11

1 Description

The braking system is of four-wheel hydraulic type, with discs at the front and drums at the rear.

A dual hydraulic circuit is used. A pressure regulating valve is incorporated to restrict the pressure to the rear brakes under heavy brake applications and according to load to prevent rear wheel lock-up.

Servo assistance is provided by a vacuum servo unit mounted within the driving cab in conjunction with the hydraulic master cylinder.

The front disc brakes are self-adjusting, as are the rear brakes. The rear adjusters are accessible through holes in the brake backplates.

The handbrake operates mechanically on the rear wheels only by means of cables.

There are differences between some components used according to model. Where this affects overhaul, this is described in the appropriate Section.

2 Maintenance and adjustment

1 At the intervals specified in Routine Maintenance, check the level of fluid in the master cylinder reservoir which is located under a flap in the centre of the facia panel (photo). Topping up should only be required rarely, and then in small quantities, to make up for the slight drop in fluid level caused by normal wear in friction linings and pads. As the linings wear, the pistons move out slightly to effectively increase the fluid capacity of the hydraulic circuits.

2 A rapid drop in fluid level and the need for frequent topping up are indicative of a leak in the system, which must be found and rectified immediately.

3 Also at the specified intervals, check the disc pad and shoe lining wear (Sections 3 and 7).

4 Periodically check the condition of the hydraulic pipes and flexible hoses as described in Section 11.

5 The small air filter in the servo unit must be renewed at the intervals specified in Routine Maintenance. The simplest way to do

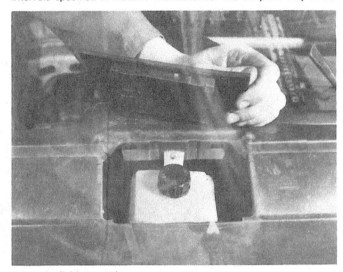

2.1 Brake fluid reservoir

this and so avoid disturbing the pushrod setting is to pull back the dust-excluding bellows at the pushrod, prise out the retaining ring and then cut the damping ring from its outer edge to the pushrod and remove it.

6 Remove the air filter using the same method.

7 Cut the new components and install them, but make sure that the cuts are not in alignment when filter and damping ring are installed.

8 If it is wished to renew the components without cutting them, then the pushrod must be disconnected and the clevis fork removed. When reassembling, refer to Section 13 for details of pushrod adjustment.

9 At three-yearly intervals, the hydraulic fluid should be renewed as by this time it will have absorbed moisture from the atmosphere and will be likely to corrode the internal parts of the hydraulic assemblies.

10 The absorption of moisture into the hydraulic fluid also reduces the boiling point of the fluid, which under heavy or prolonged braking action can cause the formation of vapour bubbles and in turn reduce braking efficiency or loss of braking altogether.

11 To renew the fluid, simply bleed a quantity from each nipple and at the same time introduce new fluid into the master cylinder reservoir. The volume of fluid drawn off should match the suggested quantities in the following table to ensure a complete change of fluid.

Caliper/wheel cylinder bleed nipple	Volume of fluid expelled (cc)
Right rear	500
Left rear	500
Right front (lower)	500
Right front (upper)	100
Left front (lower)	500
Left front (upper)	100
Pressure regulating valve	100

12 Remember that hydraulic fluid is an effective paint stripper. Wash any spillage off the vehicle bodywork immediately with plenty of cold water. It is also poisonous, so avoid syphoning by mouth, and wash after skin contact has occurred.

3 Disc pads – inspection and renewal

1 Raise the front of the vehicle and remove the front roadwheels.

2 Inspect the thickness of the pad friction material. If it has worn down to the specified minimum, the disc pads must be renewed as an axle set (four pads).

3 Should the pads be removed for any reason other than renewal, mark them so that they can be refitted in their original positions,

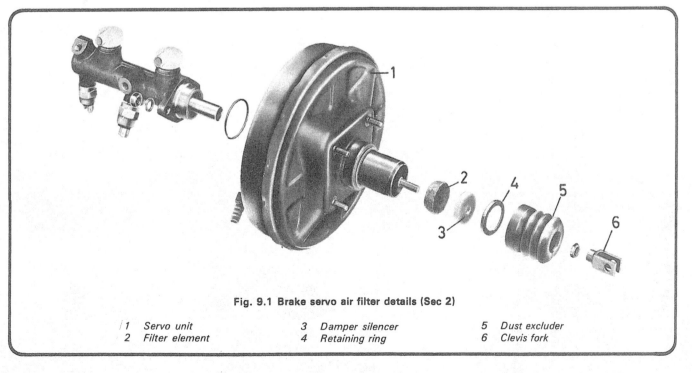

Fig. 9.1 Brake servo air filter details (Sec 2)

1 Servo unit	3 Damper silencer	5 Dust excluder
2 Filter element	4 Retaining ring	6 Clevis fork

otherwise braking will be uneven.

4   Using a thin drift, drive out the pad retaining pins towards the centre-line of the vehicle (photo).

5   Remove the cross-shaped anti-rattle spring.

6   Pull the pads out of the caliper using a pair of pliers (photo).

7   Brush dust and dirt from the caliper jaws, taking care not to inhale the dust.

8   Before the new thicker pads can be installed, the pistons must be pushed back into their bores. A piece of flat metal such as a tyre lever is ideal for this purpose. As the pistons are depressed, the fluid will be displaced from the caliper cylinders and cause the level in the master cylinder reservoir to rise. Anticipate this by syphoning some fluid out of the reservoir. An old hydrometer or poultry baster is useful for this job.

9   Slide the pads into position, making quite sure that the friction side is against the disc.

10 Fit the retainer plate. You will find that the lugs on the plate engage in the recess in the piston rim (photo).

11 Fit the cross-shaped spring and then refit the pad retaining pins (photo).

12 Apply the brake pedal hard several times.

13 Repeat the operations on the opposite side, fit the roadwheels,

lower the vehicle and top up the fluid reservoir to the level marked.

4   Hydraulic components – overhaul (general)

1   Whenever a component of the hydraulic system is overhauled, it must be emphasised that the following five essential requirements are absolutely vital to the maintenance of a safe braking system.

2   Observe scrupulous cleanliness. Never allow grit or dirt to contaminate the hydraulic internal components.

3   When cleaning hydraulic components, use only clean hydraulic fluid, methylated spirit or the brake manufacturer's cleaning solvent – nothing else.

4   Never allow anything but specified hydraulic fluid or rubber grease to enter the system.

5   Always store hydraulic fluid in a closed container as it is hygroscopic (absorbs moisture), which could cause corrosion of the internal parts of a hydraulic assembly.

6   Do not shake the fluid container for the 24 hours prior to its use. Always discard old fluid bled from the system, or if it is retained, use it solely for bleed jar purposes.

3.4 Removing a pad retaining pin

3.6 Withdrawing disc pads

3.10 Fitting disc pad and retainer plate

3.11 Inserting a disc pad retaining pin

5 Caliper – removal, overhaul and refitting

1   Raise the vehicle and remove the front roadwheel.
2   Unscrew and remove the two caliper mounting bolts. On LT 40 and 45 models, the two steering arm bolts must also be removed.
3   Slide the caliper off the disc.

Fig. 9.2 Caliper mounting bolts (LT 28, 31, 35) (Sec 5)

4   Release the flexible hose from the caliper by turning it not more than one quarter of a turn and then, holding the hose to prevent it twisting, unscrew the caliper from it.
5   Quickly cap the hose to prevent loss of fluid.
6   Brush away all external dirt and remove the pads as described in Section 3.
7   From one piston prise out the dust excluder.
8   Using air pressure applied to the hose connecting hole on the caliper, blow the piston out of its cylinder. Only low air pressure should be required to do this, such as produced by a foot or hand-operated tyre pump. Retain the opposing piston with wire, not with the fingers or they may get trapped.
9   Using a sharp pointed instrument, extract the rubber seal from the groove in the cylinder, but take care not to scratch the cylinder bore.
10   Examine the piston and bore for signs of scoring or metal-to-metal rubbed areas. If evident, renew the caliper complete.
11   Where the components are in good condition, obtain a repair kit which will contain all the necessary seals and other renewable items.
12   Install the new piston seal in its groove, using the fingers only to manipulate it into position.
13   Apply rubber grease or clean brake fluid to the piston and cylinder bore. Insert the piston squarely, but only up to the edge of the piston seal.
14   Now turn the piston so that its rim recess is aligned to match the lugs on the retainer plate. The retainer plate can in fact be used to set the piston position and then used to depress the piston into its bore and to prevent the piston from rotating. Special pliers are available to rotate a piston after installation.
15   Do not fully insert the piston until the dust excluder has been fitted. Once this is done, push the piston fully home.

Fig. 9.3 Caliper mounting bolts (LT 40, 45) (Sec 5)

Fig. 9.4 Levering out piston dust excluder (Sec 5)

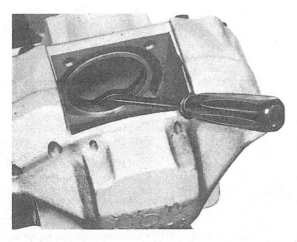

Fig. 9.5 Extracting caliper piston seal (Sec 5)

Fig. 9.6 Caliper piston alignment (rim recess arrowed) (Sec 5)

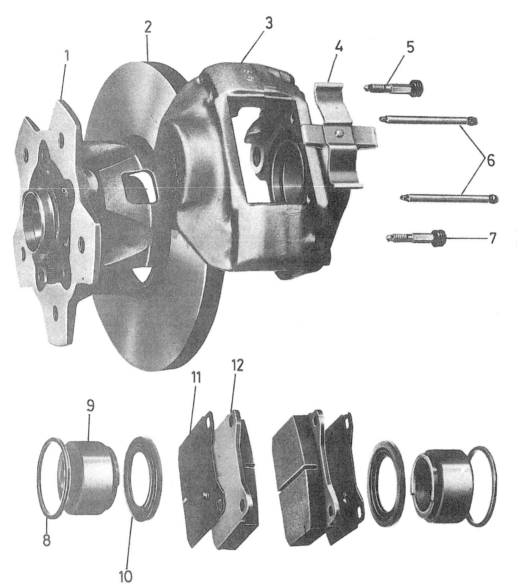

Fig. 9.7 Exploded view of typical
brake caliper (Sec 5)

1   Hub
2   Disc
3   Caliper
4   Pad anti-rattle spring
5   Upper bleed screw
6   Pad retaining pins
7   Lower bleed screw
8   Piston seal
9   Piston
10 Dust excluder
11 Piston retainer
12 Disc pad

Fig. 9.8 Fitting caliper piston dust excluder (Sec 5)

16 Repeat the operations on the opposing piston.
17 Reconnect the flexible hose, screwing the caliper onto the hose
end fitting to prevent the hose being twisted.
18 Install the caliper, tightening the mounting bolts to the specified
torque.
19 Refit the pads.
20 Bleed the hydraulic circuit as described in Section 12.

6   Disc – inspection and renewal

1   Whenever the disc pads are being checked for wear, inspect the
discs themselves. If they are deeply grooved or have an excessive wear
'lip' on the periphery of the disc then the disc must be renewed, or
refinished within the specified limits.
2   If uneven braking or judder has been occurring, suspect an out-of-
true disc and check it using a dial gauge or feeler blades against a fixed
point as the disc is rotated.
3   To remove the disc, raise the vehicle and remove the roadwheel.
4   Unbolt the caliper mounting bolts, also the steering arm bolts on
LT 40 and 45 models.
5   On all models except LT 40 and 45, pull out the hose anchor clip.
6   Slide the caliper off the disc and hang it up with a length of wire

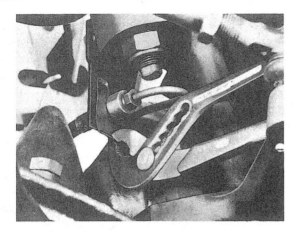

Fig. 9.9 Extracting brake hose anchor clip (Sec 6)

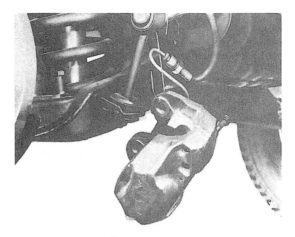

Fig. 9.10 Caliper supported on a piece of wire (Sec 6)

Fig. 9.11 Prising off hub cap (Sec 6)

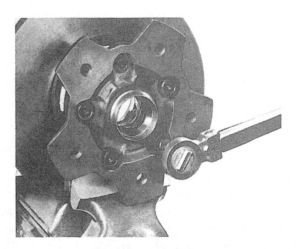

Fig. 9.12 Unscrewing a disc/hub connecting bolt (Sec 6)

Fig. 9.13 Hub nut staked (Sec 6)

Fig. 9.14 Location of front brake hose protective sleeve (Sec 6)

to prevent strain on the hose.

7 Using a hooked tool, prise off the hub grease cap. On LT 40 and 45 models this is deeply recessed in the hub.

8 Unscrew the staked hub nut, remove the thrust washer and pull the hub/disc assembly from the stub axle. Take care that the outer bearing does not drop out.

9 Secure the disc in the jaws of a vice fitted with soft metal protection.

10 Unscrew the hexagonal socket-headed screws which hold the disc and hub together.

11 Refitting is a reversal of removal. Tighten all bolts to the specified torque. Clean off any protective grease from the new disc.

12 The hub bearings must be adjusted by spinning the hub and tightening the nut with the fingers. The final setting of the nut should be made with the fingers until endfloat just disappears.

13 Stake the nut into the axle groove without disturbing the adjustment.

14 If the nut has been used twice before, use a new nut.

15 Check that the protective sleeve on the flexible hose is positioned as shown in Fig. 9.14.

16 Refit the roadwheel and lower the vehicle.

7   Rear shoe linings – inspection and renewal

1   Wear in the rear shoe linings can be observed by looking through the small hole in the brake backplate (photo).
2   If the friction material has worn down to the specified minimum or less, the shoes must be renewed as an axle set (four shoes).
3   Raise the rear of the vehicle and remove the roadwheel.
4   Remove the drum securing screw and pull off the drum. If it is tight, insert a screwdriver into the hole in the backplate and rotate the star wheel adjuster downwards (photo).
5   Slacken the handbrake cable by unscrewing the nuts at the equaliser.
6   If the drum is still stuck, use three M10 x 40 bolts screwed into the tapped holes in the drum to force it off.
7   Unhook the handbrake lever from the brake shoe (photo).
8   Remove the shoe steady cups and springs (photo). Do this by depressing the cup with a pair of pliers and twisting it through 90°. Take off the cup, spring and steady pin.
9   Disconnect the shoe lower spring and pull the shoe lower ends forward out of their anchorages (photo).
10 Unhook the spring from the automatic adjuster lever (photo).
11 Pull the lower ends of the shoes out of the anchorage block.
12 Disconnect the shoe upper return springs.
13 Make a note of which way round the shoes are fitted with reference to leading and trailing ends (area of shoe not covered by lining).

7.1 Removing shoe inspection hole plug

7.4 Removing drum retaining screw

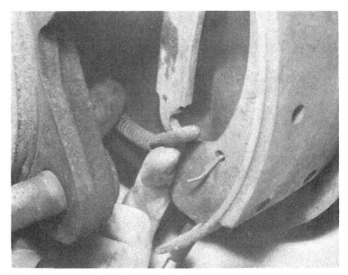

7.7 Disconnecting handbrake cable from shoe lever

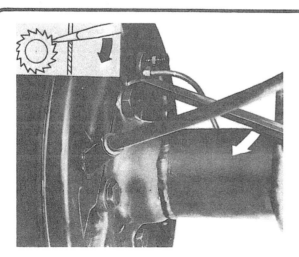

Fig. 9.15 Slackening rear brake shoe adjuster (Sec 7)

Fig. 9.16 Using bolts to remove a brake drum (Sec 7)

7.8 Removing a shoe steady spring cup

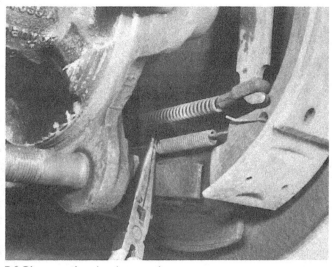

7.9 Disconnecting shoe lower spring

7.10 Disconnecting automatic adjuster lever spring

Fig. 9.17 Shoe upper return springs (Sec 7)

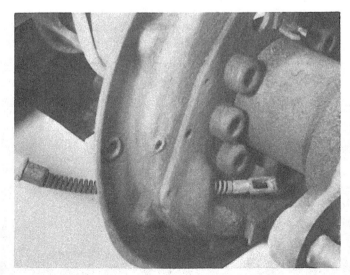

7.16 Rear brake backplate with shoes removed

14 Remove the shoes and the automatic adjuster.

15 Once the shoes are removed, on no account depress the brake pedal or the pistons may be ejected. It is advisable to retain the pistons in the cylinder using a rubber band in case they should drop out.

16 Brush away all dust and dirt from the brake backplate and from the inside of the drum. Take care not to inhale the dust as it contains asbestos which is injurious to health (photo).

17 The new shoes will be supplied complete with new levers already riveted in position. Transfer the original automatic adjuster strut to the new shoes (photos).

18 Refit by engaging the upper ends of the shoes with the pistons.

19 Refit the upper return springs, making sure that the springs are secure on the ends of the backplate lugs. It is important to note how the engagement of the spring ends varies according to vehicle model. Note also that LT 40 and 45 versions have an additional upper spring (refer to the appropriate illustration).

20 Engage the shoe lower ends in the anchorage block.

21 Clean the automatic adjuster pushrod and lightly grease the threads.

22 Lever the shoes apart with a screwdriver and insert the adjuster pushrod between the two shoes (photo).

23 Connect the lower return spring to the shoes.

24 Refit the shoe steady springs and cups.

25 Reconnect the handbrake cable to the shoe lever and check that the cable is really slack.

26 Reconnect the spring to the automatic adjuster lever.

27 Visually centralise the shoes in relation to the brake backplate and

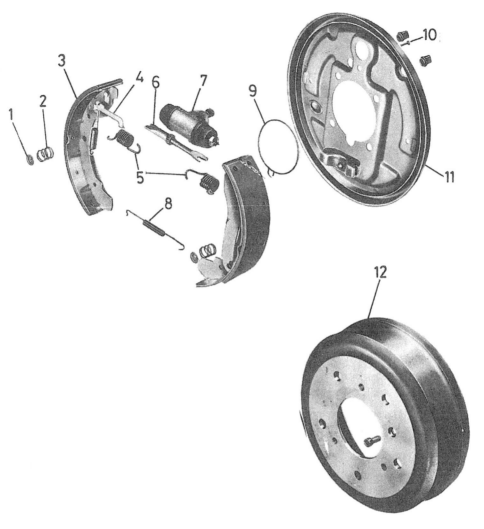

Fig. 9.18 Rear brake components (LT 28, 31 models) (Sec 7)

1	Shoe steady cupped washer	4	Automatic adjuster lever
2	Shoe steady spring	5	Upper return springs
3	Shoe	6	Automatic adjuster strut

7	Wheel cylinder	10	Shoe steady pin
8	Lower return spring	11	Backplate
9	O-ring seal	12	Drum

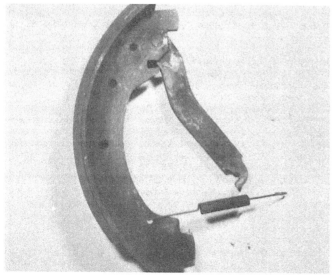

7.17a Shoe with handbrake lever

7.17b Shoe with automatic adjuster lever

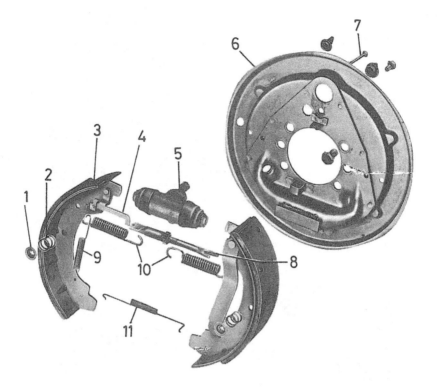

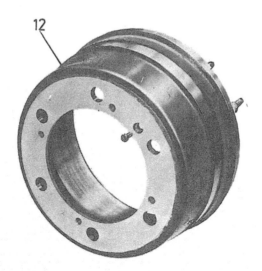

Fig. 9.19 Rear brake components (LT 35) (Sec 7)

1 Shoe steady cupped
   washer
2 Shoe steady spring
3 Shoe

4 Automatic adjuster lever
5 Wheel cylinder
6 Backplate
7 Shoe steady pin

8 Automatic adjuster strut
9 Automatic adjuster lever
   spring

10 Shoe upper return springs
11 Shoe lower return spring
12 Drum

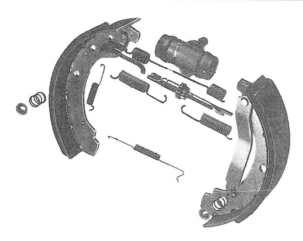

Fig. 9.20 Different shoe return springs and automatic adjuster on
LT 40, 45 models (Sec 7)

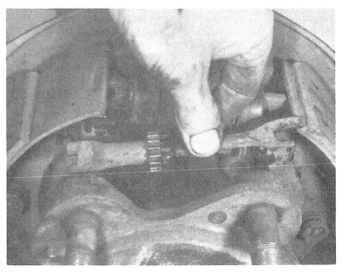

7.22 Refitting the automatic adjuster strut

Fig. 9.21 Connecting shoe lower return spring (Sec 7)

then turn the star wheel on the adjuster pushrod until the brake drum
will only just pass over the shoes.
28 Remove the drum again and adjust the handbrake cable at the
equaliser until play is eliminated between the shoe lever and the
adjuster pushrod.
29 Fit the drum and the roadwheel and lower the vehicle to the
ground.
30 Apply the handbrake several times to fully adjust the shoes to the
drum by means of the automatic adjuster mechanism.

---

8   Rear wheel cylinder – removal, overhaul and refitting

---

1   Remove the brake shoes as described in the preceding Section.
2   Disconnect the hydraulic pipe union at the rear of the cylinder and
quickly cap the end of the pipe to prevent loss of fluid. A bleed nipple
cap is useful for this purpose.
3   Unscrew the bolt which secures the wheel cylinder to the brake
backplate and remove the cylinder.
4   Brush away external dirt and take off the dust excluders.
5   Pull out the pistons. If they are difficult to remove, apply low air
pressure to the fluid inlet union hole.
6   With the pistons removed, take out the coil spring and then
examine the cylinder bore and the piston for scoring or metal-to-metal
rubbed areas. If these are evident, renew the wheel cylinder complete.
7   Where the components are in good condition, discard the old seals
and obtain a repair kit which will contain all the necessary seals and
other renewable components.
8   Use the fingers only to manipulate the new seals into position,
apply rubber grease or clean hydraulic fluid to the pistons and insert

Fig. 9.22 Handbrake cable adjusted to eliminate clearance at point
arrowed (Sec 7)

them squarely into the cylinder with the coil spring between them.
9   Fit the dust excluders.
10 Bolt the cylinder to the backplate and reconnect the hydraulic
pipeline.
11 Fit and adjust the brake shoes as described in Section 7. Refit the
brake drum.
12 Bleed the rear hydraulic circuit (Section 12).

---

9   Master cylinder – removal, overhaul and refitting

---

1   Working under the facia panel, disconnect the leads from the
brake stop-lamp switches.
2   Prise out the reservoir supply pipe unions from the master cylinder
and allow the fluid to drain into a suitable container (photos).
3   Disconnect the pipeline unions from the master cylinder and cap
the open ends of the pipes.
4   Unscrew the two nuts which hold the master cylinder to the face
of the servo unit and withdraw it. Retain the sealing ring.
5   Clean away external dirt.
6   Unscrew and remove the stop screw from the master cylinder.
7   From the end of the master cylinder, extract the circlip, remove the
plastic bush, cup, seal, washer and primary piston.
8   Remove the support ring, the cup seal, support ring, coil spring,
stop sleeve and the stroke limiting screw.

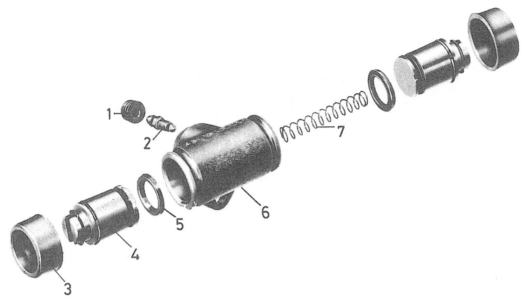

Fig. 9.23 Exploded view of rear wheel cylinder (Sec 8)

1 Dust cap	3 Dust-excluding boot	5 Piston seal	7 Spring
2 Bleed screw	4 Piston	6 Cylinder body	

9.2a Brake master cylinder and servo unit

9.2b Brake fluid reservoir pipe connections

9 Now remove the secondary piston components. These include the cup seals, piston, washer, cup seal, support ring and coil spring.
10 Examine the surfaces of the cylinder bore and the pistons for signs of scoring or metal-to-metal rubbed areas. If evident, renew the cylinder complete.
11 Where the components are in good condition, discard the old seals and obtain a repair kit which will contain all the necessary new seals and other renewable components.
12 Manipulate the seals into position using the fingers only. Take particular care that the seal lips are facing the correct way as shown in Fig. 9.24.
13 When inserting the components into the cylinder, either apply rubber grease or dip them in clean hydraulic fluid, and take care not to trap the seal lips.
14 Depress the primary piston pushrod while screwing in the stop screw.
15 The rubber plugs for the fluid inlet unions are smeared with brake fluid and simply pressed into position.
16 Refit the master cylinder to the servo using a new sealing ring.
17 Bleed the complete hydraulic system (both circuits) as described in Section 12.

10 Pressure regulating valve – testing and adjustment

1 This valve is located on a side-member and actuated from the rear axle through a lever and spring.
2 To confirm that the valve is operating, have an assistant depress the brake pedal and then release it quickly. The lever of the regulator valve should be seen to move.
3 Without a pressure gauge it is impossible to check or adjust the valve, the only rough guide to its setting being the quantity of brake lining dust present in the rear drums. A small quantity indicates that the hydraulic pressure to the rear shoes is being reduced during brake applications.
4 Where a suitable gauge is available, connect it to a rear wheel cylinder in place of the bleed nipple.
5 Have the vehicle standing on its roadwheels in an unladen condition.

LT 28 and 31

6 Measure the distance (a) between the lower face of the rear axle tube and the base of the chassis side-member (Fig. 9.25). Find the

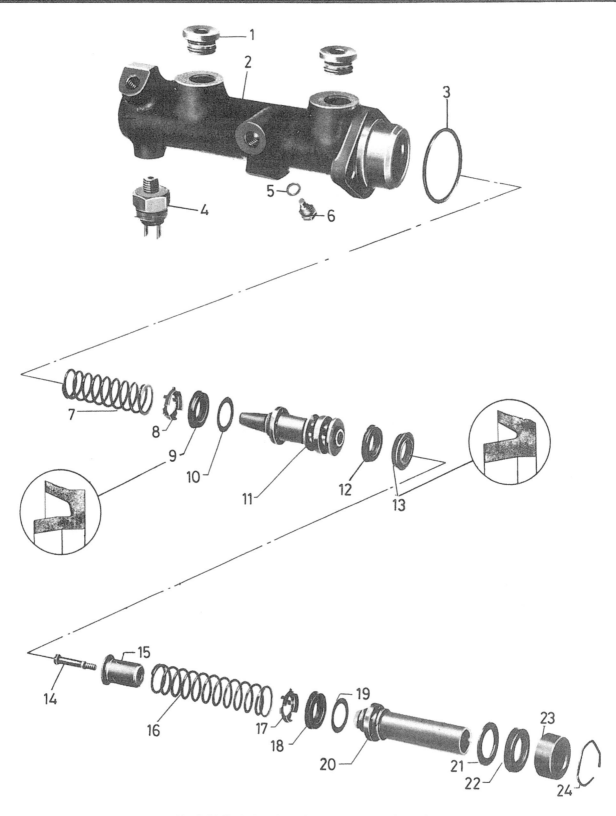

Fig. 9.24 Exploded view of master cylinder (Sec 9)

1 Sealing plug	7 Spring	13 Seal	19 Washer
2 Cylinder body	8 Support ring	14 Stroke limiting screw	20 Primary piston/pushrod
3 O-ring seal	9 Seal	15 Stop sleeve	21 Washer
4 Stop-lamp switch	10 Washer	16 Spring	22 Seal
5 Washer	11 Secondary piston	17 Support ring	23 Dust-excluding boot
6 Stop screw	12 Seal	18 Seal	24 Circlip

Fig. 9.25 Rear axle clearance diagram for setting brake pressure regulator (LT 28, 31) (Sec 10)

For (a) see text

Fig. 9.26 Rear suspension clearance diagram for setting brake pressure regulator (LT 35, 40, 45) (Sec 10)

For (a) see text

Fig. 9.27 Brake pressure regulating valve adjustment (Sec 10)

dimension in the table which follows and read off the pressure. Now compare the pressure figure from the table with the reading on the gauge which will be indicated while the brake foot pedal is held depressed.

Distance a (mm)	Pressures (bar)	
	Pick-up and Van	High roofed Van
300 to 295	7 to 10	4 to 7
295 to 290	10 to 12	7 to 9
290 to 285	12 to 14	9 to 11
285 to 280	14 to 16	11 to 13
280 to 275	16 to 18	13 to 15
275 to 270	18 to 20	15 to 17
270 to 265	20 to 22	17 to 19
265 to 260	22 to 24	19 to 21
260 to 255	24 to 26	21 to 23
255 to 250	26 to 29	23 to 26
250 to 245	29 to 32	26 to 29
245 to 240	32 to 35	29 to 32
240 to 235	35 to 38	32 to 35
235 to 230	38 to 41	35 to 38
230 to 225	41 to 44	38 to 41

LT 35, 40 and 45

7   Measure the distance (a) between the centres of the shock absorber upper and lower mounting bolts (Fig. 9.26). Find the dimension in the table which follows and read off the pressure according to model. Now compare the pressure figure from the table with the reading on the gauge which will be indicated while the brake foot pedal is held depressed.

Distance a (mm)	Pressures (bar)		
	LT 35	LT 40	LT 45
425 to 420	24 to 30	27 to 33	–
420 to 415	30 to 35	33 to 38	–
415 to 410	35 to 41	38 to 44	–
410 to 405	41 to 48	44 to 51	–
405 to 400	48 to 55	51 to 58	–
400 to 395	55 to 62	58 to 66	58 to 65
395 to 390	62 to 69	66 to 74	65 to 72
390 to 385	69 to 77	74 to 82	72 to 80
385 to 380	77 to 84	82 to 91	80 to 88
380 to 375	84 to 93	91 to 101	88 to 97
375 to 370	93 to 101	101 to 111	97 to 107
370 to 365	101 to 110	111 to 120	107 to 116
365 to 360	110 to 119	120 to 131	116 to 128

All models

8   If the pressure is found to be too high, decrease the spring tension (shorten the effective length of the spring) by moving the bolt within the limits of the elongated hole in the anchor plate (Fig. 9.27).
9   If the pressure is too low, increase the spring tension by lengthening the effective length of the spring.
10 Never attempt to adjust the valve while the foot pedal is depressed. Recheck the circuit pressure on completion of the adjustment.

11 Flexible and rigid hydraulic lines – inspection and renewal

1   Inspect the condition of the flexible hydraulic hoses leading to each of the front brake calipers and also the one at the rear axle. If they are swollen, damaged or chafed, they must be renewed.
2   Wipe the top of the brake master cylinder reservoir cap and stick a piece of adhesive tape over the vent hole. This is to stop hydraulic fluid syphoning out during subsequent operations.
3   To remove a front flexible hose, wipe the union and brackets free of dust, and undo the union nuts from the metal pipe ends.
4   Pull out the spring retaining plate and release the flexible hose from its bracket.
5   Remove the rear flexible hose by unscrewing the union nut at the pressure regulating valve and then unscrewing the hose from the three-way connector.
6   Refitting in both cases is the reverse sequence to removal. It will

be necessary to bleed the brake hydraulic system as described in Section 12. If one hose has been removed it is only necessary to bleed either the front or rear brake hydraulic system. Remember to remove the adhesive tape from the reservoir cap.

7   On completion of the work, check that the tyres do not rub against the hoses at full steering lock positions. If they do, the hoses can be given a 'set' in either direction by releasing the spring retainer plate and union nut at the bracket and twisting the hose by not more than one quarter of a turn.

8   At regular intervals wipe the steel brake pipes clean and examine them for signs of rust or denting caused by flying stones.

9   Examine the securing clips, which should be insulated to prevent wear to the pipe surface. Bend the tongues of the clips if necessary to ensure that they hold the brake pipes securely without letting them rattle or vibrate.

10 Check that the pipes are not touching any adjacent components or rubbing against any part of the vehicle. Where this is observed, bend the pipe gently away to clear.

11 Although the pipes are plated any section of pipe may become rusty through chafing and should be renewed. Brake pipes are available to the correct length and fitted with end unions from most VW dealers and can be made to pattern by many accessory suppliers. When installing the new pipes use the old pipes as a guide to bending and do not make any bends sharper than is necessary.

12 The complete system will, of course, have to be bled when the pipes have been reconnected.

12 Hydraulic system – bleeding

1   Two independent hydraulic circuits are used, one for the front brakes, one for the rear.

2   If the master cylinder has been disconnected then the complete system must be bled, but if only a component of one circuit has been disturbed, then only that particular circuit need be bled.

3   Observe the following sequence for bleeding:

Rear circuit
RHD – *Left wheel, right wheel, pressure regulator*
LHD – *Right wheel, left wheel, pressure regulator*
Front circuit
RHD – *Left wheel, right wheel*
LHD – *Right wheel, left wheel*

Bleeding – two-man method

4   Gather together a clean glass jar and a suitable length of flexible tubing which is a tight fit over the bleed screw. Engage the help of an assistant.

5   Before commencing the bleeding operation, check that all rigid pipes and flexible hoses are in good condition and that all hydraulic unions are tight. Take great care not to allow hydraulic fluid to come into contact with the vehicle paintwork, otherwise the finish will be seriously damaged. Wash off any spilled fluid immediately with cold water.

6   If hydraulic fluid has been lost from the master cylinder, due to a leak in the system, ensure that the cause is traced and rectified before proceeding further or a serious malfunction of the braking system may occur.

7   To bleed the system, clean the area around the bleed screw at the wheel cylinder to be bled. If the hydraulic system has only been partially disconnected and suitable precautions were taken to prevent further loss of fluid, it should only be necessary to bleed that part of the system. However, if the entire system is to be bled, start at the wheel furthest away from the master cylinder.

8   Remove the master cylinder filler cap and top up the reservoir. Periodically check the fluid level during the bleeding operation and top up as necessary.

9   Destroy the vacuum in the servo by giving several applications of the brake pedal in rapid succession.

10 Connect one end of the flexible tubing to the bleed screw and immerse the other end in the glass jar containing sufficient clean hydraulic fluid to keep the end of the tube submerged. Open the bleed screw half a turn and have your assistant depress the brake pedal to the floor and then slowly release it. Tighten the bleed screw at the end of each downstroke to prevent expelled air and fluid from being drawn back into the system. Repeat this operation until clean hydraulic fluid,

free from air bubbles, can be seen coming through the tube. Now tighten the bleed screw and remove the flexible tube.

Bleeding – using one-way valve kit

11 There are a number of one-man, do-it-yourself, brake bleeding kits currently available from motor accessory shops. It is recommended that one of these kits should be used wherever possible as they greatly simplify the bleeding operation and also reduce the risk of expelled air and fluid being drawn back into the system.

12 If a one-man brake bleeding kit is being used, connect the outlet tube to the bleed screw and then open the screw half a turn. If possible position the unit so that it can be viewed from the car, then depress the brake pedal to the floor and slowly release it. The one-way valve in the kit will prevent dispelled air from returning to the system at the end of each stroke. Repeat this operation until clean hydraulic fluid, free from air bubbles, can be seen coming through the tube. Now tighten the bleed screw and remove the outlet tube.

Bleeding – using a pressure bleeding kit

13 These too are available from motor accessory shops and are usually operated by air pressure from the spare tyre.

14 By connecting a pressurised container to the master cylinder fluid reservoir, bleeding is then carried out by simply opening each bleed nipple in turn and allowing the fluid to run out, rather like turning on a tap, until no air is visible in the fluid.

15 Using this system, the large reserve of hydraulic fluid provides a safeguard against air being drawn into the master cylinder during the bleeding operation.

16 This method is particularly effective when bleeding 'difficult' systems and when bleeding the entire system at time of routine fluid renewal.

All systems

17 If the entire system is being bled the foregoing procedures should now be repeated at each wheel, finishing at the wheel nearest to the master cylinder. Do not forget to recheck the fluid level in the master cylinder at regular intervals and top up as necessary.

18 When completed, recheck the fluid level in the master cylinder, top up if necessary and refit the cap. Check the 'feel' of the brake pedal which should be firm and free from any 'sponginess' which would indicate air still present in the system.

19 Discard any expelled hydraulic fluid as it is likely to be contaminated with moisture, air and dirt which makes it unsuitable for further use.

13 Vacuum servo unit – removal and refitting

1   The vacuum servo unit is mounted in conjunction with the brake master cylinder. It is recommended that the master cylinder attachments are disconnected as described in Section 9 and the master cylinder and servo removed together as one assembly.

2   Disconnect the servo pushrod from the pedal arm by pulling out the split pin and the clevis pin.

3   Disconnect the vacuum hose from the servo unit.

4   Unscrew and remove the nuts which hold the servo to the mounting bracket and withdraw the servo/master cylinder as an assembly.

5   Unbolt the brake master cylinder from the servo unit.

6   If the servo unit has proved faulty or inoperative, it should be renewed complete as apart from an air filter repair kit (see Section 2), spare parts are not available.

7   If a new servo unit is being installed, check the pushrod projection as shown in Fig. 9.29 and adjust its length where necessary by releasing the clevis fork locknut and turning the fork.

8   Refitting is a reversal of removal. Bleed the complete hydraulic system on completion as described in Section 12.

14 Handbrake – adjustment

1   Under normal circumstances, the handbrake is automatically adjusted by the action of the rear shoe automatic adjusters. However, due to stretching of the cable it is occasionally necessary to eliminate some slackness in the cable if the handbrake lever has to be pulled

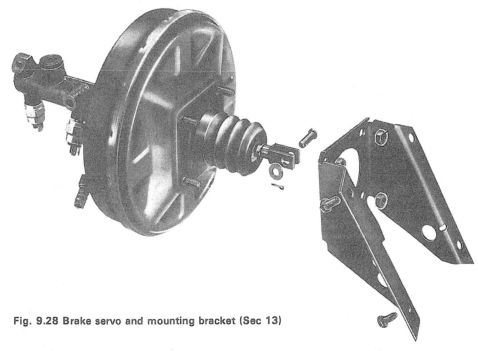

Fig. 9.28 Brake servo and mounting bracket (Sec 13)

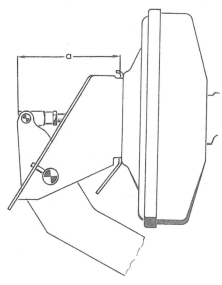

Fig. 9.29 Servo pushrod adjusting diagram (Sec 13)

a = 114.0 mm (4.49 in)

Fig. 9.30 Handbrake adjuster nut and locknut (Sec 14)

over more than four notches in order to lock the wheels.

2   With the handbrake fully off, slacken the nuts at the equaliser until the cable is quite slack.

3   Pull the handbrake lever on two notches.

4   Adjust the cable by means of the equaliser nuts until the rear roadwheels can still be turned freely by hand, but any further tightening of the nut would cause the brake shoes to bind.

5   Tighten the locknut at the equaliser.

15 Handbrake cable – renewal

1   Release the handbrake and then disconnect the primary rod from the equaliser by unscrewing and removing the locknut and adjuster nut.

2   Slip the cable end fitting from the equaliser.

3   Remove the rear roadwheels and brake drums as described earlier in this Chapter and release the cable ends from the shoe levers.

4   Drive the cable sleeves out of the brake backplate and release the cables from their guides.

5   The handbrake lever, the intermediate lever and the intermediate or primary rods can be dismantled once the pivot pins have been withdrawn. These are retained by circlips.

6   Reassembly and refitting of the new cables are reversals of removal. Adjust on completion as described in Section 14 and lubricate all pivot points.

16 Brake pedal – removal and refitting

1   The pendant type foot pedal is connected directly to the servo pushrod and pivots on a cross-shaft locked into the support bracket.

2   Disconnect the pedal arm from the pushrod clevis fork by pulling out the split pin and the clevis pin.

3   Using a screwdriver, release the pedal return spring hooked ends.

4   Drive out the roll pin from the end of the cross-shaft, withdraw the shaft and take off the pedal. Removing the roll pin is a difficult job and it is sometimes found easier to unbolt and remove the complete pedal bracket to provide better access to the roll pin.

5   The pedal arm bush can be removed by driving it out, and a new one installed by pressing it in using a vice.

6   Refitting is a reversal of removal. Apply grease to the bush and to all pivot points.

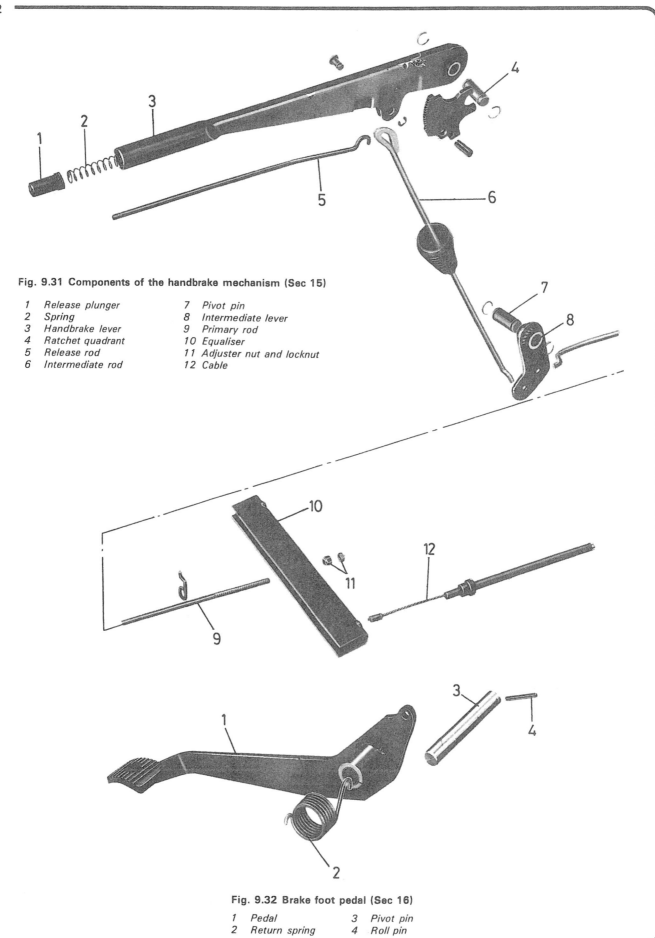

Fig. 9.31 Components of the handbrake mechanism (Sec 15)

1	Release plunger	7	Pivot pin
2	Spring	8	Intermediate lever
3	Handbrake lever	9	Primary rod
4	Ratchet quadrant	10	Equaliser
5	Release rod	11	Adjuster nut and locknut
6	Intermediate rod	12	Cable

Fig. 9.32 Brake foot pedal (Sec 16)

1	Pedal	3	Pivot pin
2	Return spring	4	Roll pin

17 Fault diagnosis – braking system

Symptom	Reason(s)
Pedal travels almost to floor before brakes operate	Brake fluid too low Caliper leaking Master cylinder leaking (bubbles in master cylinder fluid) Brake flexible hose leaking Brake line fractured Brake system unions loose Rear automatic adjusters seized
Brake pedal feels springy	New linings not yet bedded-in Brake discs or drums badly worn or cracked Master cylinder securing nuts loose
Brake pedal feels spongy and soggy	Caliper or wheel cylinder leaking Master cylinder leaking (bubbles in master cylinder reservoir) Brake pipe, line or flexible hose leaking Unions in brake system loose Air in hydraulic system
Excessive effort required to brake car	Pad or shoe linings badly worn New pads or shoes recently fitted – not yet bedded-in Harder linings fitted than standard causing increase in pedal pressure Lining and brake drums contaminated with oil, grease or hydraulic fluid Servo unit inoperative or faulty
Brakes uneven and pulling to one side	Linings and discs or drums contaminated with oil, grease or hydraulic fluid Tyre pressures unequal Brake caliper loose Brake pads or shoes fitted incorrectly Different type of linings fitted at each wheel Anchorages for front suspension or rear suspension loose Brake discs or drums badly worn, cracked or distorted
Brakes tend to bind, drag or lock-on	Air in hydraulic system Wheel cylinders seized Handbrake cables too tight
Rear wheels lock prematurely in heavy braking	Pressure regulating valve defective or maladjusted

Chapter 10 Electrical system

For modifications, and information applicable to later models, see Supplement at end of manual

Contents

Specifications

General

System type	12V negative earth, battery, alternator and pre-engaged starter motor
Battery capacity (typical)	45 Ah

Alternator

Make	Bosch or Motorola
Load current	25A at 3000 rpm
Regulating voltage	12.5 to 14.5
Brush length:	
New	10.0 mm (0.39 in)
Wear limit	5.0 mm (0.20 in)

Starter motor

Type and make	Pre-engaged, Bosch
Brush wear limit	13.0 mm (0.5 in)
Commutator minimum diameter	33.5 mm (1.32 in)
Commutator endfloat	0.1 to 0.3 mm (0.004 to 0.012 in)
Pinion clearance (pinion to stop)	0.25 to 0.50 mm (0.010 to 0.020 in)

Fuses (typical)

Number	Circuit protected	Rating
1	Dipped headlamp (LH)	8A
2	Dipped headlamp (RH)	8A
3	Main beam headlamp (LH)	8A
4	Main beam headlamp (RH)	8A
5	Heated rear window (element)	16A
6	Stop lamp, hazard warning lamps	8A
7	Interior lamps	8A
8	Direction indicator and warning lamps	8A
9	Horn, fuel cut-off valve, reversing lamps	8A
10	Heater blower	8A
11	Windscreen wipers, washers, heated rear window (switch)	8A
12	Spare	
13	Front parking, rear lamp (RH)	8A
14	Front parking, rear lamp (LH), rear number plate, rear fog lamps	8A
15	Fuel pump	8A

Torque wrench settings

	Nm	lbf ft
Alternator mounting bolt	25	18
Alternator mounting bracket bolts	40	30
Alternator adjuster link bolts	20	15
Alternator pulley nut	45	33
Starter motor mounting bolts	75	55
Wiper arm-to-spindle nut	6	4

1 Description

The electrical system is of 12 volt, negative earth type and includes a battery, a belt-driven alternator and a pre-engaged starter.

All electrical circuits are fused, and certain components and accessories operate via relays.

There are variations in the design and location of some lamps, depending upon body type. Van versions are equipped with reversing lamps.

2 Battery – care and maintenance

1   The battery on modern vehicles requires very little attention, but regular visual inspection is essential.
2   At weekly intervals, check the electrolyte level in each cell. Where necessary, add distilled or demineralised water only to bring the level up to the mark on the casing (translucent type), or to 6.0 mm ($\frac{1}{4}$ in) above the plate separators on solid casing types.
3   Topping up should only be needed at infrequent intervals. Otherwise, suspect a leak in the battery casing, or overcharging caused by a faulty regulator in the alternator.
4   Keep the battery terminals clean and tight and smear them with petroleum jelly.
5   Keep the top of the battery clean. If any corrosion is found either on the battery terminals or the battery mounting platform, neutralise it with household ammonia or a solution of sodium bicarbonate (baking powder).
6   Battery condition can only be satisfactorily checked with a battery tester but if a hydrometer is available, this can be used to check the specific gravity of each cell and the readings compared with the following table.
7   Any deviation in the specified reading for one or more cells may indicate that the battery is failing and a replacement is required.

Fully discharged	Electrolyte temperature	Fully charged
1.098	38°C (100°F)	1.268
1.102	32°C (90°F)	1.272
1.106	27°C (80°F)	1.276
1.110	21°C (70°F)	1.280
1.114	16°C (60°F)	1.284
1.118	10°C (50°F)	1.288
1.122	4°C (40°F)	1.292
1.126	−1.5°C (30°F)	1.296

8   Where necessary as a means of prolonging the useful life of the battery, to charge it after the vehicle has remained unused for a long period, or to offset excessive current drain caused by unusual use of electrical accessories or insufficient charging mileage, the battery can be charged from an external source.
9   Before connecting a mains charger, disconnect the vehicle battery leads.

3 Battery – precautions

1   Never smoke or bring a naked flame near the battery vent plugs, or an explosion could result.
2   Never disconnect a battery terminal as a means of stopping the engine.
3   To isolate the electrical system, disconnect the negative battery lead. When disconnecting both leads, always disconnect the negative lead first.
4   When jump starting from the battery of another vehicle, make sure that the cables from one battery to the other are connected negative to negative and positive to positive.

4.3 Battery is clamped at base

5   Always disconnect the battery before carrying out any electric welding on the vehicle.
6   Remember that battery electrolyte is corrosive. Neutralise any spillage immediately. Wash electrolyte off skin or clothes with plenty of cold water; seek medical advice immediately if electrolyte is splashed into the eyes or swallowed.

4 Battery – removal and refitting

1   The battery is located behind the left-hand seat in the driving cab. Remove the plastic cover.
2   Disconnect the battery terminals.
3   Unbolt and remove the clamp plate which secures the battery to the floor (photo).
4   Remove the battery, keeping it level as it is lifted out.
5   Refitting is a reversal of removal. Do not overtighten the clamp plate.

5 Alternator – maintenance

1   The only regular maintenance required is to check the tension of the drivebelt and to adjust as necessary. This is described in Chapter 2, Section 6.
2   Occasionally wipe away dirt and grease from the alternator body, and check that the wiring plug at its rear face is secure.

6 Alternator – removal, overhaul and refitting

1   Disconnect the battery leads.
2   Note the terminal connections at the rear of the alternator and disconnect the plug or multi-pin connector.
3   Undo and remove the alternator adjustment arm bolt, slacken the alternator mounting bolts and push the alternator inwards towards the engine. Lift away the drivebelt from the pulley.
4   Remove the remaining mounting bolt and carefully lift the alternator away.
5   Take care not to knock or drop the alternator otherwise this can cause irreparable damage.

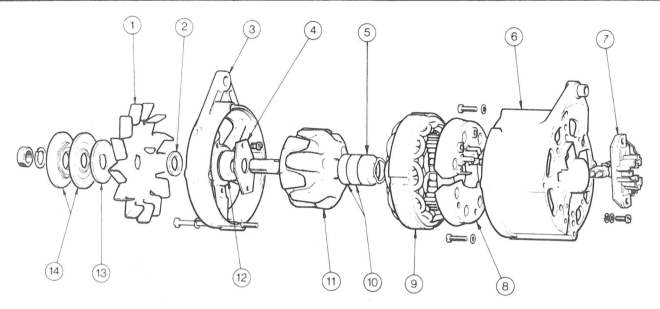

Fig. 10.1 Exploded view of Bosch alternator (Sec 6)

1 Fan	5 Slip ring end bearing	9 Stator	12 Drive end bearing
2 Spacer	6 Slip ring end housing	10 Slip rings	13 Spacer
3 Drive end housing	7 Brush box and regulator	11 Rotor	14 Pulley
4 Thrust plate	8 Rectifier (diode) pack		

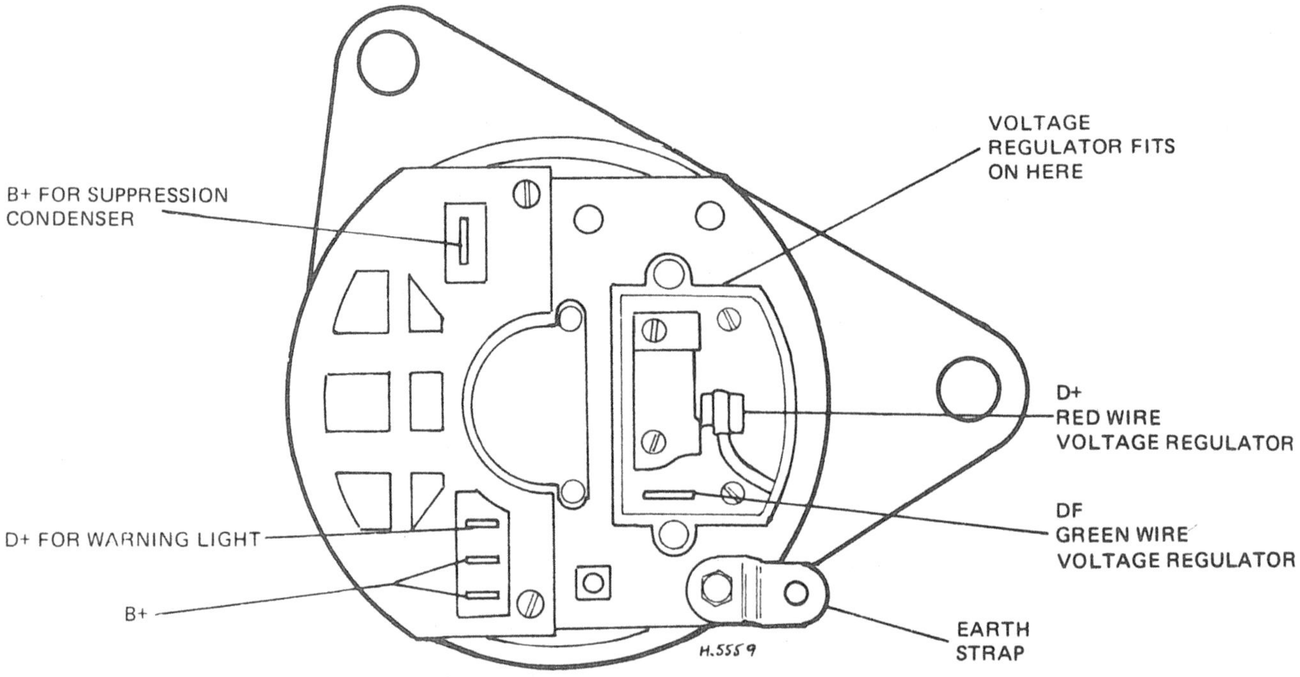

Fig. 10.2 Rear view of Motorola alternator with regulator removed, showing connections (Sec 6)

6   Due to the specialist knowledge and equipment required to test or service an alternator it is recommended that if the performance is suspect, the car be taken to an automobile electrician who will have the facilities for such work. Because of this recommendation, information is limited to the inspection and renewal of the brushes. Should the alternator not charge or the system be suspect, the following points may be checked before seeking further assistance:

(a)   Check the drivebelt tension, as described in Chapter 2
(b)   Check the battery, as described in Section 2
(c)   Check all electrical cable connections for cleanliness and security

7   Undo and remove the two screws, spring and plain washers that secure the brush box to the rear of the brush end housing. Lift away the brush box.
8   Check that the carbon brushes are able to slide smoothly in their guides without any sign of binding.
9   Measure the length of the brushes. If they have worn down to the specified minimum or less, they must be renewed.
10   Hold the brush wire with a pair of pliers and unsolder it from the brush box. Lift away the two brushes.
11   Insert the new brushes and check to make sure that they are free to move in their guides. If they bind, lightly polish with a very fine file.
12   Solder the brush wire ends to the brush box.

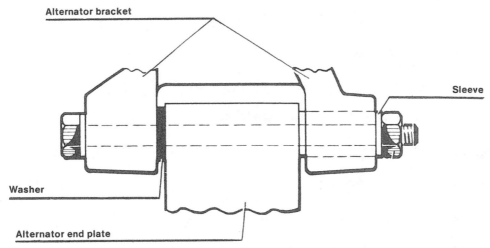

Fig. 10.3 Alternator mounting pivot bolt (Sec 6)

13 Whenever new brushes are fitted new springs should be fitted also.

14 Refitting the brush box is the reverse sequence to removal.

15 On some models a Motorola alternator is fitted instead of the Bosch type. Overhaul operations are similar.

16 Refitting the alternator is a reversal of removal, but note the location of the washer and sleeve on the mounting bolt. Tension the drivebelt as described in Chapter 2.

---

7 Starter motor – description and testing in vehicle

1 The starter motor is of pre-engaged type and is operated by the combined ignition/starter switch mounted on the side of the steering column.

2 If the starter motor operates but only very slowly, this is probably due to a discharged battery, or to loose or corroded battery or starter motor terminals.

3 If the starter motor does not turn the engine although the battery is well charged, suspect sticking or worn brushes or a dirty or scored commutator.

---

8 Starter motor – removal and refitting

1 Disconnect the battery negative terminal.

2 Unbolt the exhaust downpipe from the manifold flange.

3 Disconnect the leads from the starter motor and solenoid terminals.

4 Unscrew and remove the mounting bolts and withdraw the starter motor from the clutch bellhousing.

5 Refitting is a reversal of removal. Tighten the mounting bolts to the specified torque.

---

9 Starter motor – overhaul

1 Servicing operations should be limited to renewal of brushes, renewal of the solenoid, the overhaul of the starter drive gear and cleaning the commutator.

2 The major components of the starter should normally last the life of the unit and in the event of failure, a factory exchange replacement should be obtained.

3 Disconnect the motor field winding lead from the solenoid terminal.

4 Remove the screws which secure the solenoid to the drive end cover and withdraw the solenoid, at the same time unhooking it from the drive engagement lever.

5 Remove the dust cap from the end of the starter motor and extract the lockwasher, shims and gasket.

6 Unscrew the tie-bolts and withdraw the motor end cover.

7 Extract the carbon brushes from their holders and remove the

Fig. 10.4 Ignition key positions (Sec 7)

1 Off   3 Start
2 Ignition on (steering
   unlocked)

Fig. 10.5 Starter solenoid terminals (Sec 8)

1 From battery   3 From ignition/starter switch
2 To ignition coil   4 Field winding connection

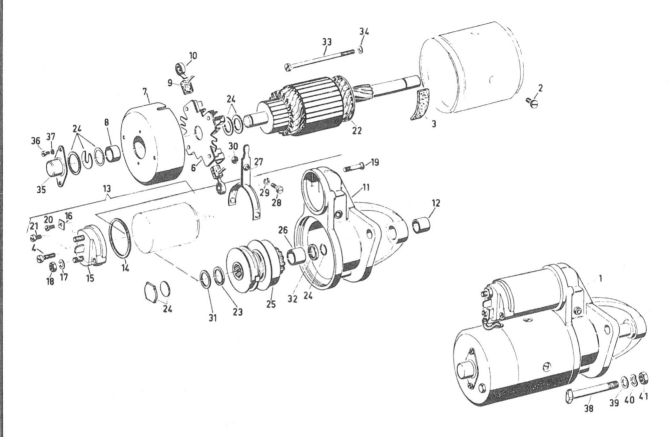

Fig. 10.6 Exploded view of Bosch starter motor (Sec 9)

1	Starter	12	Bearing sleeve	22	Armature
2	Screw	13	Solenoid	23	Seal
3	Insulator	14	Shim	24	Shims and C-washer
4	Screw with lockwasher	15	Solenoid cover	25	Gear
6	Brush holder	16	Clamp	26	Bearing sleeve
7	Commutator bearing	17	Lockwasher	27	Operating lever
8	Bearing sleeve	18	Nut	28	Screw
9	Carbon brush	19	Screw	29	Lockwasher
10	Spring	20	Screw	30	Hex nut
11	Drive bearing	21	Screw	31	Stop washer

32	Stop ring
33	Screw
34	Plain washer
35	Dust cap
36	Screw
37	Washer
38	Hammer head screw
39	Plain washer
40	Washer
41	Hex nut

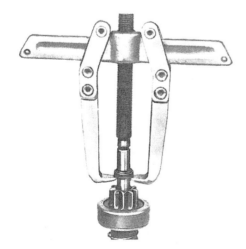

Fig. 10.7 Drawing starter drive stop ring over jump ring (Sec 9)

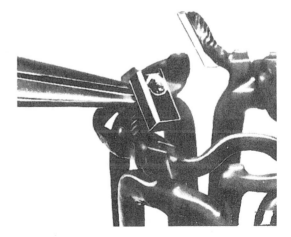

Fig. 10.8 Starter motor brush lead ready for soldering (Sec 9)

brush mounting plate.

8   Withdraw the yoke from the drive end cover and then unscrew the engagement lever pivot bolt.

9   Withdraw the armature complete with engagement lever.

10   Using a piece of suitable tubing, drive the stop ring back up the armature shaft to expose the jump ring. Extract the jump ring and pull off the starter drive components.

11   Measure the overall length of each of the two brushes and where they are worn below the minimum recommended (see Specifications) renew them by unsoldering and resoldering. Ensure that each brush slides freely in its holder. If necessary, rub with a fine file and clean any accumulated carbon dust or grease from the holder with a fuel moistened rag.

12   Normally, the commutator may be cleaned by holding a piece of non-fluffy rag moistened with fuel against it as it is rotated by hand. If on inspection, the mica separators are level with the copper segments then they must be undercut by between 0.020 and 0.032 in (0.5 and 0.8 mm). Undercut the mica separators of the commutator using an old hacksaw blade ground to suit. The commutator may be polished with a piece of very fine glass paper – never use emery cloth as the carborundum particles will become embedded in the copper surfaces.

13   Wash the components of the drive gear in paraffin and inspect for wear or damage, particularly to the pinion teeth, and renew as appropriate.

14   Reassembly is a reversal of dismantling. Oil the sliding surfaces of the pinion assembly with a light oil, applied sparingly.

15   Note that the endfloat of the armature must be within the specified limits. This can be adjusted by varying the shims located under the starter motor dust cap.

10 Fuses and relays – general

1   The fuse box is located under the left-hand side of the instrument panel (photo).

2   The circuits protected are marked on the lid of the fuse block. Refer also to the Specifications at the beginning of this Chapter.

3   Always renew a blown fuse with one of similar capacity and if it blows again immediately, trace the fault – very often faulty wiring insulation.

4   Certain electrical accessories are operated via relays which are mounted on the fuse block. These relays include the following circuits:

Headlamp dimmer and flasher
Heated rear window
Hazard warning lamps and direction indicators

5   Never be tempted to bypass a persistently blowing fuse with silver foil or wire. Serious damage or fire could result.

11 Direction and hazard warning units – general

1   These are included in the relays referred to in the preceding Section.

2   Failure of a lamp in one of these circuits may be due to the bulb itself but if the frequency of flashing increases then it may be due to a poor earth connection at one of the bulb holders or lamp units.

10.1 Fuses and relays

3   Once these two items have been checked, the relay must be suspected. Renew it by pulling the original unit from its sockets and pushing the new one into position. No repair is possible.

12 Steering column switch – removal and refitting

1   Remove the steering wheel as described in Chapter 11.

2   Disconnect the lead from the battery negative terminal.

3   Unscrew and remove the four screws which retain the steering column combination switch (photo).

4   Disconnect the switch wiring at the connector plug on the side of the steering column (photo).

5   Withdraw the switch assembly from the steering column (photo).

6   Refitting is a reversal of removal.

13 Steering column lock – removal and refitting

1   Disconnect the battery.

2   Remove the steering wheel as described in Chapter 11.

3   Remove the steering column switches as described in the preceding Section.

4   Pull the column upper shroud upwards and remove it.

5   On early models (up to October 1975), pull out the locking plate with a pair of pliers.

6   Extract the cross-head screw which secures the centre shroud section to the steering lock and remove the shroud.

7   Pull the lower shroud section off against the tension of its fixing clip.

8   Drill out the shear head bolt which holds the column lock. If the

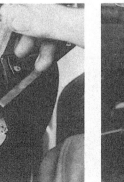

12.3 Removing steering column switch screws

12.4 Disconnecting steering column switch wiring

12.5 Removing steering column switch

bolt is drilled to take an 'easy out' bolt extractor or similar, the shear head bolt can be removed without difficulty.

9   Disconnect the ignition switch wiring harness and remove the lock/switch assembly.

10 Refitting is a reversal of removal. Use a new shear head bolt, but do not shear off the bolt head until the gap between the top edge of the upper shroud and the lower edge of the steering wheel hub has been adjusted to between 2.0 and 3.0 mm (0.079 and 0.118 in). Do this by altering the position of the lock. Check the lock for correct operation and then tighten the bolt until its head shears off.

14 Ignition/starter switch – removal and refitting

Up to 1975
1   Remove the steering column lock as described in the preceding Section.

2   Extract the small screw and remove the ignition/starter switch from the lock assembly.

October 1975 on
3   To remove the ignition/starter switch on these later models, drill a 3.0 mm (0.118 in) diameter hole in the lock housing in accordance with Fig. 10.11. The hole will provide access to the cylinder retaining plunger which should be depressed to release the lock cylinder.

All models
4   Refitting the switch is a reversal of removal.

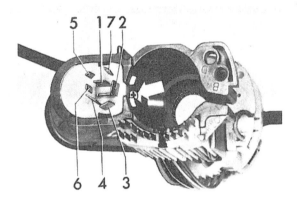

Fig. 10.10 Ignition/starter switch viewed from below (refer to Wiring Diagram). Securing screw is arrowed (Sec 14)

1	Terminal 30	5	Terminal P
2	Terminal 50	6	Terminal S
3	Terminal 15	7	Terminal G
4	Terminal X		

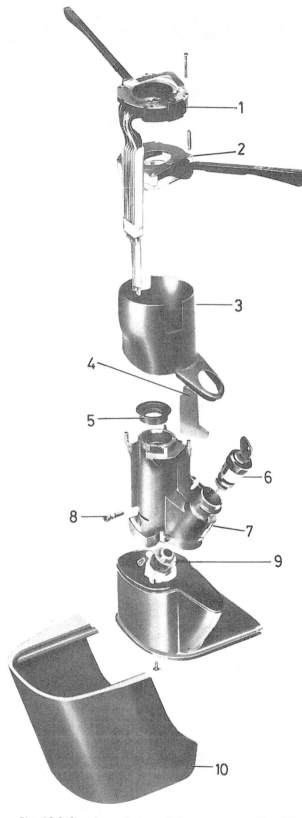

Fig. 10.9 Steering column switch arrangement (Sec 13)

1	Direction indicator switch	6	Lock cylinder
2	Windscreen wiper/washer switch	7	Steering lock housing
3	Column upper shroud	8	Shear bolt
4	Locking plate (early models)	9	Column centre shroud
5	Bearing (non-removable)	10	Column lower shroud

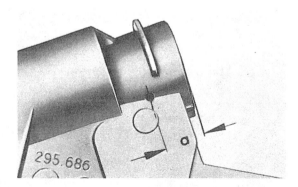

Fig. 10.11 Ignition switch removal drilling diagram (later models) (Sec 14)

a = 13.0 mm (0.5 in)

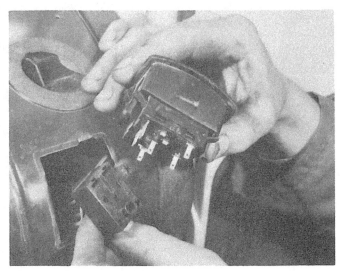

15.3 Removing lighting switch

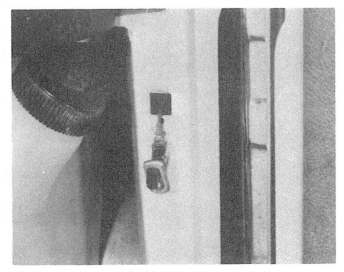

15.6 Courtesy lamp switch withdrawn

15 Switches (general) – removal and refitting

1   The facia-mounted switches can be removed once the battery has been disconnected.

2   Depress the switch retaining tabs, accessible by reaching up behind the facia panel, and pull the switch forward.
3   Identify the wiring connections and disconnect the leads or connecting plug (photo).
4   Refitting is a reversal of removal.
5   The horn switch located in the centre of the steering wheel is described in Chapter 11.
6   The courtesy lamp switches are secured by a single screw (photo).

16 Headlamp bulb – renewal

1   Extract the headlamp panel screws and remove the panel (photo).
2   Extract the headlamp rim retaining screws and withdraw the headlamp. Do not move the headlamp beam adjusting screws (photo).

Standard bulb
3   From the rear of the headlamp, pull off the wiring plug and take off the rubber dust cap (photo).
4   Twist the bulb holder retaining ring and remove it.
5   Withdraw the bulb holder/bulb assembly (photo).
6   Install the new bulb holder so that its alignment lug is correctly engaged.
7   Refit the retainer, plug and dust cap.
8   Refit the lamp unit and screw in the retaining ring screws.

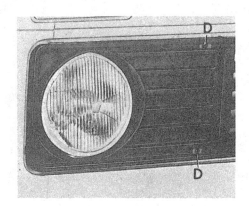

Fig. 10.12 Headlamp panel securing screws (D) (Sec 16)

16.1 Removing headlamp trim panel

Fig. 10.13 Headlamp screws (Sec 16)

A   Adjustment screw   C   Securing screws
B   Adjustment screw

16.2 Removing headlamp retaining ring screw

Fig. 10.14 Bulb (standard type) components (Sec 16)

1   Dust cap   3   Bulb/holder
2   Retaining ring   4   Front parking bulb

16.3 Removing headlamp dust-excluding cap

16.5 Removing headlamp bulb

Halogen bulb

9   From the rear of the headlamp, pull off the wiring plug and take off the rubber dust cap.

10   Squeeze the ends of the retaining clip together and swivel the clip away.

11   Lower the bulb with a piece of paper and remove it. Fit the new bulb, again holding it with paper, and making sure that the centre tab is uppermost. Never finger a halogen type bulb as the grease deposited on the glass will shorten the life of the bulb. If the bulb is accidentally touched, clean it with methylated spirit.

12   Swivel the retaining clip over the bulb holder, squeeze the ends of the clip together and engage them in their retaining lugs.

13   Install the rubber dust cap and connect the wiring plug.

14   Refit the lamp unit and screw in the retaining screws.

17   Headlamps – beam alignment

1   Where possible, have the headlamp beams checked annually in the autumn by your dealer on optical beam setting equipment. Where this is not possible, or in an emergency, the following procedure may be used.

2   Position the vehicle on level ground, 10 ft (3 m) from a wall or screen during the hours of darkness. The wall or board must be at

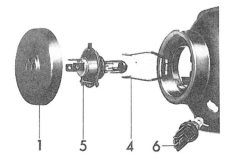

Fig. 10.15 Bulb (Halogen type) components (Sec 16)

1   Dust cap   5   Bulb/holder
4   Retaining spring clip   6   Front parking bulb

right-angles to the centre-line of the vehicle.

3   Draw a vertical line on the board in line with the centre-line of the vehicle.

4   Bounce the vehicle on its suspension to ensure correct settlement, then measure the height beteeen the ground and the centre of the lamps.

5   Measure the distance between the centres of the lamps to be adjusted, and mark the board.

6   With the lamps switched to main beam, turn the adjusting screws to bring the brightest points of light over the intersections of the two lines drawn on the board.

18 Exterior lamps – bulb renewal

Front parking

1   The bulb holders are located in the headlamp reflectors. Remove the headlamp as described in Section 16 and twist the parking lamp bulb from the reflector in an anti-clockwise direction (photo).

2   Refitting is a reversal of removal.

Front direction indicator

3   Extract the screws and remove the lens (photo).

4   The bulb is of bayonet fitting type.

5   When refitting the lens, do not overtighten the screws and make sure that the gasket is in good condition.

Rear lamp cluster

6   The operations are similar to those just described for the front direction indicator lamp (photo).

19 Interior lamps – bulb renewal

Driving cab lamp

1   Insert a thin screwdriver between the lamp lens and the panel at the end opposite to the switch. Prise out the lamp (photo).

2   Remove the bulb and then fit the new one.

3   Press the lamp into the panel until the clip engages.

Instrument panel warning and indicator lamps

4   Some of these bulbs are accessible by reaching up under the instrument panel.

5   Other bulbs can only be reached if the instrument panel is first released and pulled forward as described in the next Section.

6   Pull the plastic bulb holder out of the panel and withdraw the wedge base bulb from the holder (photo).

7   Refitting is a reversal of removal.

20 Instrument panel – removal and refitting

1   Disconnect the battery and pull off the heater control lever knobs (photo).

2   Unscrew and remove the four fixing screws from the instrument panel (photo).

3   Pull the panel forward until the back of the speedometer can be reached and the cable disconnected from it. If a tachograph is fitted, disconnect this in a similar way (photo).

4   Disconnect the multi-pin plugs and earthing tags.

5   Remove the instrument panel (photo).

6   The individual instruments may now be removed as necessary. Take great care not to damage the printed circuit (photo).

7   Refitting is a reversal of removal.

21 Speedometer and tachograph cables – renewal

Speedometer

1   Withdraw the instrument panel and disconnect the speedometer drive cable from the speedometer head.

2   Working under the vehicle at the gearbox, release the locking plate and pull the cable from its recess.

3   Release the cable clips and draw it through the various rubber grommets.

4   The new inner cable may be greased with multi-purpose grease before installation, but do not apply any grease to the last few inches before it enters the speedometer head, or lubricant may find its way into the instrument and cause sticking of the internal parts.

18.1 Removing parking lamp bulb from headlamp

18.3 Removing direction indicator lamp lens

18.6 Removing rear lamp lens

19.1 Interior lamp withdrawn

19.6 Instrument panel bulb withdrawn

20.1 Pulling off heater control lever knob

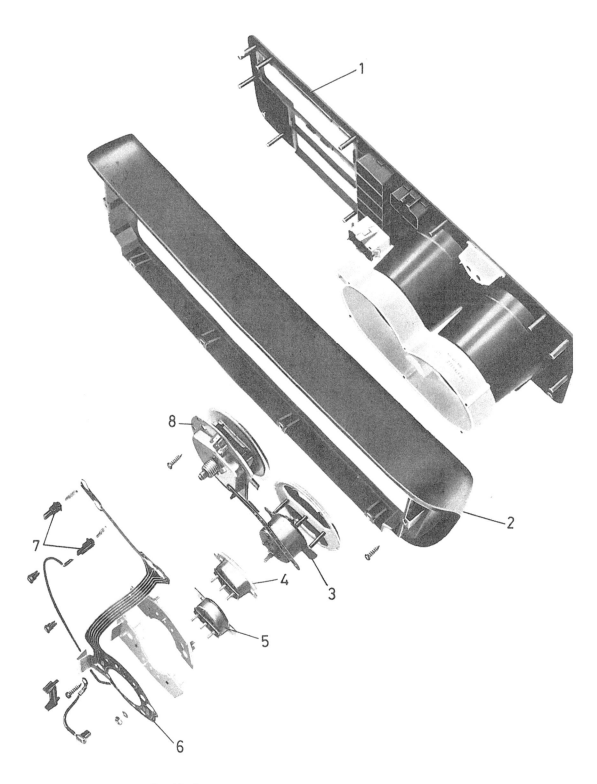

Fig. 10.16 Instrument panel without tachograph (Sec 20)

1 Panel	3 Clock	5 Fuel contents gauge	7 Bulbs
2 Crash pad	4 Coolant temperature gauge	6 Printed circuit board	8 Speedometer

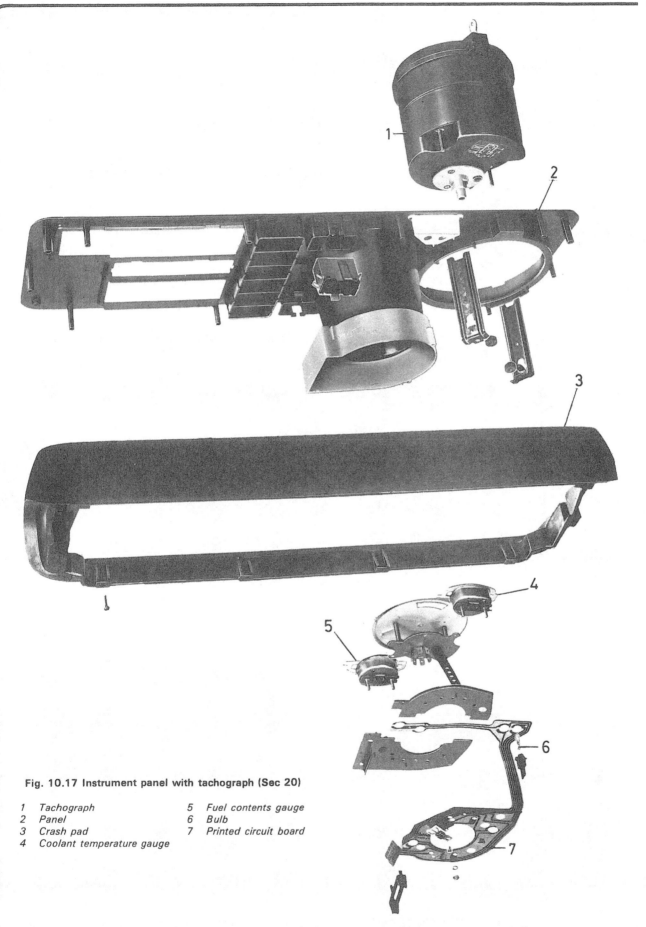

Fig. 10.17 Instrument panel with tachograph (Sec 20)

1	Tachograph	5	Fuel contents gauge
2	Panel	6	Bulb
3	Crash pad	7	Printed circuit board
4	Coolant temperature gauge		

20.2 Extracting an instrument panel screw

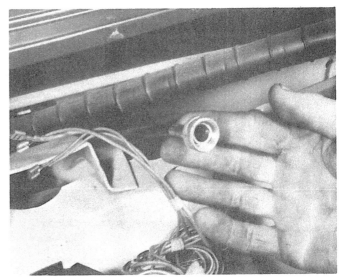

20.3 Speedometer cable disconnected from head

20.5 Instrument panel removed

20.6 Rear view of instrument panel

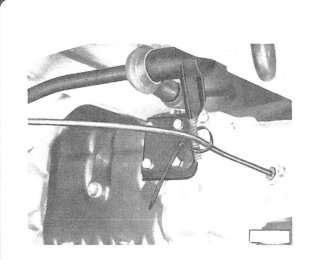
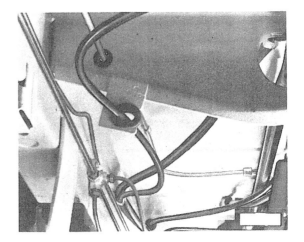

Fig. 10.18 Speedometer cable routing (Sec 21)

Fig. 10.19 Tachograph cable routing (Sec 21)

5   Install the new cable, making sure that the bends are not too sharp and there are no kinks or twists.

Tachograph

6   The tachograph drive cable is sealed to its drivegear adaptor box, otherwise renewal of the cable is carried out in a similar way to that described for the speedometer cable.

7   Note that if any seals on the tachograph are broken, regulations may require the fitting of new seals by an authorised dealer or service engineer.

22 Fuel contents and coolant temperature gauges – general

1   Reference should also be made to Chapter 2, Section 7, and to Chapter 3, Section 6.

2   Until July 1976, these instruments were fitted with integral individual voltage stabiliser units. After this date, a combined voltage stabiliser was fitted to serve both instruments.

3   It is important not to interchange the voltage stabilisers or printed circuits between one type and another. The printed circuits are identified as shown in Figs. 10.22 and 10.23.

23 Windscreen wiper blades and arms – removal and refitting

1   To remove a wiper blade, pull the arm from the glass until it locks.

2   Squeeze together the ends of the plastic insert at the U-shaped end of the arm and slide the blade out of the arm, twisting it to unhook it (photos).

3   To remove an arm, take out the plastic plug and unscrew the retaining nut (photo).

4   Withdraw the arm from the splined driving spindle (photo).

5   Refitting is a reversal of removal, but when installing the arm make sure that the wiper motor is in the parked position having been switched off by the wiper switch, and set the arm in accordance with Fig. 10.24.

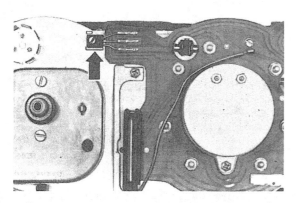

Fig. 10.20 Typical instrument voltage stabiliser (Sec 22)

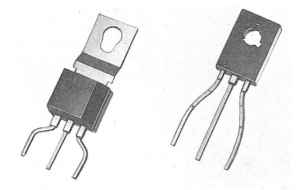

Fig. 10.21 Different types of instrument voltage stabiliser (Sec 22)

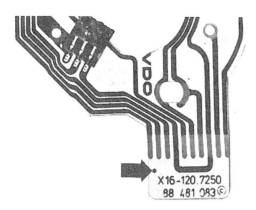

Fig. 10.22 Printed circuit marking (early models) (Sec 22)

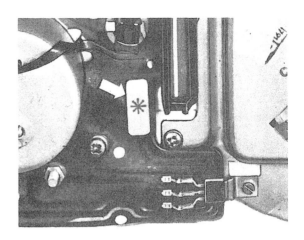

Fig. 10.23 Printed circuit marking (later models) (Sec 22)

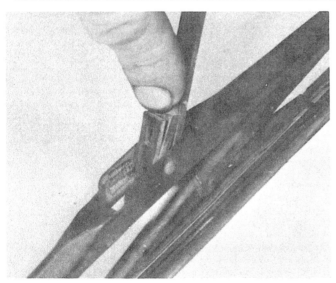

23.2a Releasing wiper blade from arm

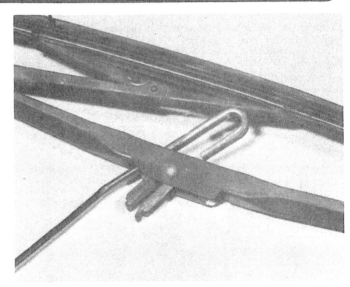

23.2b Removing wiper blade from arm

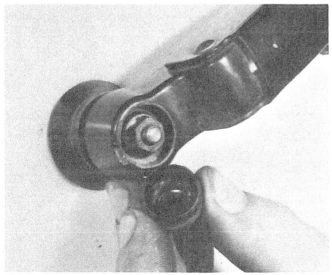

23.3 Removing wiper arm plug

23.4 Wiper arm removed

24 Windscreen wiper motor and linkage – removal and refitting

1   The wiper motor and linkage are accessible from under the facia panel within the driving cab (photo).
2   Disconnect the battery.
3   Remove the wiper blades and arms as described in the preceding Section.
4   Unscrew the three wiper motor mounting bolts and withdraw the motor far enough to be able to unscrew the nut from the crankarm and to pull it from the drive spindle splines.
5   Disconnect the wiring multi-plug and lift the motor assembly away.

6   Unbolt the wheelboxes and carefully withdraw the linkage from behind the heater.
7   Refitting is a reversal of removal, but the drive crankarm must be fitted to the motor spindle in accordance with Fig. 10.25 or 10.26 with the motor in the parked position, having been switched off by the wiper switch, not the ignition switch.

25 Windscreen and headlamp washers – maintenance

1   All later models are equipped with an electrically-operated windscreen washer, while some versions also have headlamp washers. Early models have a foot-operated pump.

Fig. 10.24 Wiper blade setting and washer jet pattern (Sec 23)

a   120.0 mm (4.72 in)
b   130.0 mm (5.1 in)
c   Manual 50.0 mm (1.96 in),
    electric 200.0 mm (7.9 in)

d   Manual 50.0 mm (1.96 in),
    electric 200.0 mm (7.9 in)
f   540.0 mm (21.3 in)

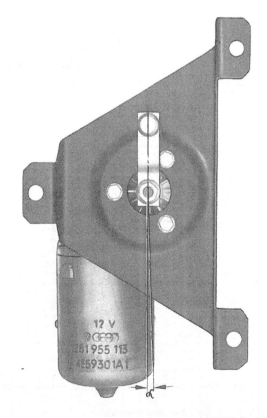

24.1 Location of wiper motor

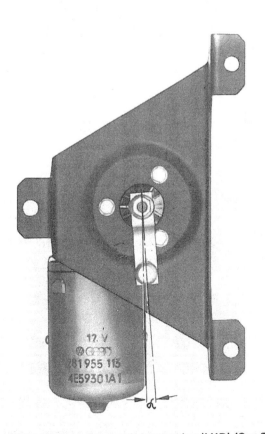

Fig. 10.25 Wiper motor crank arm setting (LHD) (Sec 24)

a   8° tolerance

Fig. 10.26 Wiper motor crank arm setting (RHD) (Sec 24)

a   8° tolerance

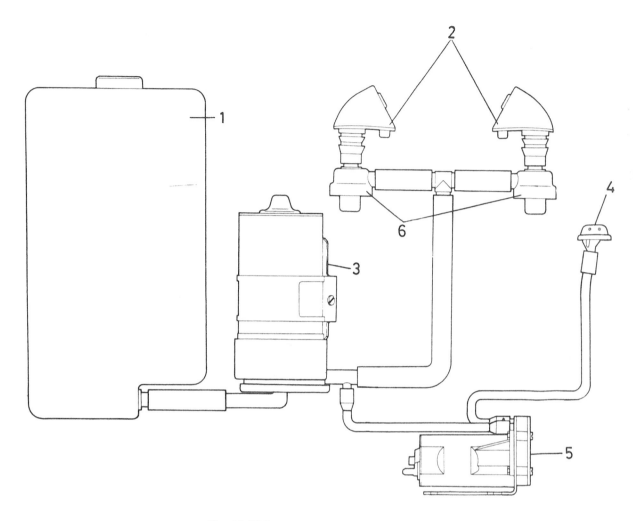

Fig. 10.27 Electric washer system (Sec 25)

1 Fluid container
2 Headlamp washer jets
3 Headlamp washer pump
4 Windscreen washer jets
5 Windscreen washer pump
6 Pressure regulating valves

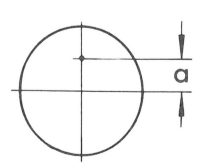

Fig. 10.28 Headlamp washer jet adjusting diagram (Sec 25)

a = 24.0 mm (0.94 in)

Fig. 10.29 Foot operated windscreen washer pump (Sec 25)

Fig. 10.30 Chassis side-member earthing point (Sec 27)

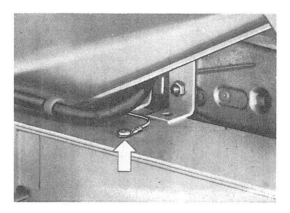

Fig. 10.31 Rear crossmember earthing point (Sec 27)

2 Maintain the level in the fluid container and use washer cleaning liquid or some methylated spirit added to the water, particularly in freezing temperatures. Never add cooling system antifreeze to the washer fluid.
3 Check the security of the washer tubes and the electrical leads from time to time and keep the jets adjusted so that the jet pattern is as shown in Figs. 10.24 and 10.28.

26 Heated rear window – description and precautions

1 Certain models are fitted with a driving cab heated rear window.
2 A relay for this electrical accessory is located on the fuse block.
3 Avoid scratching the heater element and do not stick labels over it.
4 Clean the glass with water only, and rub in the direction of the element heater strips.
5 It may be possible to repair a break in the heater element by using a proprietary conductive paint. Failing this, renewal of the window will be necessary.

27 Earthing points – location and function

1 Satisfactory operation of the starter motor, ignition system and electrical accessories depends upon the provision of clean, secure earth bonds.
2 Check these regularly. They include:

Gearbox extension housing to side-member
Battery negative lead to body
Steering column flexible coupling
Rear left-hand chassis side-member
Rear chassis crossmember

28 Fault diagnosis – electrical system

Symptom	Reason(s)
Starter motor fails to turn engine	Battery discharged
	Battery defective internally
	Battery terminal leads loose or earth lead not securely attached to body
	Loose or broken connections in starter motor circuit
	Starter motor switch or solenoid faulty
	Starter brushes badly worn, sticking, or brush wires loose
	Commutator dirty, worn or burnt
	Starter motor armature faulty
	Field coils earthed
Starter motor turns engine very slowly	Battery in discharged condition
	Starter brushes badly worn, sticking, or brush wires loose
	Loose wires in starter motor circuit
Starter motor operates without turning engine	Pinion or flywheel gear teeth broken or worn
Starter motor noisy or has an excessively rough engagement	Pinion or flywheel gear teeth broken or worn
	Starter motor retaining bolts loose
Battery will not hold charge for more than a few days	Battery defective internally
	Electrolyte level too low or electrolyte too weak due to leakage
	Plate separators no longer fully effective
	Battery plates severely sulphated
	Alternator belt slipping
	Battery terminal connections loose or corroded
	Alternator not charging properly
	Short in lighting circuit causing continual battery drain

Symptom	Reason(s)
Ignition light fails to go out; battery runs flat in a few days	Alternator belt loose and slipping or broken Alternator faulty

Failure of individual electrical equipment to function correctly is dealt with alphabetically, item-by-item, as follows:

Symptom	Reason(s)
Fuel gauge gives no reading	Fuel tank empty! Cable between tank sender unit and gauge earthed or loose Fuel gauge case not earthed Fuel gauge supply cable interrupted Fuel gauge unit broken
Fuel guage registers full all the time	Faulty instrument voltage stabilizer Cable between tank unit and gauge broken or disconnected
Horn operates all the time	Horn push either earthed or stuck down Horn cable to horn push earthed
Horn fails to operate	Blown fuse Cable or cable connection loose, broken or disconnected Horn has an internal fault
Horn emits intermittent or unsatisfactory noise	Cable connections loose
Lights do not come on	Blown fuse If engine not running, battery discharged Light bulb filament burnt out or bulbs broken Wire connections loose, disconnected or broken Light switch shorting or otherwise faulty
Lights come on but fade out	If engine not running, battery discharged
Lights give very poor illumination	Lamp glasses dirty Reflector tarnished or dirty Lamps badly out of adjustment Incorrect bulb with too low wattage fitted Existing bulbs old and badly discoloured Electrical wiring too thin not allowing full current to pass
Lights work erratically – flashing on and off, especially over bumps	Battery terminals or earth connection loose Lights not earthing properly Contacts in light switch faulty
Wiper motor fails to work	Blown fuse Wire connections loose, disconnected or broken Brushes badly worn Armature worn or faulty Field coils faulty
Wiper motor works very slowly and takes excessive current	Commutator dirty, greasy or burnt Drive to wheelboxes bent or unlubricated Wheelbox spindle binding or damaged Armature bearings dry Armature badly worn or faulty
Wiper motor works slowly and takes little current	Brushes badly worn Commutator dirty, greasy or burnt Armature badly worn or faulty
Wiper motor works but wiper blades remain static	Wheelbox gear and spindle damaged or worn Wiper motor gearbox parts badly worn

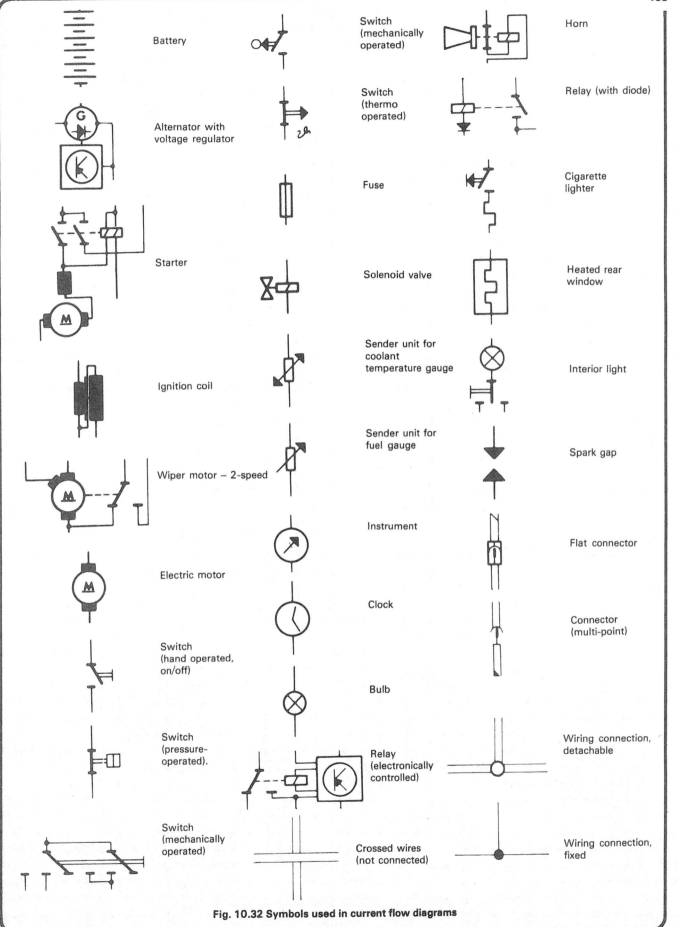

Fig. 10.32 Symbols used in current flow diagrams

154

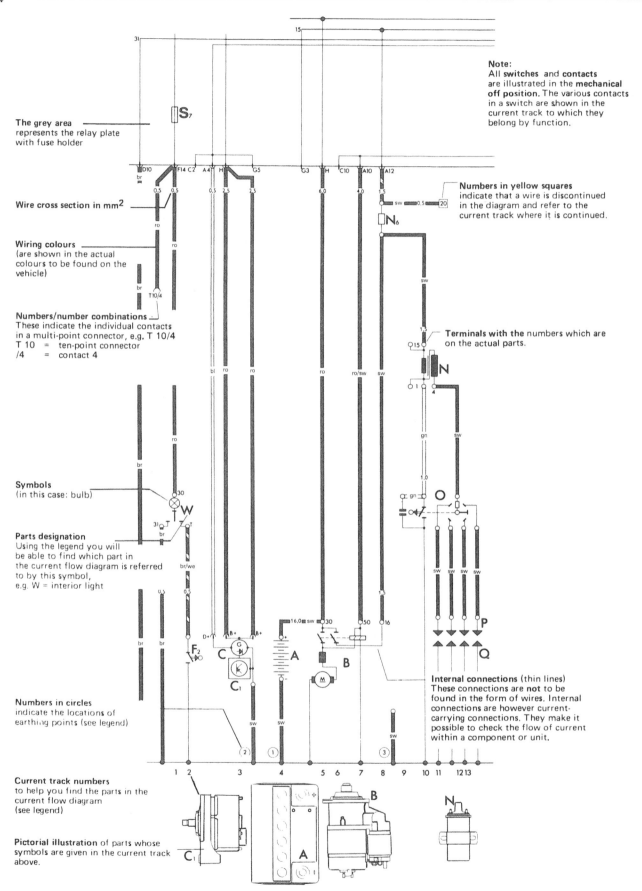

The grey area represents the relay plate with fuse holder

Note:
All **switches** and **contacts** are illustrated in the **mechanical off position**. The various contacts in a switch are shown in the current track to which they belong by function.

Wire cross section in mm²

Numbers in yellow squares indicate that a wire is discontinued in the diagram and refer to the current track where it is continued.

Wiring colours (are shown in the actual colours to be found on the vehicle)

Numbers/number combinations
These indicate the individual contacts in a multi-point connector, e.g. T 10/4
T 10 = ten-point connector
/4 = contact 4

Terminals with the numbers which are on the actual parts.

Symbols (in this case: bulb)

Parts designation
Using the legend you will be able to find which part in the current flow diagram is referred to by this symbol,
e.g. W = interior light

Numbers in circles
indicate the locations of earthing points (see legend)

Internal connections (thin lines)
These connections are **not** to be found in the form of wires. Internal connections are however current-carrying connections. They make it possible to check the flow of current within a component or unit.

Current track numbers
to help you find the parts in the current flow diagram (see legend)

Pictorial illustration of parts whose symbols are given in the current track above.

Fig. 10.33 Sample current flow diagram with explanatory notes

Key to Fig. 10.34

Designation			In current track
A	–	Battery	4
B	–	Starter	5,6,7,8
C	–	Alternator	3
C1	–	Voltage regulator	3
D	–	Ignition/starter switch	15,16,17
E	–	Windscreen wiper switch	57,58
E1	–	Light switch	19,20
E2	–	Turn signal switch	41
E3	–	Emergency light switch	38,39,40,42
E4	–	Headlight dip and flasher switch	28
E9	–	Blower switch	52,53
E15	–	Heated rear window switch	48,49
F	–	Brake light switch	34
F1	–	Oil pressure switch	64
F2	–	Door contact switch	2
F4	–	Reversing light switch	67
G	–	Fuel gauge sender unit	66
G1	–	Fuel gauge	59
G2	–	Coolant temperature sender unit	65
G3	–	Coolant temperature gauge	60
H	–	Horn button	70
H1	–	Horn	69
J	–	Headlight dip and flasher relay	28,29,30,31
J2	–	Emergency light relay	39,41
J9	–	Heated rear window relay	50,51
J31	–	Holder for wash/wipe intermittent relay	55,57
K1	–	High beam warning lamp	33
K2	–	Alternator warning lamp	62
K3	–	Oil pressure warning lamp	61
K5	–	Turn signal warning lamp	63
K6	–	Emergency light warning lamp	43
K8	–	Blower warning lamp	54
K10	–	Heated rear window warning lamp	49
L1	–	Twin filament bulb, left headlight	29,31
L2	–	Twin filament bulb, right headlight	30,32
L10	–	Instrument panel insert bulb	18,19,20

Designation			In current track
M1	–	Side light bulb, left	26
M2	–	Tail light bulb, right	25
M3	–	Side light bulb, right	24
M4	–	Tail light bulb, left	27
M5	–	Bulb, front turn signal, left	45
M6	–	Bulb, rear turn signal left	44
M7	–	Bulb, front turn signal, right	47
M8	–	Bulb, rear turn signal, right	46
M9	–	Brake light bulb, left	36
M10	–	Brake light bulb, right	35
M16	–	Bulb, reversing lights	68
N	–	Ignition coil	10,11
N1	–	Automatic choke	13
N3	–	Bypass air cutoff valve	14
N6	–	Series resistance (for ignition coil)	8
O	–	Distributor	10,11,12,13
P	–	Spark plug connectors	11,12,13
Q	–	Spark plugs	11,12,13
S1 to S15 – Fuses in fusebox			
T	–	Cable adaptor, behind instrument panel	
T1	–	Flat connector, single, near electric fuel pump	
T3	–	Connector, three-point, behind instrument panel	
T10	–	Connector, ten-point, on instrument panel insert	
V	–	Windscreen wiper motor	55,56
V2	–	Blower motor	52,53
V14	–	Electric fuel pump	72
W	–	Interior light	
Z1	–	Heated rear window	51
①	–	Earthing strap, battery/body	
②	–	Earthing strap, alternator/engine	
③	–	Earthing strap, gearbox/chassis	
⑩	–	Earthing point, instrument panel insert	

Colour code

bl	Blue	gn	Green	ro	Red
br	Brown	gr	Grey	sw	Black
ge	Yellow	li	Lilac	ws	White

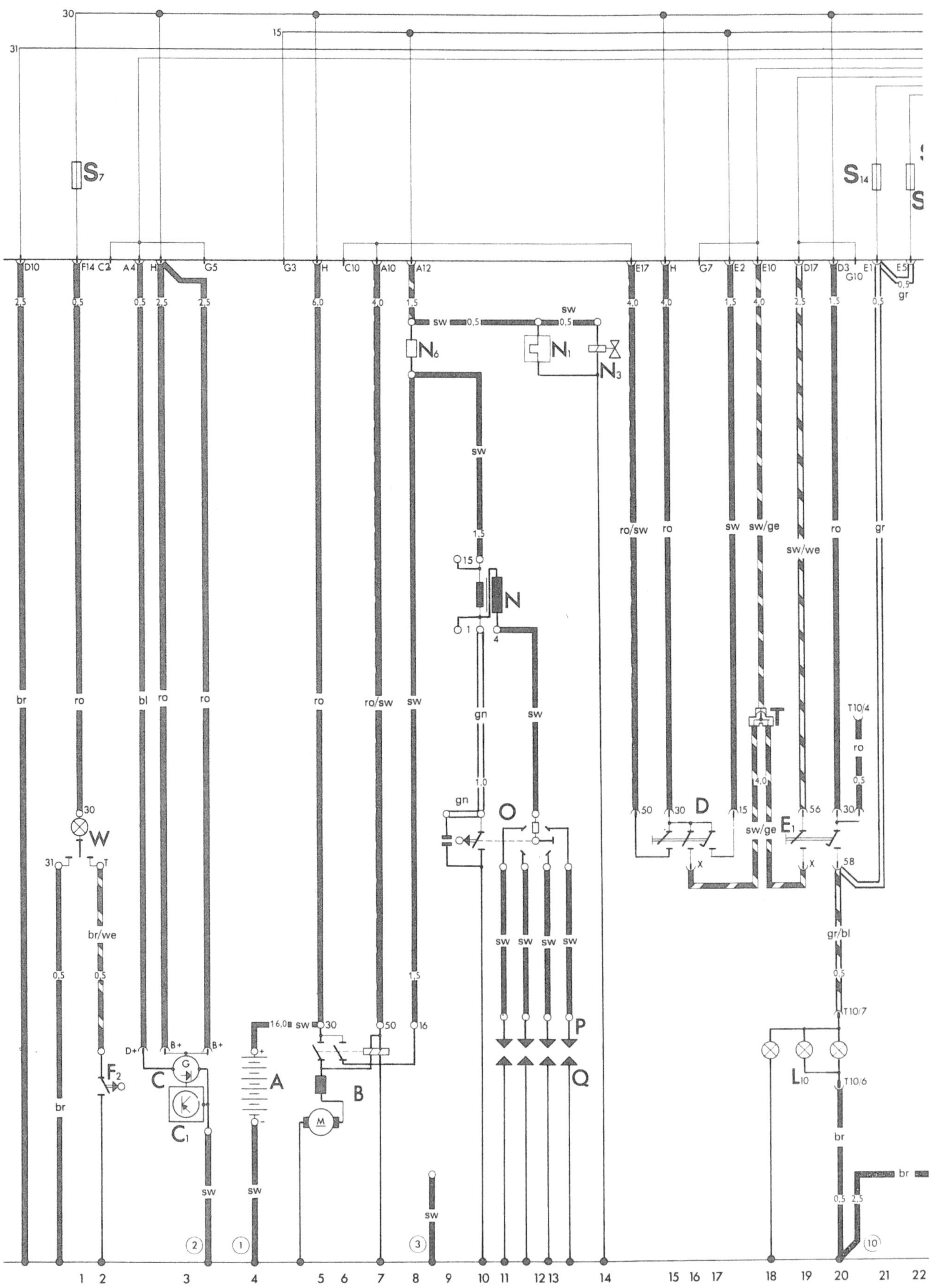

Fig. 10.34 Main current flow diagram for VW LT Series vehicles

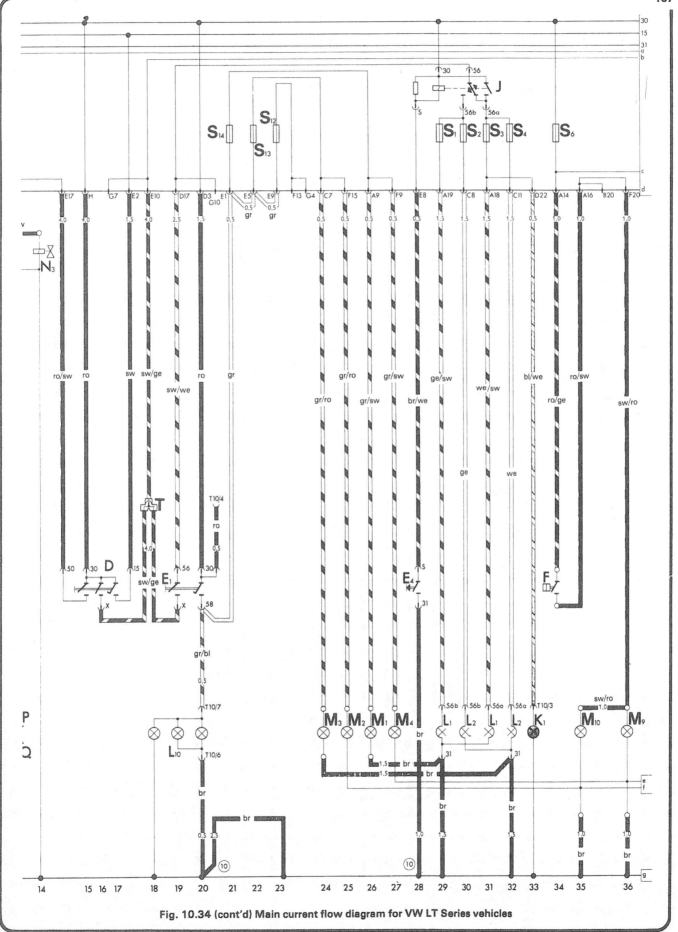

Fig. 10.34 (cont'd) Main current flow diagram for VW LT Series vehicles

Fig. 10.34 (cont'd) Main current flow diagram for VW LT Series vehicles

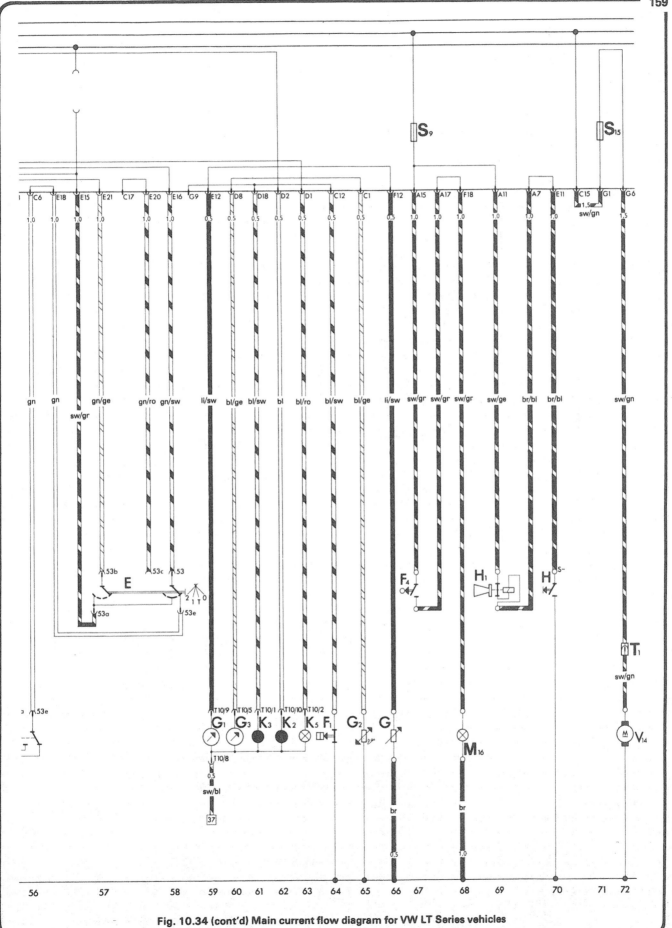

Fig. 10.34 (cont'd) Main current flow diagram for VW LT Series vehicles

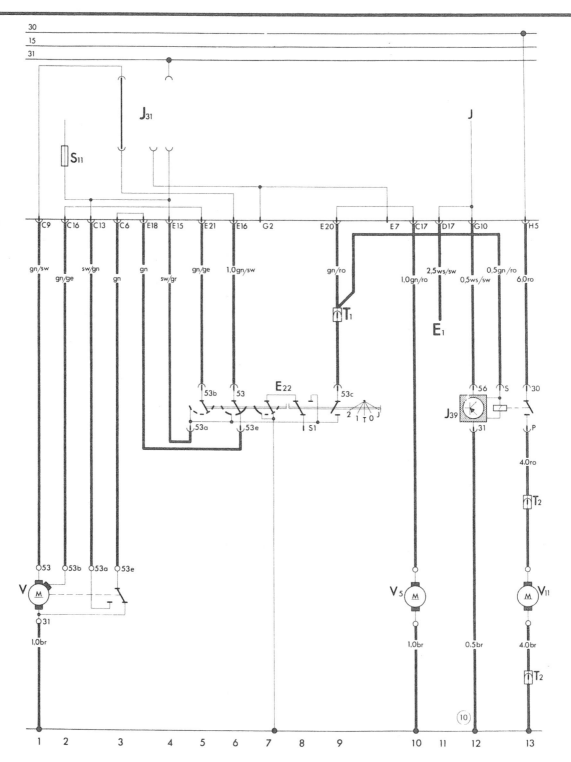

Fig. 10.35 Supplementary wiring diagram for headlight washers

Key to Fig. 10.35. For colour code see main diagram key

Designation		In current track	Designation		In current track
E1	– To light switch, terminal 56	11	T1	– Connector, single, near fusebox	
E22	– Windscreen wiper switch for intermittent operation	5-9	T2	– Connector, two-point, near headlight washer pump	
J	– To headlight dip and flasher relay, terminal 56	12	V	– Windscreen wiper motor	1-3
J31	– Reserved for intermittent wash/wipe relay	3,4	V5	– Windscreen washer pump	10
J39	– Headlight washer relay	12,13	V11	– Headlight washer pump	13
S11	– Fuse in fusebox	2	⑩	– Earthing point, instrument panel	

Chapter 11 Suspension and steering

For modifications, and information applicable to later models, see Supplement at end of manual

Contents

Specifications

Front suspension and steering
Type:

LT 28, 31, 35 .... Independent, coil springs with upper and lower arms, telescopic shock absorbers and anti-roll bar (not LT 35 van)

LT 40, 45 .... Solid beam axle with leaf springs, telescopic shock absorbers and anti-roll bar

Wheelbase .... 2500 or 2950 mm (98.4 or 116.1 in) according to model
Track (unladen) .... 1735 mm (68.3 in)
Turning circle (between kerbs):

LT 28, 31 .... 10.75 m (35.3 ft)
LT 35 .... 11.9 m (39.0 ft)
LT 40, 45 .... 11.6 m (38.1 ft)

Wheel lock angle:

	Inner	Outer
LT 28, 31, 35	37° 0'	32° 30'
LT 40, 45	40° 0'	34° 0'

Number of turns of steering wheel, lock to lock:

LT 28, 31, 35 .... 5.2
LT 40, 45 .... 4.9
Steering ratio (overall) .... 24 : 1
Steering box ratio:

LT 28, 31, 35 .... 20 : 1
LT 40, 45 .... 23.7 : 1
Ground clearance (fully loaded) .... 150 or 180 mm (5.9 or 7.1 in) according to model
Steering box oil capacity (LT 40, 45 only) .... 0.5 l (0.88 pints)

Front wheel alignment
Camber angle:

LT 28, 31, 35 .... 0° 40' ± 20' positive
LT 40, 45 .... 0° 35' ± 35' positive
Maximum variation between sides .... 0° 30'
Castor angle:

LT 28, 31, 35 .... 2° 20' ± 30'
LT 40, 45 .... 3° 25' ± 20'
Toe in:

LT 28, 31, 35 .... 2.2 to 6.6 mm (0.087 to 0.260 in)
LT 40, 45 .... 5.0 to 8.0 mm (0.197 to 0.315 in)

Rear suspension

Type:

LT 28, 31, 35	Semi-elliptic leaf springs with telescopic shock absorbers. Panhard rod on LT 28, 31, anti-roll bar on LT 35
LT 40, 45	Semi-elliptic leaf springs with telescopic shock absorbers and anti-roll bar

Track:

Single wheels	1720 mm (67.7 in)
Twin wheels	1500 mm (59.1 in)

Roadwheels

Type	Pressed steel

Wheel size:

LT 28, 31	6J x 14
LT 35	5J x 14
LT 40, 45	5J x 14

Tyres

Sizes:

LT 28	185 R14 C8PR or 195 R14 C8PR
LT 31	205 R14 C8PR
LT 35	185 R14 C6PR
LT 40, 45	185 R14 C8PR or 195 R14 C8 PR

Pressures*:	Front	Rear
LT 28	3.8 bar (55 lbf/in^2)	4.5 bar (65 lbf/in^2)
LT 31 (Van)	3.2 bar (46 lbf/in^2)	4.5 bar (65 lbf/in^2)
LT 31 (Pick-up, Chassis Cab)	2.9 bar (42 lbf/in^2)	4.5 bar (65 lbf/in^2)
LT 35 (Twin wheels)	3.7 bar (54 lbf/in^2)	3.0 bar (43 lbf/in^2)
LT 40	4.2 bar (60 lbf/in^2)	3.0 bar (43 lbf/in^2)
LT 45	4.2 bar (60 lbf/in^2)	3.7 bar (54 lbf/in^2)

*Representative selection only. Consult operator's handbook or tyre specialist

Torque wrench settings

	Nm	lbf ft
Front suspension (LT 28, 31, 35)		
Balljoint to upper wishbone	60	44
Upper wishbone pivot bolt	100	74
Balljoint to lower wishbone	60	44
Lower wishbone pivot bolt	100	74
Balljoints to stub axle carrier	125	92
Shock absorber upper mounting	25	18
Shock absorber lower mounting	20	15
Radius rod to wishbone	100	74
Radius rod to frame bracket	230	170
Crossmember bolts (M14)	125	92
Crossmember bolts (M10)	60	44
Anti-roll bar clamp bolts	30	22
Anti-roll bar to wishbone	25	18
Front suspension (LT 40, 45)		
Spring eyebolt to shackle	90	66
Spring shackle to frame member	90	66
Spring eyebolt to frame bracket	140	104
Spring U-bolts	160	118
Shock absorber mounting bolts	90	66
Anti-roll bar end link bolts	90	66
Anti-roll bar clamp bolts	45	33
Rear suspension (LT 28, 31)		
Spring eyebolt to shackle	80	59
Spring shackle to frame member	80	59
Spring eyebolt to frame bracket	130	96
Spring U-bolts	120	89
Panhard rod mounting bolts	60	44
Shock absorber upper mounting bolt	45	33
Shock absorber lower mounting bolt	75	55
Rear suspension (LT 35)		
As for LT 28, 31 except for:		
Shock absorber mounting bolts	60	44
Anti-roll bar end link bolts	60	44
Anti-roll bar clamp bolts	40	30

Rear suspension (LT 40,45)	Nm	lbf ft
Leaf spring nuts and bolts	170	125
Shock absorber mounting bolts	170	125
Anti-roll bar and link bolts	80	59
Anti-roll bar clamp bolts	45	33

Steering (LT 28, 31, 35)		
Drop arm to shaft nut	70	52
Idler arm pinch-bolt	75	55
Idler bellcrank lever pinch-bolt	75	55
Steering box mounting bolts	60	44
Drag link balljoint to drop arm	30	22
Drag link balljoint to bellcrank	30	22
Tie-rod balljoint to steering arm	30	22
Tie-rod balljoint to bellcrank	30	22
Tie-rod end locknut	40	30
Steering wheel nut	50	37
Steering shaft splined coupling pinch-bolts	25	18

Steering (LT 40, 45)		
Drag link balljoint to drop arm	35	26
Drag link balljoint to bellcrank	35	26
Tie-rod balljoint to steering arm	35	26
Tie-rod clamp bolts	40	30
Drop arm to shaft nut	250	185
Steering box mounting bolts	100	74
Steering wheel nut	50	37
Steering shaft splined coupling pinch-bolt	25	18

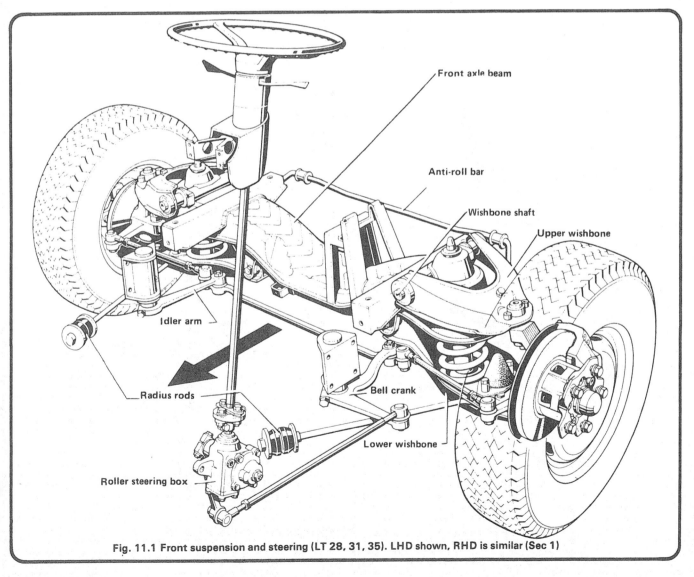

Fig. 11.1 Front suspension and steering (LT 28, 31, 35). LHD shown, RHD is similar (Sec 1)

1 Description

Front suspension

On LT 28, 31 and 35 models, the front suspension is of independent type with coil springs and telescopic shock absorbers (photo).

Upper and lower wishbones are used in conjunction with radius rods and an anti-roll bar (except LT 35 Van).

On LT 40 and 45 models, a solid beam axle is used together with leaf springs, telescopic shock absorbers and an anti-roll bar.

Rear suspension

The rear suspension on all models is similar, having semi-elliptic leaf springs, rubber helpers and telescopic shock absorbers. On LT 28 and 31 models a Panhard type transverse rod is fitted, but on other versions, an anti-roll bar is used (photo).

Steering

The steering gear is of worm and roller type, with a collapsible steering column.

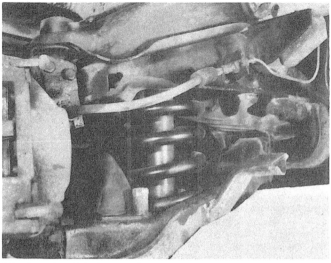

1.0a One side of the independent front suspension

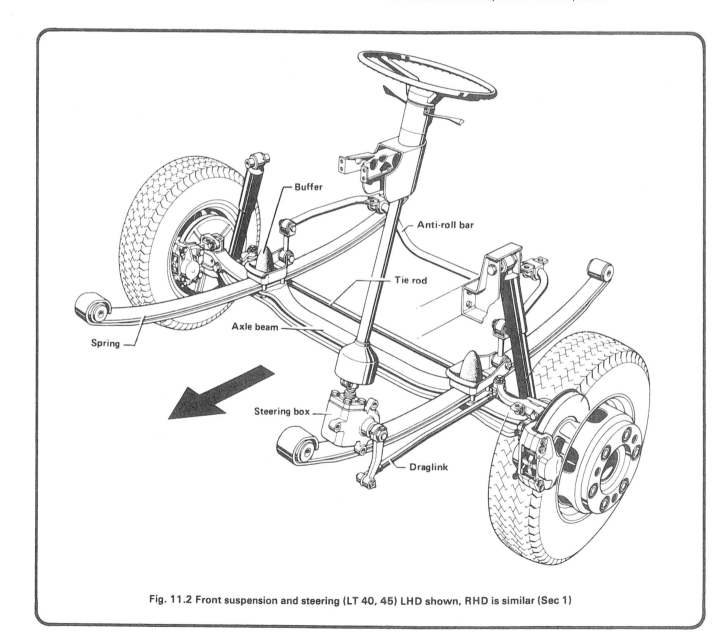

Fig. 11.2 Front suspension and steering (LT 40, 45) LHD shown, RHD is similar (Sec 1)

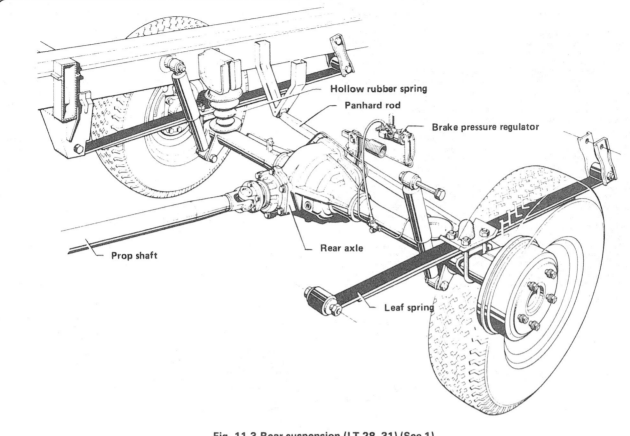

Hollow rubber spring

Panhard rod

Brake pressure regulator

Prop shaft

Rear axle

Leaf spring

Fig. 11.3 Rear suspension (LT 28, 31) (Sec 1)

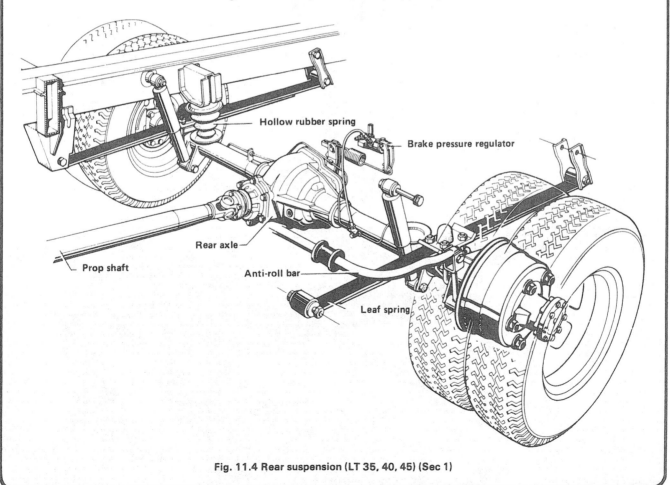

Hollow rubber spring

Brake pressure regulator

Rear axle

Prop shaft

Anti-roll bar

Leaf spring

Fig. 11.4 Rear suspension (LT 35, 40, 45) (Sec 1)

1.0b Rear suspension rubber helper springs

2   Maintenance and inspection

Lubrication

1   On LT 28, 31 and 35 models, no lubrication is required to the steering or suspension joints and the steering box is a sealed unit.
2   On LT 40 and 45 models, the upper and lower kingpin bushes require lubrication at the intervals specified in Routine Maintenance, using the grease nipples provided.
3   The steering box on LT 40 and 45 models can be topped up. New steering boxes are supplied 'dry' and must be filled before use with the correct grade and quantity of oil.

Inspection

4   Regularly inspect the gaiters on the steering balljoints for splits. Where evident, renew the assembly as described in Section 23.
5   At the intervals specified in Routine Maintenance, raise the front of the vehicle. While an assistant turns the steering wheel rapidly first in one direction and then in the other through about 10°, observe the tie-rod balljoints for wear and movement of the ball socket in relation to the steering arm. If evident, renew the balljoint assembly.
6   On LT 40 and 45 models, grip each front roadwheel and move it in and out and up and down to check for wear in the kingpin bushes. Renew if necessary as described in Section 16.
7   The steering gear is considered worn if the steering wheel has a free movement when turned exceeding 15.0 mm (0.59 in) measured

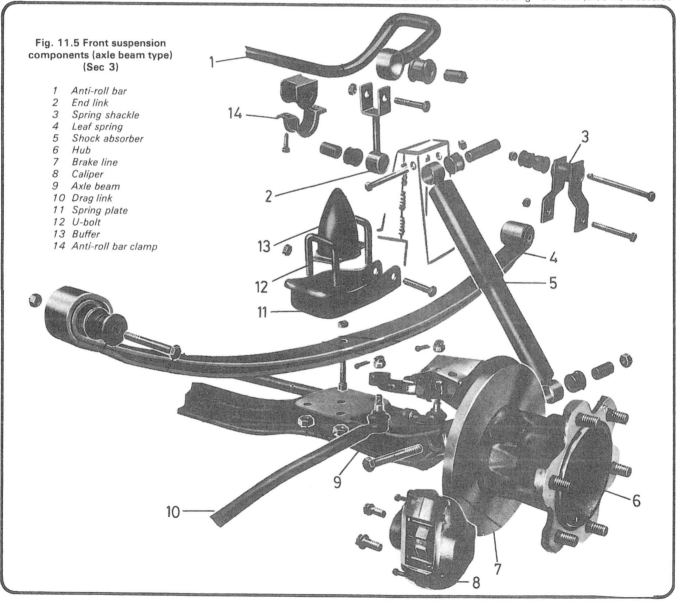

Fig. 11.5 Front suspension components (axle beam type) (Sec 3)

1   Anti-roll bar
2   End link
3   Spring shackle
4   Leaf spring
5   Shock absorber
6   Hub
7   Brake line
8   Caliper
9   Axle beam
10   Drag link
11   Spring plate
12   U-bolt
13   Buffer
14   Anti-roll bar clamp

at the steering wheel rim.

8   Check all suspension flexible mounting bushes for wear. If their holes have become elongated, renew then.

9   Examine the shock absorber casings for signs of fluid leakage and test them periodically as described in Section 3.

10   Remember that the roadwheels and tyres play a vital part in the behaviour of the suspension and steering, refer to Section 29.

11   Check the front wheel alignment and the steering angles regularly, and immediately where uneven tyre wear is noticed. Refer to Section 30 for details.

3   Front shock absorber – removal, testing and refitting

1   On LT 28, 31 and 35 models, unscrew the self-locking nut from the top of the shock absorber piston rod. Flats are provided to hold the piston rod from turning while the nut is unscrewed.

2   Draw the piston rod down and take off the top mounting cushions, spacer and cup washers.

3   Unscrew and remove the two lower mounting bolts and draw the shock absorber downwards out of the suspension coil spring and wishbones.

4   On LT 40 and 45 models, unscrew and remove the upper and lower pivot mounting bolts and remove the shock absorber complete with flexible bushes and sleeves.

5   To test a shock absorber, grip the lower mounting in the jaws of a vice so that the unit is held vertically. Now fully extend and contract the shock absorber several times. Any tendency to jerky or erratic action, or lack of resistance or seizure, will indicate the need for renewal.

6   Refitting is a reversal of removal. Make sure that the flexible bushes are in good condition, otherwise renew them.

7   Tighten all bolts and nuts to the specified torque.

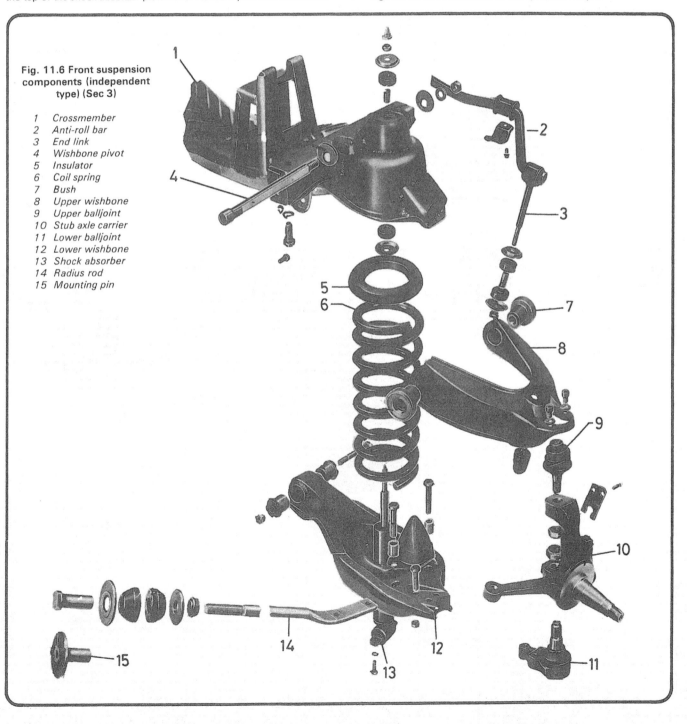

Fig. 11.6 Front suspension components (independent type) (Sec 3)

1   Crossmember
2   Anti-roll bar
3   End link
4   Wishbone pivot
5   Insulator
6   Coil spring
7   Bush
8   Upper wishbone
9   Upper balljoint
10 Stub axle carrier
11 Lower balljoint
12 Lower wishbone
13 Shock absorber
14 Radius rod
15 Mounting pin

4　Front hub bearings – adjustment and renewal

1　If when a front roadwheel is raised and the tyre gripped, the wheel can be rocked or endfloat is evident, the bearings are in need of adjustment.

2　To adjust, prise off the hub grease cap and unscrew the nut to relieve the staking.

3　Screw in the nut finger tight, at the same time turning the roadwheel until endfloat just disappears.

4　Without moving the position of the nut, stake it into the shaft groove. If the nut has been staked twice before, use a new nut.

5　Failure to eliminate endfloat or 'rocking' will indicate worn bearings. Do not be tempted to overtighten the nut in an effort to solve the problem.

6　To renew the front hub bearings, remove the brake caliper as described in Chapter 9 and hang it up out of the way. There is no need to disconnect the hydraulic hose.

7　Remove the staked nut and take out the thrust washer.

8　Pull the hub/disc slightly forward and then push it back. This will enable the outer bearing to be removed.

9　Remove the hub completely from the stub axle.

10　Examine the bearing and oil seal running surfaces on the stub axle. If they are rough or scored, the stub axle will have to be renewed.

11　Prise out the oil seal from the hub and then remove the inner bearing cage.

12　The bearing outer tracks may be removed from the hub by pressing them out or by careful use of a drift.

13　Take care not to mix up the components of the separate bearing sets if both front hubs are being dismantled at the same time.

14　Install the new bearing outer tracks and the inner roller cage. Tap a new oil seal into the hub until it is flush.

15　Half fill the interior of the hub with wheel bearing grease and smear some onto the oil seal lips.

16　Install the hub/disc to the stub axle. Fit the outer roller cage and the thrust washer and screw on the nut.

17　Adjust the bearings as described in paragraphs 3 and 4 of this Section.

18　Tap the grease cap into position, fit the brake caliper, tightening the bolts to the specified torque, and install the roadwheel.

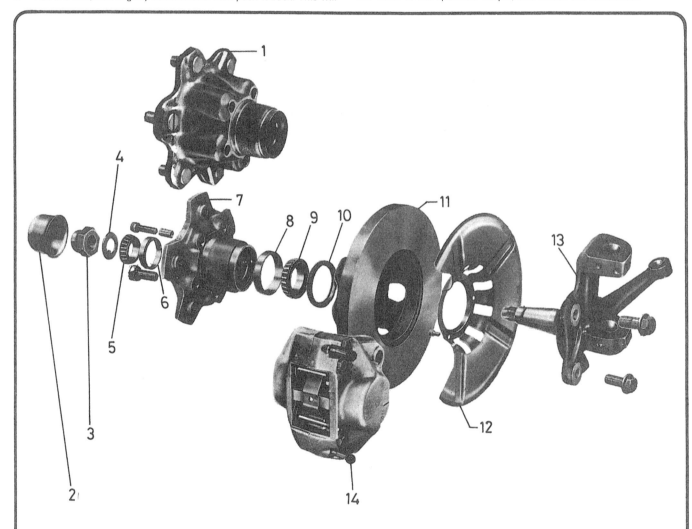

Fig. 11.7 Front hub components (LT 28, 31, 35) (Sec 4)

1　Hub (LT 35)	5　Outer roller cage	9　Inner roller cage	12　Disc shield
2　Grease cap	6　Bearing outer track	10　Oil seal	13　Stub axle carrier
3　Hub nut	7　Hub (LT 28, 31)	11　Brake disc	14　Caliper
4　Thrust washer	8　Bearing track		

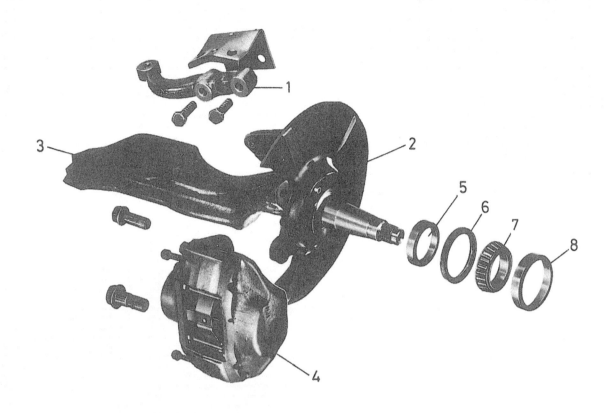

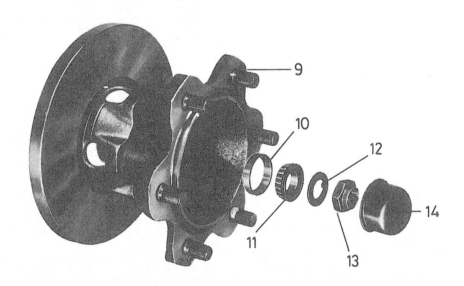

Fig. 11.8 Front hub components (LT 40, 45) (Sec 4)

1 Steering arm	5 Inner bearing track	9 Hub	12 Thrust washer
2 Disc shield	6 Oil seal	10 Bearing track	13 Hub nut
3 Axle	7 Inner bearing roller cage	11 Outer bearing roller cage	14 Grease cap
4 Caliper	8 Bearing track		

Fig. 11.9 Hub nut correctly staked (Sec 4)

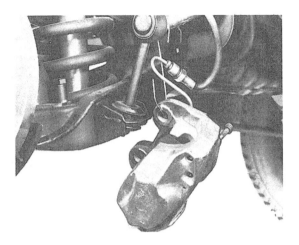

Fig. 11.10 Brake disc caliper supported on wire (Sec 4)

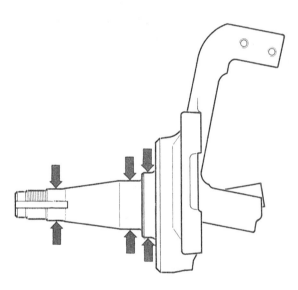

Fig. 11.11 Stub axle checking surfaces (Sec 4)

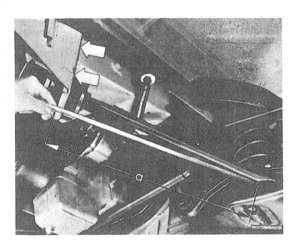

Fig. 11.12 Radius rod setting diagram (Sec 6)

Measure distance a from inner edge of wishbone cut-out to edge of angled lip of bracket (arrowed)

5 Front anti-roll bar – removal and refitting

1   This is simply a matter of disconnecting the end links from the lower wishbone and then removing the clamps from the body.
2   Renew the rubber bushes as necessary by pressing or by using a bolt, nut, distance pieces and thrust washers to draw out the bushes or to refit them. Dipping the bushes in brake hydraulic fluid will make the job easier.
3   When refitting, tighten bolts and nuts to the specified torque.

6 Radius rod (LT 28, 31, 35) – removal and refitting

1   Slacken the shouldered nut on the threaded part of the radius rod and unscrew and remove the mounting pin from the end of the rod.
2   Unbolt the rear end of the radius rod from the suspension wishbone and remove the rod, noting the location of the flexible cushions and washers.
3   Refit by reversing the removal operations. As the setting of the rods affects the castor angle (see Section 30), its effective length must be adjusted by turning the mounting pin until dimension 'a' shown in Fig. 11.12 is between 532 and 537 mm (20.9 and 21.1 in). Tighten the shouldered nut when the dimension is correct.

7 Front suspension coil spring and suspension lower arm (LT 28, 31, 35) – removal and refitting

1   Raise the front of the vehicle and support it securely under the chassis numbers.
2   Remove the roadwheel.
3   Pull out the securing clip which retains the hydraulic hose to its bracket.
4   Remove the brake caliper and hang it up out of the way.
5   Disconnect the anti-roll bar from the suspension arm.
6   Remove the shock absorber as described in Section 3.
7   Unscrew the nut on the suspension lower swivel joint two or three turns. Using a suitable balljoint extractor, release the balljoint taper pin from the stub axle carrier. If a suitable extractor is not available, place a thick bolt and nut between the ends of both swivel joint taper pins and unscrew the nut on the bolt to increase its length and so release the lower taper pin. Make sure that the nut is not removed from the taper pin during the operation or the threads will be damaged, and also the suspension arm would fly downwards with possibly dangerous consequences.
8   Do not remove the nut from the now released lower swivel joint until the suspension arm has been securely supported and slightly raised on a jack.
9   Now remove the nut and gently lower the jack.
10 Once the suspension lower arm has been lowered sufficiently, the

Fig. 11.13 Lower swivel joint nut unscrewed (Sec 7)

Fig. 11.14 Lowering suspension lower arm with a jack (Sec 7)

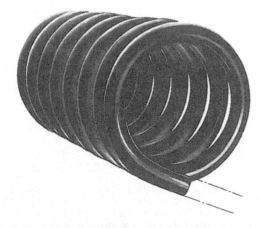

Fig. 11.15 Coil spring straight end (Sec 7)

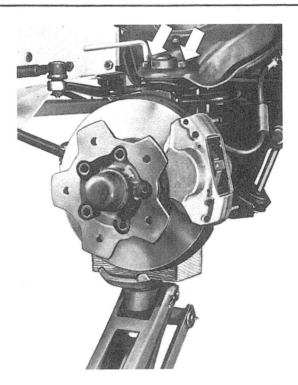

Fig. 11.16 Wishbone balljoint mounting bolts (Sec 8)

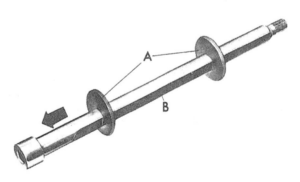

Fig. 11.17 Upper wishbone pivot shaft (Sec 8)

A   Eccentric washers   B   Shaft

coil spring can be removed.

11 The suspension arm can be removed by unbolting and removing the pivot bolt from its inboard end and disconnecting the radius rod.

12 Refitting is a reversal of removal, but observe the following points. When installing the coil spring, make sure that the straighter end of the spring coil is at the bottom. Set the radius rod as described in Section 6.

8   Front suspension upper arm (LT 28, 31, 35) – removal and refitting

1   Raise the vehicle and support it securely under the chassis members.

2   Remove the roadwheel and place a jack under the suspension lower arm to support it.

3   Using an Allen key, remove the two bolts which hold the upper swivel joint to the suspension arm.

4   Lower the jack under the suspension arm. As the stub axle carrier moves downwards, the upper swivel joint will be released from the upper arm.

5   Mark the position of the eccentric washers on either side of the suspension arm pivot in relation to the arm, using quick-drying paint or scratch marks. This is important to maintain the approximate setting of

the camber angle (see Section 30).

6   Unscrew the nut from the pivot shaft at the inboard end of the suspension upper arm.

7   Withdraw the pivot shaft and remove the suspension arm.

8   Refitting is a reversal of removal, but make sure that the end of the pivot shaft which has the hexagon socket is towards the front of the vehicle, and that the eccentric washers and the arm are in their originally marked alignment.

9   It is recommended that the camber angle is checked on completion as described in Section 31.

---

9 Front suspension upper and lower swivel joints (LT 28, 31, 35) – removal and refitting

Lower joint

1   Raise and support the front of the vehicle under the chassis members.

2   Remove the roadwheel.

3   Pull out the clip which secures the brake flexible hose to the support bracket, unbolt the caliper and tie it up out of the way.

4   Release the lower swivel joint from the stub axle carrier as described in Section 7, paragraphs 7, 8 and 9.

5   Unbolt the swivel joint from the lower arm and remove it.

Upper joint

6   Release the upper joint from the stub axle carrier as described for the lower joint. Unbolt and remove it from the suspension upper arm.

All joints

7   Refitting the new joints is a reversal of removal, but make sure that the socket-headed retaining bolts have their heads on the top surface of the suspension arms.

8   Tighten all nuts and bolts to the specified torque.

---

10 Front suspension arm bushes – renewal

This is a job best left to your dealer, as the bush locating collars are welded in position and the welds will have to be ground off to remove them. A press will be needed to remove and install the bushes and the collars spot welded in position again to complete the work.

---

11 Stub axle carrier (LT 28, 31, 35) – removal and refitting

1   Raise the front of the vehicle and support it securely under the chassis members.

2   Remove the roadwheel and then place a jack under the suspension lower arm.

3   Pull out the spring clip which retains the brake flexible hose to its support bracket, unbolt the caliper and tie it up out of the way.

4   Remove the hub/disc assembly as described in Section 4.

5   Unbolt and remove the disc shield.

6   Unscrew the nuts on the upper and lower swivel joints by two or three turns.

7   Release the two swivel joint taper pins from the stub axle carrier as described in Section 9.

8   Using a suitable balljoint separator, disconnect the tie-rod ends from the steering eye.

9   Making sure that the jack is in contact with the underside of the suspension lower arm, unscrew and remove both swivel joint nuts. Lower the jack and remove the stub axle carrier.

10   Refitting is a reversal of removal. Tighten all nuts and bolts to the specified torque.

---

12 Front axle assembly (LT 28, 31, 35) – removal and refitting

1   This operation is rarely required except where major body or chassis member repairs are being carried out.

2   Raise the front of the vehicle, support it securely and remove the front roadwheels.

3   Remove the front seats and the engine cover.

4   Use a crossbar, supported on pillars standing on the floor of the

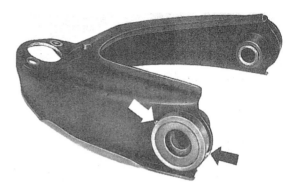

Fig. 11.18 Suspension arm bush collar weld points (Sec 10)

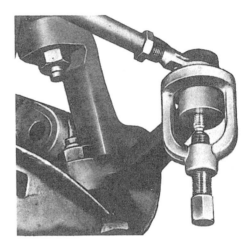

Fig. 11.19 Disconnecting a tie-rod end balljoint (Sec 11)

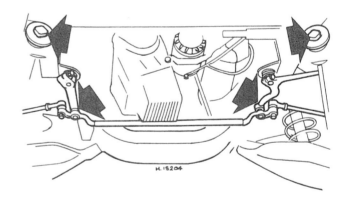

Fig. 11.20 Radius rod and relay rod connections (Sec 12)

driving cab and connected to the engine lifting hooks, to support the weight of the engine.

5   Working under the vehicle, disconnect the ends of the radius rods from their body anchor brackets.

6   Disconnect the centre relay section of the steering tie-rod from the steering box drop arm and the idler arm. Use a balljoint separator to do this.

7   Unbolt the anti-roll bar clamps from the body and then disconnect the brake pipeline at the four-way union of the side-member. Plug the end of the pipe quickly to prevent loss of fluid.

8   Working from under the engine mounting brackets, unscrew and remove the engine mounting nuts.

9   Support the axle crossmember on a trolley jack and then unbolt it

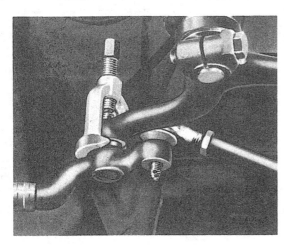

Fig. 11.21 Releasing relay rod balljoint (Sec 12)

Fig. 11.22 Anti-roll bar clamps and brake pipeline four-way union (Sec 12)

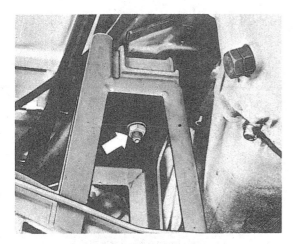

Fig. 11.23 Engine mounting nut (Sec 12)

Fig. 11.24 Crossmember-to-side-member bolts (Sec 12)

Fig. 11.25 Removing independent type front suspension (Sec 12)

from the side members.

10 Lower the jack slowly and with the help of at least one assistant, remove the suspension assembly from the vehicle.

11 Refitting is a reversal of removal. Tighten all nuts and bolts to the specified torque and set the radius rod lengths as described in Section 6.

12 Bleed the brake front hydraulic circuit as described in Chapter 9.

13 Front leaf spring (LT 40, 45) – removal and refitting

1 Slightly raise the front of the vehicle and support it securely under the chassis members.

2 Place a jack under the axle beam at the end from which the leaf spring is to be removed.

3 Disconnect the anti-roll bar drop link from the spring plate.

4 Disconnect the shock absorber lower mounting and contract it fully.

5 Unscrew and remove the bolts from the eyes of the leaf spring.

6 Unscrew the nuts from the spring U-bolts.

7 Remove the spring plate and the leaf spring.

8 Refitting is a reversal of removal, but make sure that the spring eye which is not completely closed is located at the front. Use the jack under the axle to raise or lower the spring to align the spring and side-member bolt holes.

14 Front suspension flexible bushes (LT 40, 45) – renewal

The flexible bushes which are used in the leaf spring eyes and shackle, and in the anti-roll bar drop links, are best removed with a press. If one is not available however, use a bolt, nut, washers and distance pieces to draw out the old bushes and to install the new ones. Use a little brake hydraulic fluid as a lubricating medium to ease installation of the bushes.

15 Stub axle carrier (LT 40, 45) – removal and refitting

1   Raise the front of the vehicle and support it accurately under the chassis members and the axle beam.

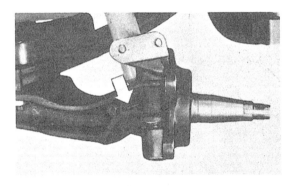

Fig. 11.26 Kingpin retaining roll pin (Sec 15)

2   Remove the hub/disc and bearings as described in Section 4.
3   Unscrew the nut from the tie-rod end balljoint taper pin, and then disconnect it from the steering arm of the stub axle carrier using a suitable extractor.
4   Drive out the securing roll pin.
5   From the top of the stub axle carrier, extract the circlip, the end cap with grease nipple and the two O-rings.
6   Using a depth gauge, measure the distance from the top of the stub axle carrier to the top face of the kingpin and record it.
7   Remove the kingpin in a downward direction. If the bushes are badly worn the pin may almost fall out, but otherwise a portable press may be required to remove it. Take great care not to damage the bushes if they are not going to be renewed.
8   Remove the stub axle carrier.
9   Refitting is a reversal of removal. Grease the kingpin before pressing it in from the bottom, and note the location of the shim and thrust bearing. The maximum up-and-down movement between the axle beam and the stub axle carrier is 2.0 mm (0.079 in). Where this is found to be excessive, change the shim for a thicker one from the nine thicknesses available.
10   Set the kingpin to its originally recorded position in the stub axle carrier and drive in a new roll pin.
11   Fit the O-rings, end caps, grease nipples and circlips.
12   Fit the hub/disc and adjust the bearings.
13   Reconnect the tie-rod and the brake caliper.
14   Apply a grease gun to the kingpin nipples.

16 Kingpin bushes (LT 40, 45) – renewal

1   Without the special tools, this is a job best left to your dealer.
2   Where the tools are available, proceed as follows, having removed

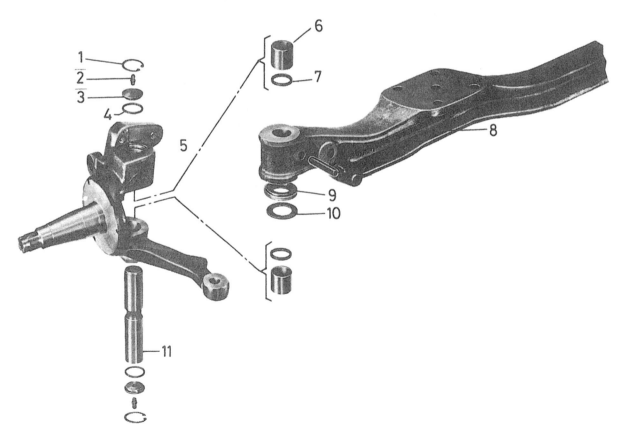

Fig. 11.27 Kingpin and associated components (LT 40, 45) (Sec 15)

1 Circlip	4 O-ring	7 Sealing ring	10 Shim
2 Grease nipple	5 Stub axle carrier	8 Axle	11 Kingpin
3 Cap	6 Bush	9 Thrust bearing	

the stub axle carrier from the vehicle (see preceding Section).
3   Press the upper and lower bushes from the stub axle carrier.
4   Press in the new bushes so that they are positioned as shown in Fig. 11.28. Use a depth gauge to measure the setting.
5   Insert a reamer (Tool No 3043) into the lower bush and ream both bushes to size at the same time. Do not ream from the wrong direction, as the reamer has two different diameter cutters.
6   Clean out any swarf and press the sealing rings in flush.

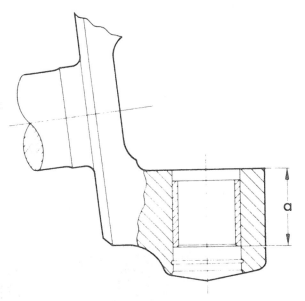

Fig. 11.28 Kingpin bush installation diagram (Sec 16)

a = 33.0 mm (1.30 in)

17 Front axle (LT 40, 45) – removal and refitting

1   Raise the front of the vehicle and support it securely under the chassis members.
2   Remove the roadwheels.
3   Remove the hub/disc and bearing assemblies as described in Section 4.
4   Disconnect the shock absorber lower mountings.
5   Disconnect the anti-roll bar drop links.
6   Disconnect the steering tie-rod end balljoints from the steering arms of the stub axle carrier.
7   Support the axle beam on a trolley jack and disconnect the leaf springs from it by unscrewing the U-bolt nuts.
8   Withdraw the axle beam from the vehicle.
9   Refitting is a reversal of removal. Tighten all nuts and bolts to the specified torque and adjust the hub bearings as described in Section 4. Make sure that the steering lock stop is located on the axle face nearer the rear of the vehicle.

Fig. 11.29 Reaming kingpin bushes (Sec 16)

18 Rear shock absorbers – removal, testing and refitting

1   Removal of the rear shock absorbers on all models is similar, simply unscrew and remove the anchor bolts from the upper and lower eyes of the shock absorber and lift it away.
2   Test with reference to the method used in Section 3.
3   Refitting is a reversal of removal. Tighten the mounting nuts to the specified torque.

19 Panhard rod (LT 28, 31) – removal and refitting

1   Unscrew and remove the pivot bolts which attach one end of the Panhard rod to the bracket on the chassis side-member and the other end to the rear axle casing. Remove the rod.
2   Refit by reversing the removal operations. Tighten the bolts to the specified torque.

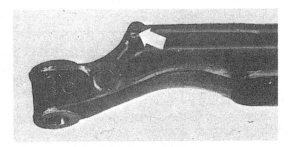

Fig. 11.30 Axle beam lock stop (arrowed) (Sec 17)

20 Rear anti-roll bar (LT 35, 40, 45) – removal and refitting

1   Unbolt the end drop links.
2   Unbolt the clamps which hold the anti-roll bar to the body.
3   Remove the bar.
4   Refitting is a reversal of removal. Tighten all nuts and bolts to the specified torque.

21 Rear roadspring – removal and refitting

1   Raise the rear of the vehicle slightly and support it under the chassis members.
2   Place a jack under the rear axle tube near the end from which the leaf spring is to be removed.
3   Unscrew and remove the spring eyebolts.
4   Unscrew and remove the U-bolt nuts and take off the spring plate.
5   Remove the spring.
6   Refit by reversing the removal operations, raising or lowering the axle slightly with the jack to align the spring eyebolt holes with those in the side-member front brackets. Tighten all nuts and bolts to the specified torque. **Do not** use grease on the U-bolt threads.

22 Rear suspension flexible bushes – renewal

1   The flexible bushes which are used in the leaf spring eyes and shackle, in the anti-roll bar or Panhard rod attachments and in the shock absorber mountings, are best removed and installed on a press.
2   Where a press is not available, use a bolt, nut, washers and distance pieces to draw out the old bushes and to install the new ones. Use a little brake hydraulic fluid as a lubricating medium to ease installation of the bushes.

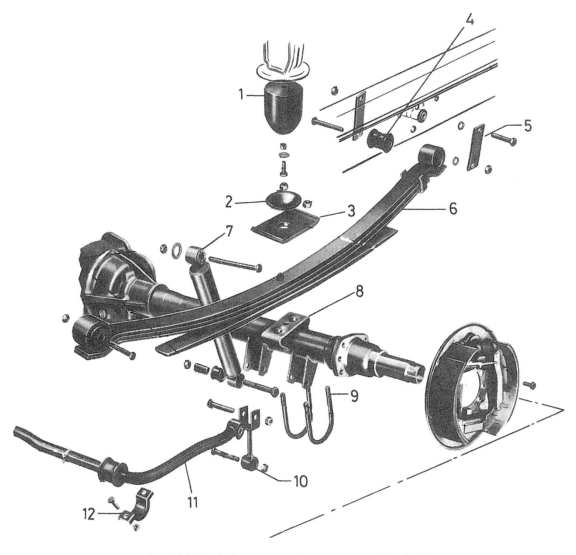

Fig. 11.31 Typical rear roadspring components (Sec 21)

1	Buffer	4	Bush	7	Shock absorber	10	End link
2	Seat	5	Shackle plate	8	Axle casing	11	Anti-roll bar
3	Spring plate	6	Leaf spring	9	U-bolt	12	Clamp

23 Steering linkage balljoints – renewal

1   Wear in the balljoints can only be rectified by renewal of the component. The tie-rod end balljoints can be removed from one tie-rod but not from the other one, where they must be renewed as an assembly together with the rod.
2   The drag link has one removable balljoint, the other one being permanently attached to the rod.
3   The removal of all steering link rods is similar. Unscrew and remove the nut from the balljoint taper pin and then use a balljoint extractor to separate the taper pin from the eye.
4   Where the balljoint is detachable from the rod, always mark the setting (number of threads exposed) before removing it. In this way the new balljoint may be installed in the same relative position as the original.
5   Never apply grease to a taper pin before connecting it to its eye.
6   Sometimes it is found that it is impossible to unscrew the taper pin nut as the ball turns in its socket at the same time. Where this occurs, place a jack under the socket and apply pressure to force the taper pin into tighter contact with the eye.
7   However careful you may have been to adjust the position of the new balljoint assemblies to their original positions, always check the

front end alignment as described in Section 30.
8   Always use new split pins in the castellated nuts.

24 Idler arm (LT 28, 31, 35) – removal, overhaul and refitting

1   Two idler arm assemblies are used in the steering linkage. The following operations may be required to one or both if wear in the idler arm bushes has been detected.
2   Disconnect the relay rod balljoints from the idler arms, using a suitable extractor.
3   Disconnect the drag link balljoint from the idler arm.
4   Unscrew the idler bracket mounting bolts and remove the idler arm assembly from the vehicle.
5   Grip the bracket in the jaws of a vice and unscrew and remove the idler arm pinch-bolt.
6   Take off the idler arm from its spindle.
7   Remove the idler spindle bushes by drifting them out, or by attaching a claw type slide hammer to them.
8   Press in the new bushes or use a bolt, nut and thrust washers to draw them into position. Make sure that the top bush is flush when installed. The bottom bush should stand proud by between 0.8 and 1.1 mm (0.03 and 0.04 in).

Fig. 11.32 Steering linkage (independent
suspension) (Sec 23)

1 Cap
2 Idler spindle
3 Steering box
4 Dished washer
5 Bush
6 Idler
7 Drag link
8 Drop arm
9 Bush
10 Special washer (plastic
side to ball wrench)
11 Seal
12 Support ring
13 O-ring
14 Bellcrank lever
15 Relay rod (part)
16 Adjustable tie-rod
17 Stub axle carrier
18 Idler
19 Idler arm
20 Relay rod (part)
21 Non-adjustable tie-rod
22 Stub axle

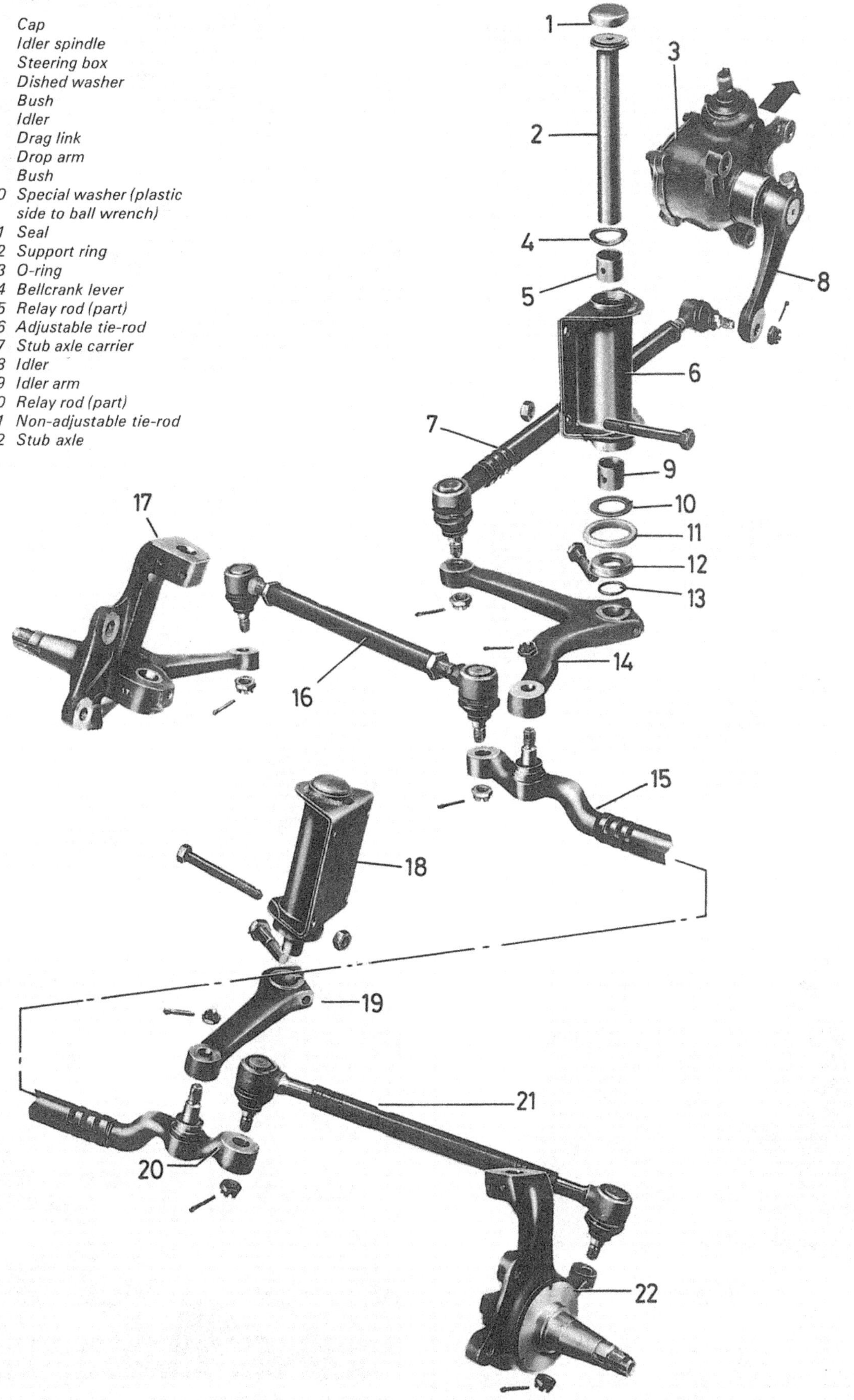

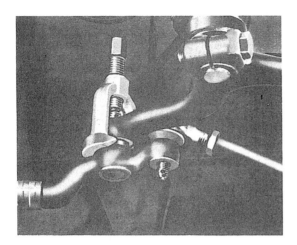

Fig. 11.33 Disconnecting relay rod balljoint (Sec 24)

25.2 Removing the horn button

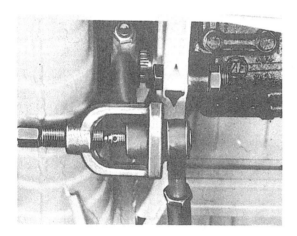

Fig. 11.34 Disconnecting drag link balljoint (Sec 24)

25.3 Horn switch earthing wire

Fig. 11.35 Idler lower bush projection (Sec 24)

a = 0.8 to 1.1 mm (0.03 to 0.04 in)

25.5 Removing the steering wheel

9   Refitting is a reversal of removal. Apply grease generously to the inside of the idler spindle housing and bushes, and observe carefully the sequence of washers and seals fitted above the idler arm.

25 Steering wheel – removal and refitting

1   Centre the steering wheel so that the front roadwheels are in the straight-ahead position.
2   From the centre of the steering wheel hub, prise out the plastic horn button. Do not attempt to lever under the metal bezel (photo).
3   Pull off the earthing wire (photo).
4   Unscrew the steering wheel nut.
5   Pull the wheel from the splined shaft (photo).
6   Take out the small coil spring.
7   Refitting is a reversal of removal, but if there is any doubt as to the centralising of the steering gear, check that the notch on the pinion shaft dust excluder is in alignment with the projection on the steering box (Fig. 11.36).
8   Tighten the steering wheel nut to the specified torque.

26 Steering column – removal and refitting

1   Disconnect the battery.
2   Set the steering to its centre position as described in the preceding Section.
3   Pull the steering column shroud from the column by giving it a sharp jerk to overcome the resistance of its securing clip (photo).
4   Disconnect the steering column switch wiring harness at the connector plugs.
5   At the base of the column, unscrew and remove the splined coupling pinch-bolt.
6   At the upper end of the column, disconnect the support bracket.
7   Withdraw the column/shaft assembly complete with steering wheel from the driving cab.
8   The column switches may be removed, also the steering column lock; refer to Chapter 10.
9   The flexible and splined coupling components may be removed (note the earth bond), but further dismantling is not recommended. Rather, obtain a complete column/shaft assembly.
10 Refitting is a reversal of removal.

27 Steering gear – adjustment

1   If slight knocking is evident when traversing uneven surfaces, then the worm and roller backlash may require adjusting.
2   To do this, raise the front of the vehicle to lift the roadwheels off the ground.
3   Set the steering to its central position as described in Section 25.
4   Using a suitable puller, remove the drag link from the drop arm on the steering box.
5   Now turn the steering wheel through one half of a turn.
6   Loosen the adjuster screw locknut on the steering box and unscrew the adjuster screw one turn.
7   Grip the drop arm. While pushing and pulling it, turn the adjuster screw in until all endfloat in the roller shaft disappears.
8   Without altering the setting of the screw, tighten the locknut.
9   If the steering wheel is turned about a further one-eighth of a turn, a slight endfloat will be detected at the drop arm. This is correct.
10 Reconnect the drag link, tighten the castellated nut to the specified torque and insert a new split pin.

28 Steering box – removal and refitting

1   Centre the steering (Section 25).
2   Using a suitable balljoint extractor, disconnect the drag link from the drop arm at the steering box.
3   Unscrew the two bolts which connect the splined pinion coupling to the flexible coupling at the base of the steering column.
4   Unbolt the steering box from the chassis side-member, and remove it from the vehicle.
5   Spare parts are not available for the repair of steering boxes; only complete assemblies are available.

Fig. 11.36 Steering centred alignment marks (Sec 25)

26.3 Removing the column centre shroud section

Fig. 11.37 Steering flexible coupling earth bond (Sec 26)

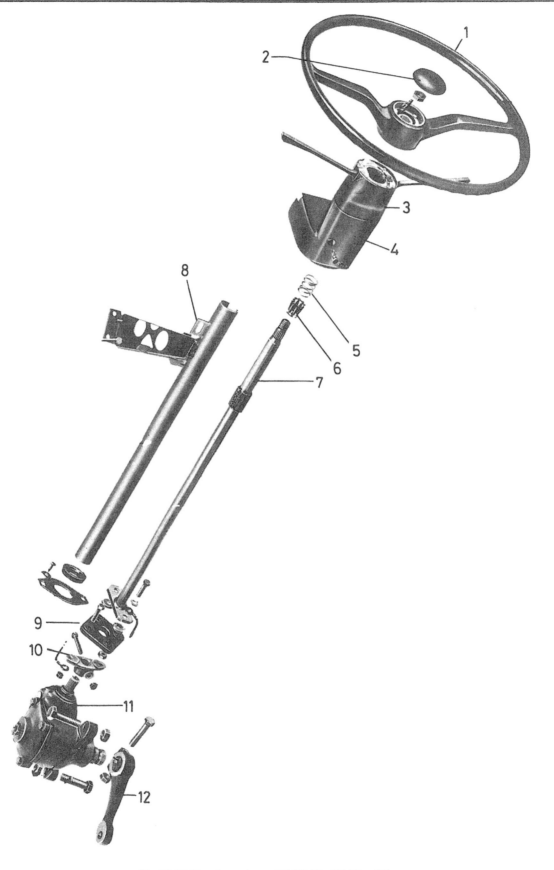

Fig. 11.38 Steering column (LT 28, 31, 35) (Sec 26)

1 Steering wheel	4 Lower shroud	7 Steering shaft	10 Splined coupling
2 Horn button	5 Spring	8 Column tube	11 Steering box
3 Upper shroud	6 Support ring	9 Flexible disc	12 Drop arm

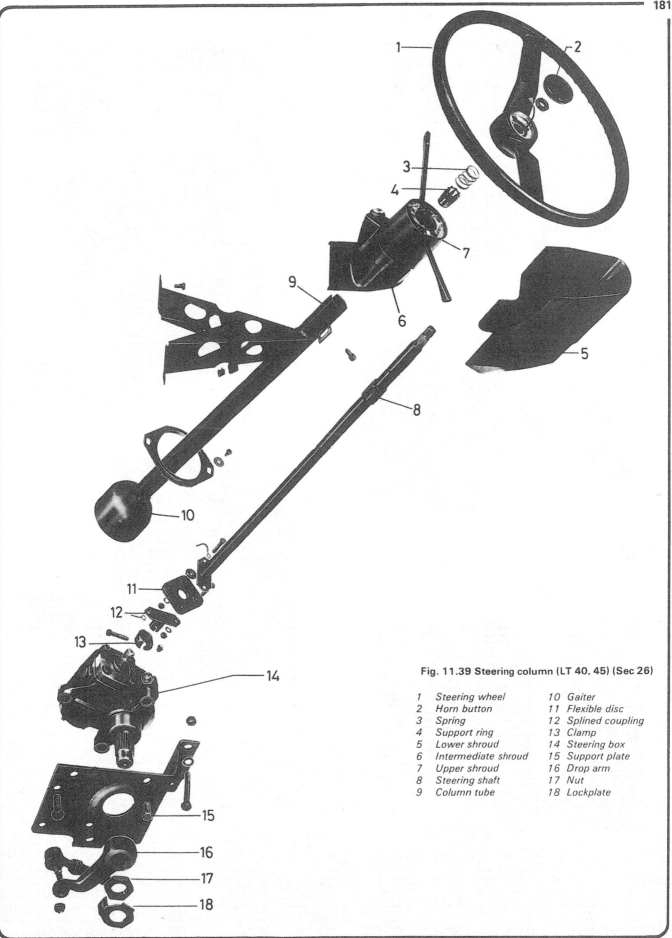

Fig. 11.39 Steering column (LT 40, 45) (Sec 26)

1	Steering wheel	10	Gaiter
2	Horn button	11	Flexible disc
3	Spring	12	Splined coupling
4	Support ring	13	Clamp
5	Lower shroud	14	Steering box
6	Intermediate shroud	15	Support plate
7	Upper shroud	16	Drop arm
8	Steering shaft	17	Nut
9	Column tube	18	Lockplate

Fig. 11.40 Steering box adjuster screw and locknut (Sec 27)

Fig. 11.42 Master spline alignment (LT 28, 31, 35) (Sec 28)

Fig. 11.44 Correct setting of drop arm offset (Sec 28)

Fig. 11.41 Steering splined coupling connecting bolts (Sec 28)

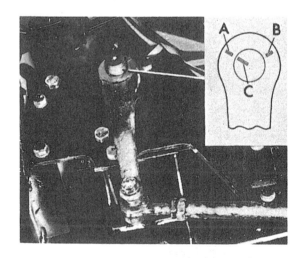

Fig. 11.43 Drop arm alignment marks (LT 40, 45) (Sec 28)

A with C for LHD vehicles   B with C for RHD vehicles

8   Refitting is a reversal of removal, but note that the drop arm and its splined shaft have a master spline and groove to ensure correct fitting on LT 28, 31 and 35 models, while on LT 40 and 45 models, the engraved marks must be aligned according to which side the steering is located (LHD or RHD) and as shown in Fig. 11.42. Make sure also that the offset in the drop arm is towards the steering box.
9   Make sure that the steering wheel and the box are correctly centralised before connecting the splined and flexible couplings, and remember the earth bond.

6   The drop arm should be removed after unscrewing its pinch-bolt for fitting to the new steering box.
7   Steering boxes for LT 28, 31 and 35 models are supplied sealed and full of lubricant. Those for LT 40 and 45 models are supplied 'dry' and must be filled after installation in the vehicle.

29 Roadwheels and tyres – general

1   The roadwheels on all models are of pressed steel construction, but vary in size and design. Never interchange different wheel types.
2   Certain models are equipped with twin rear wheels.

3   Periodically check the wheels for rim damage and rusting and take the appropriate action.

4   Steel braced radial ply tyres are fitted as standard. When new tyres are being purchased, always buy tyres of similar construction. Textile ply tyres if fitted to these vehicles will be found to wear very quickly.

5   Keep the tyres inflated to the specified pressure and examine them frequently for damage to the sidewalls, and for nails and flints embedded in the treads.

6   The roadwheels are balanced statically and dynamically off the vehicle during production. Correct balance must be maintained by having them re-balanced when new tyres are fitted, when a puncture has been repaired, and halfway through the useful life of the tyres.

7   On LT 35 Pick-up models, the spare wheel must be stowed on its bracket so that the dished side of the wheel is downwards, otherwise the fuel tank may be damaged.

30 Steering angles and front wheel alignment – explanation of terms

1   Accurate front wheel alignment is essential to good steering and for even tyre wear. Before considering the steering angles, check that the tyres are correctly inflated, that the front wheels are not buckled, the hub bearings are not worn or incorrectly adjusted and that the steering linkage is in good order, without slackness or wear at the joints.

2   Wheel alignment consists of four factors:

Camber is the angle at which the roadwheels are set from the vertical when viewed from the front or rear of the vehicle. Positive camber is the angle (in degrees) that the wheels are tilted outwards at the top from the vertical.

Castor is the angle between the steering axis and a vertical line when viewed from each side of the vehicle. Positive castor is indicated when the steering axis is inclined towards the rear of the vehicle at its upper end.

Steering axis inclination is the angle, when viewed from the front or rear of the vehicle, between the vertical and an imaginary line drawn between the upper and lower suspension swivel balljoints or kingpin centres.

Toe is the amount by which the distance between the front inside edges of the roadwheel rims differs from that between the rear inside edges. If the distance between the front edges is less than that at the rear, the wheels are said to toe-in. If the distance between the front inside edges is greater than that at the rear, the wheels toe-out.

3   Due to the need for precision gauges to measure the small angles of the steering and suspension settings, it is preferable that adjustment of camber and castor is left to a service station having the necessary equipment.

4   For information purposes however, the methods of adjusting or correcting these angles are described in the following two Sections.

31 Front wheel alignment (LT 28, 31, 35)

Camber
1   The camber angle is adjusted by releasing the suspension upper arm pivot nut and using an Allen key in the pivot bolt socket, turning the pivot bolt. This will cause the eccentric washers to rotate and so move the suspension arm relative to the side-member.

Castor
2   The castor angle is varied according to the setting of the radius rods. Provided the length indicated in Fig. 11.12 is between 529 and 540 mm (20.8 and 21.3 in), then the angle will be within tolerance. When installing a rod after dismantling or renewal, set its length to between 532 and 537 mm (20.9 and 21.1 in).

Toe-in
3   Set the front roadwheels in the straight-ahead position. The vehicle must be parked on smooth, level ground, and be normally laden.

4   Obtain a tracking gauge. These are available in various forms from accessory stores, or one can be fabricated from a length of steel tubing, suitably cranked to clear the sump and bellhousing, and having

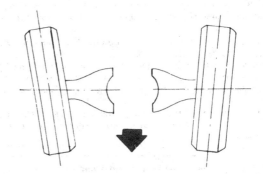

Fig. 11.45 Front wheel alignment diagram. Arrow points to front of vehicle (Sec 30)

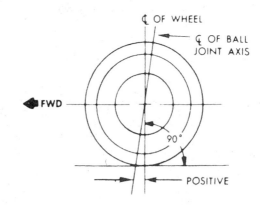

Fig. 11.46 Castor diagram (Sec 30)

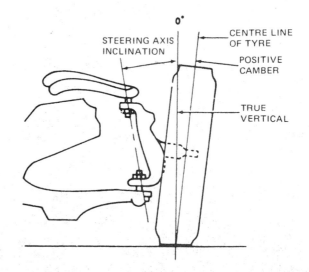

Fig. 11.47 Steering axis inclination and camber diagram (independent suspension shown) (Sec 30)

a setscrew and locknut at one end.

5   With the gauge, measure the distance between the two wheel inner rims (at hub height) at the rear of the wheel. Push the vehicle forward to rotate the wheel through 180° (half a turn) and measure the distance between the wheel inner rims, again at hub height, at the front of the wheel. The last measurement should be less than the first by the appropriate toe-in (see Specifications).

6   Where the toe-in is found to be incorrect, release the locknuts on the adjustable tie-rod and turn the rod. The opposite tie-rod is non-adjustable. Only turn the tie-rod a quarter or half a turn before

rechecking the toe-in.

7   On completion, tighten the toe-rod locknuts without disturbing their setting. Check that the balljoints are in the centres of their arcs of travel.

8   With the roadwheels still in the straight-ahead positon, check the steering centralising marks as described in Section 25. If they are found to be out of alignment, then disconnect the drag link from the steering box drop arm, release the locknut and adjust the position of the link balljoint so that with the steering now centralised, the balljoint taper pin will pass directly into the eye of the drop arm.

9   If the position of the steering wheel spokes is still incorrect, remove the steering wheel as described in Section 25 and reposition it on its splines.

32 Front wheel alignment (LT 40, 45)

Camber and castor

1   These are non-adjustable. Any deviation from the specified values will be due to a distorted axle beam or incorrectly mounted axle in relation to the roadsprings.

Toe-in

2   On these models, the tie-rod is of one-piece construction, locked to the balljoint assemblies by clamps.

3   The methods of toe adjustment and steering drag link setting are otherwise similar to those described in Section 31.

33 Fault diagnosis – suspension and steering

Symptom	Reason(s)
Steering feels vague, car wanders and floats at speed	Tyre pressures uneven Shock absorbers worn Spring broken Steering gear balljoints badly worn Suspension geometry incorrect Steering mechanism free play excessive Front suspension and rear axle pick-up points out of alignment
Stiff and heavy steering	Tyre pressures too low No oil in steering gear (LT 40, 45) Front wheel toe-in incorrect Suspension geometry incorrect Steering gear incorrectly adjusted too tightly Steering column badly misaligned
Wheel wobble and vibration	Seized balljoints or swivels Wheel nuts loose Front wheels and tyres out of balance Steering balljoints badly worn Hub bearings badly worn or incorrectly adjusted Steering gear free play excessive Front springs weak or broken

Chapter 12 Bodywork and underframe

For modifications, and information applicable to later models, see Supplement at end of manual

Contents

1 Description

All models in this light truck range are of forward control design, with a steel chassis and cab/body of semi-unitary construction.

The front chassis members are designed to permit progressive deformation in case of impact. The safety cage cab design, which incorporates a roll-over bar, also contributes to providing good occupant safety levels.

Short or long wheelbase panel vans are available. High roof panel vans, double cab chassis pick-up and chassis cab versions of various capacities complete a wide range of options.

2 Bodywork and chassis – maintenance

1 The main requirement is to keep the body and chassis clean. Mud encourages rust and should be removed regularly using a high pressure hose.

2 Accumulations of oil or grease should be removed with a steam cleaner or a water-soluble grease solvent.

3 Check the underbody protective coating periodically and make good where necessary.

4 An older vehicle may well benefit from wax injection in the chassis box sections.

5 Keep the door drain holes clear (photo).

6 Oil door hinges and locks regularly. On van versions with sliding doors, the cover must be removed from the centre guide rail before the hinge pivots are accessible for oiling. To do this extract the two cover securing screws, noting that the front screw is not accessible until the door is opened about 15 mm (0.6 in).

7 Remove the cover by tapping it gently, using a wedge inserted at its rear edge.

8 If necessary, the beading can be stuck to the cover using a suitable adhesive.

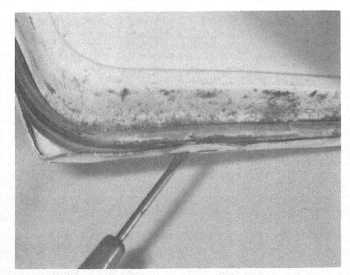

2.5 Clearing a door drain hole

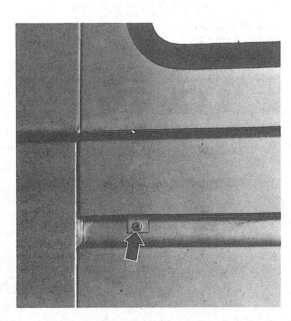

Fig. 12.1 Sliding door guide rail cover rear screw (Sec 2)

Fig. 12.2 Sliding door guide rail cover front screw (Sec 2)

Fig. 12.3 Sliding door hinge pivot oiling points (Sec 2)

3   Body damage – repair

The photographic sequences on pages 190 and 191 illustrate the operations detailed in the following sub-sections.

Repair of minor scratches in the bodywork

If the scratch is very superficial, and does not penetrate to the metal of the bodywork, repair is very simple. Lightly rub the area of the scratch with a paintwork renovator, or a very fine cutting paste, to remove loose paint from the scratch and to clear the surrounding bodywork of wax polish. Rinse the area with clean water.

Apply touch-up paint to the scratch using a fine paint brush; continue to apply fine layers of paint until the surface of the paint in the scratch is level with the surrounding paintwork. Allow the new paint at least two weeks to harden: then blend it into the surrounding paintwork by rubbing the scratch area with a paintwork renovator or a very fine cutting paste. Finally, apply wax polish.

Where the scratch has penetrated right through to the metal of the bodywork, causing the metal to rust, a different repair technique is

required. Remove any loose rust from the bottom of the scratch with a penknife, then apply rust inhibiting paint to prevent the formation of rust in the future. Using a rubber or nylon applicator fill the scratch with bodystopper paste. If required, this paste can be mixed with cellulose thinners to provide a very thin paste which is ideal for filling narrow scratches. Before the stopper-paste in the scratch hardens, wrap a piece of smooth cotton rag around the top of a finger. Dip the finger in cellulose thinners and then quickly sweep it across the surface of the stopper-paste in the scratch; this will ensure that the surface of the stopper-paste is slightly hollowed. The scratch can now be painted over as described earlier in this Section.

Repair of dents in the bodywork

When deep denting of the bodywork has taken place, the first task is to pull the dent out, until the affected bodywork almost attains its original shape. There is little point in trying to restore the original shape completely, as the metal in the damaged area will have stretched on impact and cannot be reshaped fully to its original contour. It is better to bring the level of the dent up to a point which is about $\frac{1}{8}$ in (3 mm) below the level of the surrounding bodywork. In cases where the dent is very shallow anyway, it is not worth trying to pull it out at all. If the underside of the dent is accessible, it can be hammered out gently from behind, using a mallet with a wooden or plastic head. Whilst doing this, hold a suitable block of wood firmly against the outside of the panel to absorb the impact from the hammer blows and thus prevent a large area of the bodywork from being 'belled-out'.

Should the dent be in a section of the bodywork which has a double skin or some other factor making it inaccessible from behind, a different technique is called for. Drill several small holes through the metal inside the area – particularly in the deeper section. Then screw long self-tapping screws into the holes just sufficiently for them to gain a good purchase in the metal. Now the dent can be pulled out by pulling on the protruding heads of the screws with a pair of pliers.

The next stage of the repair is the removal of the paint from the damaged area, and from an inch or so of the surrounding 'sound' bodywork. This is accomplished most easily by using a wire brush or abrasive pad on a power drill, although it can be done just as effectively by hand using sheets of abrasive paper. To complete the preparation for filling, score the surface of the bare metal with a screwdriver or the tang of a file, or alternatively, drill small holes in the affected area. This will provide a really good 'key' for the filler paste.

To complete the repair see the Section on filling and re-spraying.

Repair of rust holes or gashes in the bodywork

Remove all paint from the affected area and from an inch or so of the surrounding 'sound' bodywork, using an abrasive pad or a wire brush on a power drill. If these are not available a few sheets of abrasive paper will do the job just as effectively. With the paint removed you will be able to gauge the severity of the corrosion and therefore decide whether to renew the whole panel (if this is possible) or to repair the affected area. New body panels are not as expensive as most people think and it is often quicker and more satisfactory to fit a new panel than to attempt to repair large areas of corrosion.

Remove all fittings from the affected area except those which will act as a guide to the original shape of the damaged bodywork (eg headlamp shells etc). Then, using tin snips or a hacksaw blade, remove all loose metal and any other metal badly affected by corrosion. Hammer the edges of the hole inwards in order to create a slight depression for the filler paste.

Wire brush the affected area to remove the powdery rust from the surface of the remaining metal. Paint the affected area with rust inhibiting paint; if the back of the rusted area is accessible treat this also.

Before filling can take place it will be necessary to block the hole in some way. This can be achieved by the use of aluminium or plastic mesh, or aluminium tape.

Aluminium or plastic mesh is probably the best material to use for a large hole. Cut a piece to the approximate size and shape of the hole to be filled, then position it in the hole so that its edges are below the level of the surrounding bodywork. It can be retained in position by several blobs of filler paste around its periphery.

Aluminium tape should be used for small or very narrow holes. Pull a piece off the roll and trim it to the approximate size and shape required, then pull off the backing paper (if used) and stick the tape over the hole; it can be overlapped if the thickness of one piece is insufficient. Burnish down the edges of the tape with the handle of a screwdriver or similar, to ensure that the tape is securely attached to the metal underneath.

Bodywork repairs – filling and re-spraying

Before using this Section, see the Sections on dent, deep scratch, rust holes and gash repairs.

Many types of bodyfiller are available, but generally speaking those proprietary kits which contain a tin of filler paste and a tube of resin hardener are best for this type of repair. A wide, flexible plastic or nylon applicator will be found invaluable for imparting a smooth and well contoured finish to the surface of the filler.

Mix up a little filler on a clean piece of card or board – measure the hardener carefully (follow the maker's instructions on the pack) otherwise the filler will set too rapidly or too slowly.

Using the applicator apply the filler paste to the prepared area; draw the applicator across the surface of the filler to achieve the correct contour and to level the filler surface. As soon as a contour that approximates to the correct one is achieved, stop working the paste – if you carry on too long the paste will become sticky and begin to 'pick up' on the applicator. Continue to add thin layers of filler paste at twenty-minute intervals until the level of the filler is just proud of the surrounding bodywork.

Once the filler has hardened, excess can be removed using a metal plane or file. From then on, progressively finer grades of abrasive paper should be used, starting with a 40 grade production paper and finishing with 400 grade wet-and-dry paper. Always wrap the abrasive paper around a flat rubber, cork, or wooden block – otherwise the surface of the filler will not be completely flat. During the smoothing of the filler surface the wet-and-dry paper should be periodically rinsed in water. This will ensure that a very smooth finish is imparted to the filler at the final stage.

At this stage the 'dent' should be surrounded by a ring of bare metal, which in turn should be encircled by the finely 'feathered' edge of the good paintwork. Rinse the repair area with clean water, until all of the dust produced by the rubbing-down operation has gone.

Spray the whole repair area with a light coat of primer – this will show up any imperfections in the surface of the filler. Repair these imperfections with fresh filler paste or bodystopper, and once more smooth the surface with abrasive paper. If bodystopper is used, it can be mixed with cellulose thinners to form a really thin paste which is ideal for filling small holes. Repeat this spray and repair procedure until you are satisfied that the surface of the filler, and the feathered edge of the paintwork are perfect. Clean the repair area with clean water

and allow to dry fully.

The repair area is now ready for final spraying. Paint spraying must be carried out in a warm, dry, windless and dust free atmosphere. This condition can be created artificially if you have access to a large indoor working area, but if you are forced to work in the open, you will have to pick your day very carefully. If you are working indoors, dousing the floor in the work area with water will help to settle the dust which would otherwise be in the atmosphere. If the repair area is confined to one body panel, mask off the surrounding panels; this will help to minimise the effects of a slight mis-match in paint colours. Bodywork fittings (eg chrome strips, door handles etc) will also need to be masked off. Use genuine masking tape and several thicknesses of newspaper for the masking operations.

Before commencing to spray, agitate the aerosol can thoroughly, then spray a test area (an old tin, or similar) until the technique is mastered. Cover the repair area with a thick coat of primer; the thickness should be built up using several thin layers of paint rather than one thick one. Using 400 grade wet-and-dry paper, rub down the surface of the primer until it is really smooth. While doing this, the work area should be thoroughly doused with water, and the wet-and-dry paper periodically rinsed in water. Allow to dry before spraying on more paint.

Spray on the top coat, again building up the thickness by using several thin layers of paint. Start spraying in the centre of the repair area and then, using a circular motion, work outwards until the whole repair area and about 2 inches of the surrounding original paintwork is covered. Remove all masking material 10 to 15 minutes after spraying on the final coat of paint.

Allow the new paint at least two weeks to harden, then, using a paintwork renovator or a very fine cutting paste, blend the edges of the paint into the existing paintwork. Finally, apply wax polish.

Major damage repair

Major body or chassis damage, whether due to accident or to corrosion, should be rectified by your dealer. A check for distortion must be carried out after completing such work, using special jigs and accurate measuring equipment.

***

4 Bumpers – removal and refitting

Front bumpers
1 Unscrew and remove the number plate.
2 Prise the rubber caps from the bumper securing bolts.

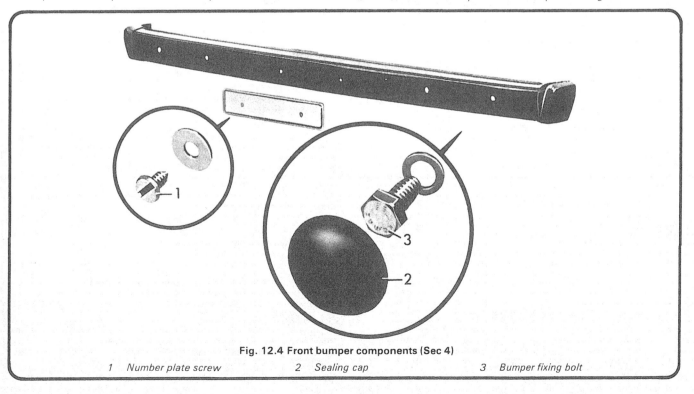

Fig. 12.4 Front bumper components (Sec 4)

1 Number plate screw	2 Sealing cap	3 Bumper fixing bolt

Fig. 12.5 Rear bumper fixing points (Sec 4)

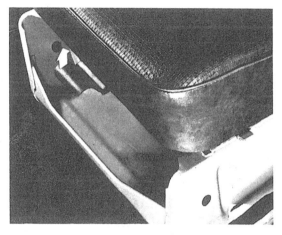

Fig. 12.6 Seat adjusting lever (Sec 6)

3   Unscrew the bolts and remove the bumper.

Rear bumper
4   Working under the rear of the vehicle, unbolt the bumper brackets from the chassis member and triangular reinforcement plate.

All bumpers
5   Refitting is the reverse of the removal procedure.

5   Radiator grille – removal and refitting

1   Extract the fixing screws, which are visible from the front of the vehicle, and withdraw the grille (photo).
2   Refitting is a reversal of removal.

6   Seats – removal and refitting

Driver's seat
1   Refer to Chapter 1, Section 4.

Passenger seat
2   Pull the seat adjusting lever and the smaller locking lever upwards. Hold them in this position while the seat is pushed towards the front of the vehicle out of its runners.
3   Refitting is the reverse of the removal procedure.

7   Facia panel – removal and refitting

1   Disconnect the battery.
2   Reach up behind the facia panel, squeeze the lighting switch retaining tangs together and remove the switch.
3   Remove the heater blower and hazard warning switches in a

5.1 Removing a grille screw

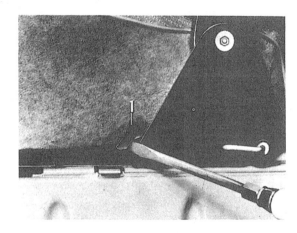

Fig. 12.7 Seat locking lever (1) (Sec 6)

similar way.
4   Again working under the facia panel, pull the convoluted air ducts from the heater casing.
5   Remove the instrument panel as described in Chapter 10.
6   Reach through the opening left by removal of the instrument panel and disconnect the fresh air ducts.
7   Remove the cover plate from the brake hydraulic fluid reservoir and release the reservoir by extracting its single securing screw.
8   Where fitted, remove the radio and speaker.
9   Unscrew and remove the fixing screws from the front edge of the facia panel just below the windscreen.
10 Unscrew and remove the facia panel end screws which attach the panel to the body pillars.
11 Working under the lower edge of the facia panel, extract the screws which hold it to the support brackets. Remove the facia panel.
12 Refitting is a reversal of removal.

8   Front door trim panel – removal and refitting

1   Open the door and unscrew and remove the door lock plunger knob (photo).
2   Peel the plastic cover from the window winder handle, extract the retaining screw and remove the handle (photos).
3   Prise out the small panels from the door pull handle and extract the fixing screws (photo).
4   Prise out the dished section from the door lock interior handle to

Fig. 12.8 Removing the lighting switch (Sec 7)

Fig. 12.11 Brake fluid reservoir (1). Screw arrowed (Sec 7)

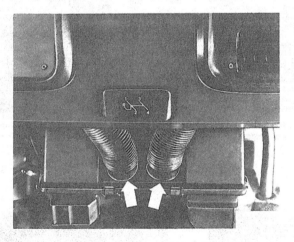

Fig. 12.9 Heater casing air ducts (Sec 7)

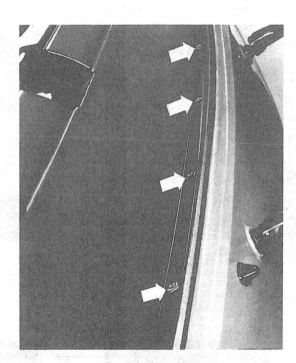

Fig. 12.12 Facia panel front securing screws (Sec 7)

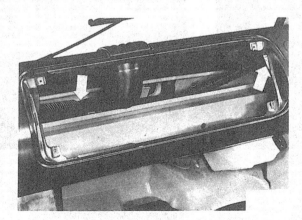

Fig. 12.10 Fresh air hoses (Sec 7)

reveal the handle fixing screw. Remove the screw and the handle (photos).

5   Insert the fingers or a broad blade between the edge of the trim panel and jerk the panel away to release the fixing clips (photo).

6   Remove the panel and carefully peel away the waterproof sheet (photo).

7   Refitting is a reversal of removal, but tuck the foil sections of the waterproof sheet into the door aperture as shown (Fig. 12.16).

9   Front door – dismantling and reassembly

1   Remove the door trim panel as previously described.

2   Wind the window down until the glass can be detached from the window lift by unscrewing the two bolts. Lower the glass.

3   Peel back the window seal channel and extract the screws which hold the upper part of the glass guide rail.

4   Unscrew the bolt which holds the lower part of the glass guide rail and withdraw the rail and seal in a downward direction.

This photographic sequence shows the steps taken to repair the dent and paintwork damage shown above. In general, the procedure for repairing a hole will be similar; where there are substantial differences, the procedure is clearly described and shown in a separate photograph.

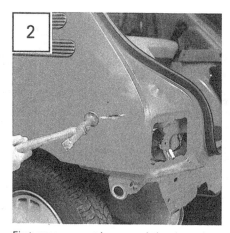

First remove any trim around the dent, then hammer out the dent where access is possible. This will minimise filling. Here, after the large dent has been hammered out, the damaged area is being made slightly concave.

Next, remove all paint from the damaged area by rubbing with coarse abrasive paper or using a power drill fitted with a wire brush or abrasive pad. 'Feather' the edge of the boundary with good paintwork using a finer grade of abrasive paper.

Where there are holes or other damage, the sheet metal should be cut away before proceeding further. The damaged area and any signs of rust should be treated with Turtle Wax Hi-Tech Rust Eater, which will also inhibit further rust formation.

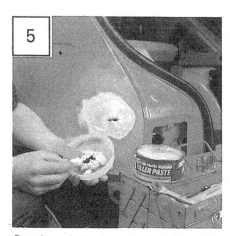

For a large dent or hole mix Holts Body Plus Resin and Hardener according to the manufacturer's instructions and apply around the edge of the repair. Press Glass Fibre Matting over the repair area and leave for 20-30 minutes to harden. Then ...

... brush more Holts Body Plus Resin and Hardener onto the matting and leave to harden. Repeat the sequence with two or three layers of matting, checking that the final layer is lower than the surrounding area. Apply Holts Body Plus Filler Paste as shown in Step 5B.

For a medium dent, mix Holts Body Plus Filler Paste and Hardener according to the manufacturer's instructions and apply it with a flexible applicator. Apply thin layers of filler at 20-minute intervals, until the filler surface is slightly proud of the surrounding bodywork.

For small dents and scratches use Holts No Mix Filler Paste straight from the tube. Apply it according to the instructions in thin layers, using the spatula provided. It will harden in minutes if applied outdoors and may then be used as its own knifing putty.

Use a plane or file for initial shaping. Then, using progressively finer grades of wet-and-dry paper, wrapped round a sanding block, and copious amounts of clean water, rub down the filler until glass smooth. 'Feather' the edges of adjoining paintwork.

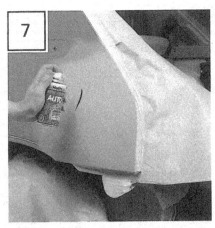

7

Protect adjoining areas before spraying the whole repair area and at least one inch of the surrounding sound paintwork with Holts Dupli-Color primer.

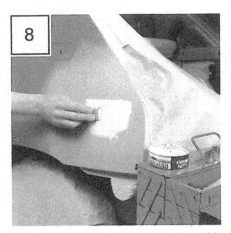

8

Fill any imperfections in the filler surface with a small amount of Holts Body Plus Knifing Putty. Using plenty of clean water, rub down the surface with a fine grade wet-and-dry paper – 400 grade is recommended – until it is really smooth.

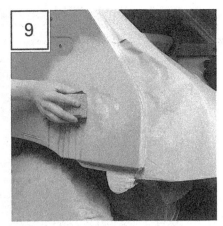

9

Carefully fill any remaining imperfections with knifing putty before applying the last coat of primer. Then rub down the surface with Holts Body Plus Rubbing Compound to ensure a really smooth surface.

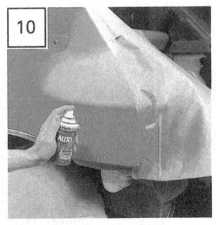

10

Protect surrounding areas from overspray before applying the topcoat in several thin layers. Agitate Holts Dupli-Color aerosol thoroughly. Start at the repair centre, spraying outwards with a side-to-side motion.

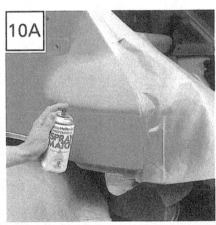

10A

If the exact colour is not available off the shelf, local Holts Professional Spraymatch Centres will custom fill an aerosol to match perfectly.

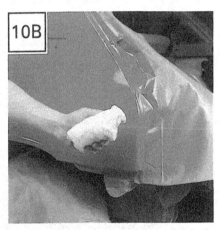

10B

To identify whether a lacquer finish is required, rub a painted unrepaired part of the body with wax and a clean cloth.

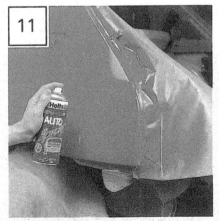

11

If *no* traces of paint appear on the cloth, spray Holts Dupli-Color clear lacquer over the repaired area to achieve the correct gloss level.

12

13

The paint will take about two weeks to harden fully. After this time it can be 'cut' with a mild cutting compound such as Turtle Wax Minute Cut prior to polishing with a final coating of Turtle Wax Extra.

14

When carrying out bodywork repairs, remember that the quality of the finished job is proportional to the time and effort expended.

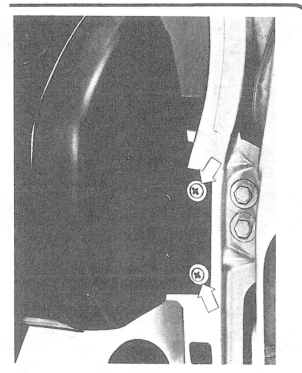

Fig. 12.13 Facia panel to pillar fixing screws (Sec 7)

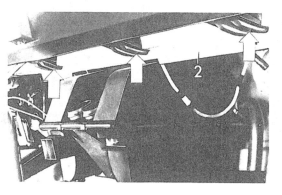

Fig. 12.14 Facia panel carrier (2). Arrows show screws (Sec 7)

8.2a Removing a window winder handle cover

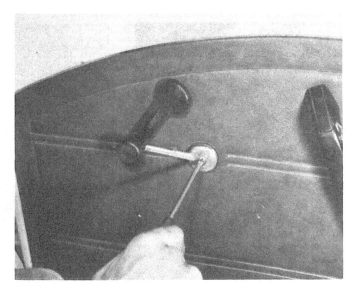

8.2b Extracting the window winder handle screw

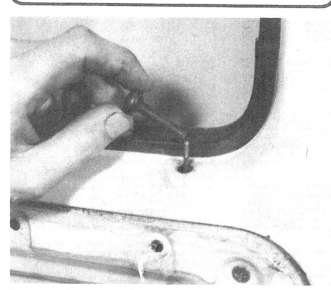

8.1 Unscrewing a door lock plunger knob

8.3 Removing a door pull screw

8.4a Remote control handle recessed cover

8.4b Extracting the remote control escutcheon screw

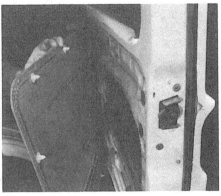

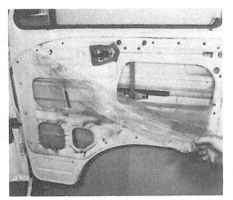

8.5 Removing the door trim

8.6 Removing the door waterproof sheet

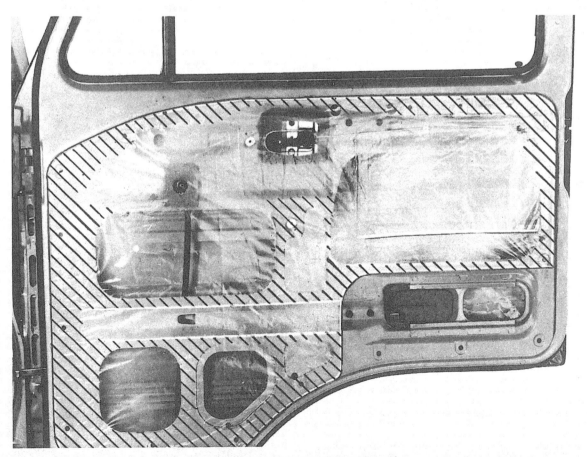

Fig. 12.15 Door waterproof sheet (Sec 8)

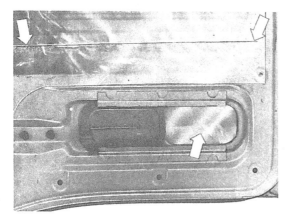

Fig. 12.16 Door foil flaps (arrowed) (Sec 8)

Fig. 12.17 Glass-to-lifter fixing bolts (Sec 9)

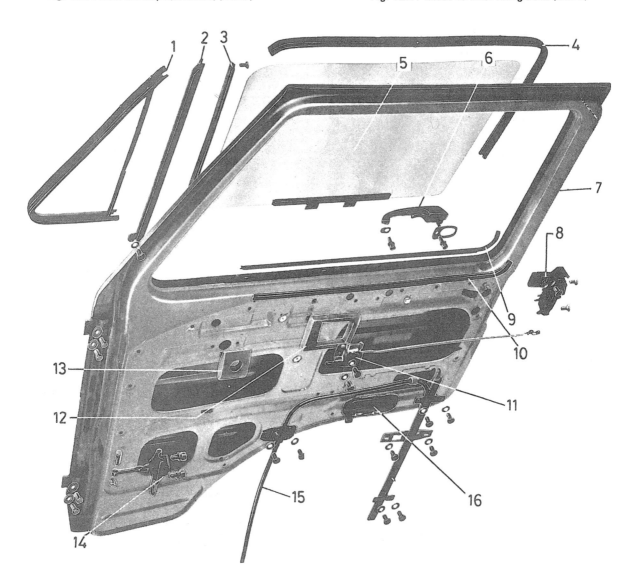

Fig. 12.18 Front door components (Sec 9)

1	Quarter window weatherseal	5	Glass	9	Outer weatherseal	13	Seal
2	Glass guide rail	6	Door exterior handle	10	Inner weatherseal	14	Door check
3	Door weatherseal	7	Door frame	11	Remote control	15	Window lifter
4	Glass guide channel	8	Door lock	12	Seal	16	Vent slide

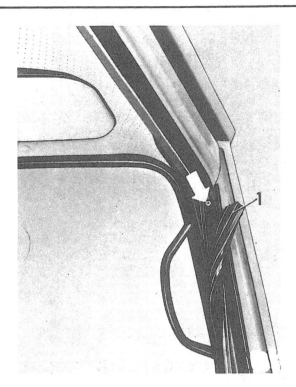

Fig. 12.19 Glass channel securing screw (arrowed) (Sec 9)

1 Seal

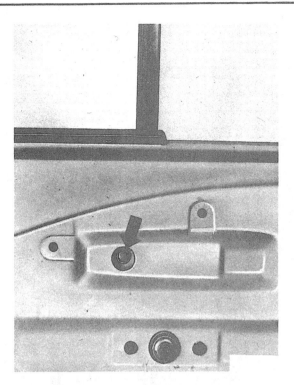

Fig. 12.20 Glass guide rail lower fixing bolt (Sec 9)

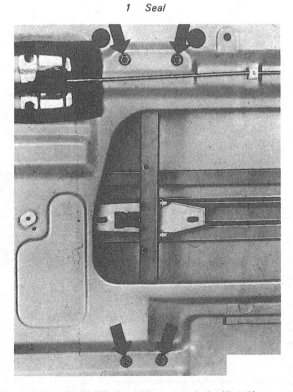

Fig. 12.21 Window lifter upper bolts (Sec 9)

Fig. 12.22 Window lifter lower bolts (arrowed) (Sec 9)

2 Sheet metal lug

5 Prise the inner and outer window slot seals from their retaining clips.

6 Remove the quarter window and its weatherseal.

7 Raise the window glass by hand and remove it from the door cavity by turning and tilting it.

8 Unbolt the window regulator, bend back the metal lug and remove the regulator (photo).

9 The door exterior handle may be removed by unscrewing the two socket-headed screws.

10 Disconnect the spring (3) from the remote control rod (4) (Fig. 12.24). Release the remote control fixing screws and remove the remote control assembly (photo).

11 The door lock can be removed from the door cavity after releasing its securing screws from the door edge (Fig. 12.25).

12 Refitting the components is a reversal of removal, but before tightening the bolts which hold the glass to the window lift, raise the

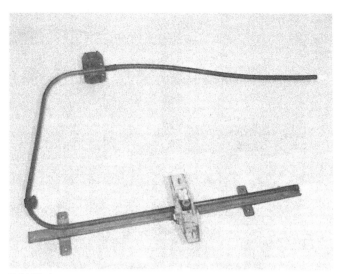

9.8 Window control mechanism

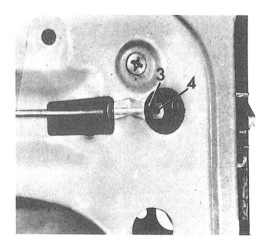

Fig. 12.24 Remote control pullrod (Sec 9)

3   *Spring*   4   *Control rod*

Fig. 12.25 Door lock fixing screws (Sec 9)

Fig. 12.23 Door exterior handle fixing screws (Sec 9)

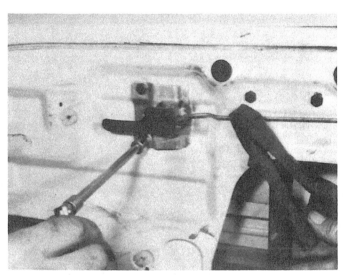

9.10 Removing a remote control handle screw

glass fully to align it in its guides. Holes are provided in the door inner panel so that the lift bolts can be fully tightened when the glass is in the fully raised position.

10 Front door – removal and refitting

1   Open the door and disconnect the door check link. To do this, extract the C-clip at the base of the link pivot pin and remove the pin (photo).
2   Support the door in the fully open position using jacks or blocks with pads of rag to prevent damage to the paintwork.
3   Unbolt the hinges from the body A pillar and lift the door from the cab (photo).
4   Refitting is a reversal of removal, but align the door so that when closed there is an equal gap all round its edge and the door panel is flush with the adjacent body panels. The door can be moved to correct

Fig. 12.26 Window lifter fixing bolts viewed through access holes (Sec 9)

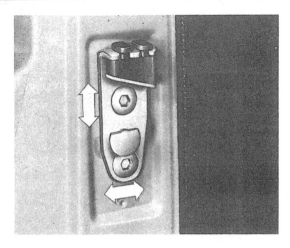

Fig. 12.27 Door striker plate. Arrows show possible adjustment (Sec 10)

alignment by slackening the hinge bolts.

5   The striker on the door pillar should be moved as necessary to provide smooth, positive closure. Release the striker by unscrewing the socket-headed screws.

11 Panel van rear doors – removal, refitting and adjustment

1   Open the door fully and remove the trim panel (if fitted) by pulling it from its retaining clips.

2   Disconnect the door check strap by pulling out the retaining pin.

3   Disconnect the leads from the rear number plate lamp by separating the plugs. Pull the leads out of the door.

4   Unscrew the hinge socket-headed bolts and withdraw the door from the pillar.

5   Refit by reversing the removal operations, but have the hinge bolts finger tight at first until the door can be closed and its alignment checked. Make sure that there is an even gap all round the door edge,

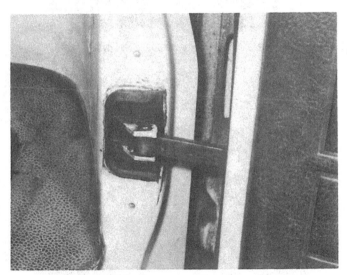

10.1 Front door check link

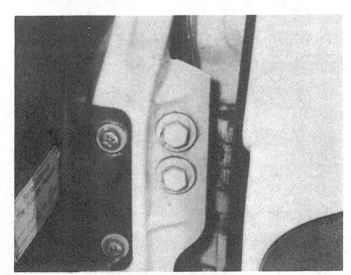

10.3 Front door hinge bolts and facia end screws

Fig. 12.28 Van rear door check strap (Sec 11)

1   Strap   2   Pin

Fig. 12.29 Van rear number plate leads (Sec 11)

3   Connectors   4   Cable clip

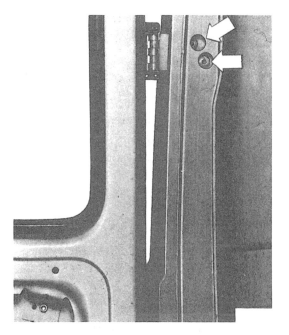

Fig. 12.30 Van rear door hinge bolts (Sec 11)

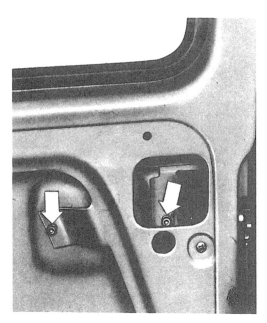

Fig. 12.31 Van rear door exterior handle fixing bolts (Sec 12)

that the waistline rib is level with the rib on the body, and that the door
panel is flush with the adjacent body panels. Move the door as
necessary within the limits of the hinge elongated bolt holes to correct
misalignment, then tighten the hinge bolts fully.
6   Check the lock adjustment as described in Section 13.

12 Panel van rear door primary lock – removal and refitting

1   Remove the door exterior handle by unscrewing the two securing
nuts from the interior of the door.
2   Extract the three screws and withdraw the lock from the door
cavity.
3   Refitting is a reversal of removal.

13 Panel van rear door secondary lock – removal, refitting and adjustment

1   Open the door. Extract the screw and remove the handle from the
door edge.
2   Reach inside the door cavity and prise off the clips from the

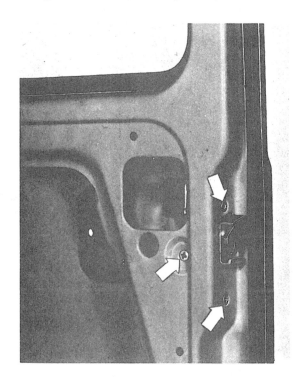

Fig. 12.32 Van rear door lock fixing screws (Sec 12)

locking rod pins.
3   Pull the rods from the pins and then withdraw them from their
guides at the top and bottom of the door.
4   Extract the screws and withdraw the lock from the door cavity.
5   Refit by reversing the removal operations, make sure to adjust the
locking rods as described in the following paragraphs before clipping
them to the pins on the lock. Have the handle in the unlocked position
before adjusting.
6   The upper locking rod is correctly adjusted when the dimension (a)
between the tip of the lug and the end face of the guide is as shown
in Fig. 12.38.
7   The lower locking rod is correctly adjusted when the tip of the lug
is level with the surface of the guide. Adjustment must be made either
by turning the locking lugs through one complete turn, or by rotating

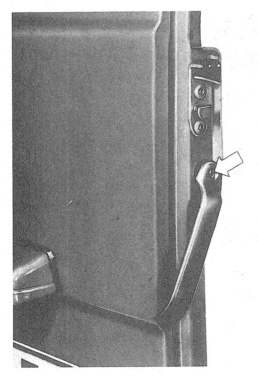

Fig. 12.33 Van rear door secondary lock handle. Fixing screw
arrowed (Sec 13)

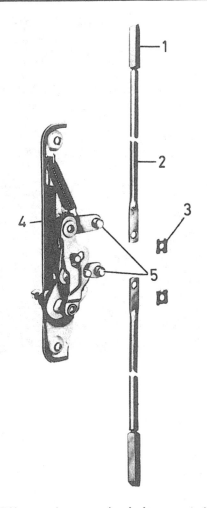

Fig. 12.35 Van rear door secondary lock components (Sec 13)

1	Lug	4	Lock
2	Locking rod	5	Lock pins
3	Clip		

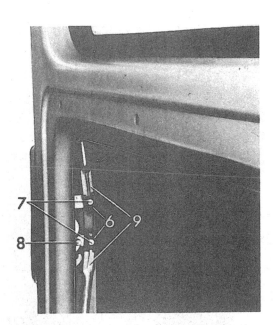

Fig. 12.34 Van rear door locking rods (Sec 13)

6	Clips	8	Lock
7	Pins	9	Locking rods

the locking rod through half a turn (180°).
8   The door striker plates may be moved as necessary to provide
smooth positive closure by releasing their socket-headed screws.
9   Close the doors and check for rattles. If evident, the cause is
insufficient pressure between the rod guide (17) and the bonded
rubber stop (16) (Fig. 12.40). This can be overcome by inserting a
shim (Part No 281 827 533A).

Fig. 12.36 Removing van rear door lower locking rod (Sec 13)

9	Locking rod	10	Rod guide

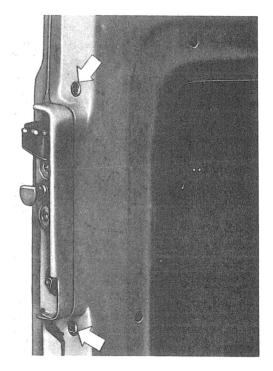

Fig. 12.37 Van rear door secondary lock fixing screws (Sec 13)

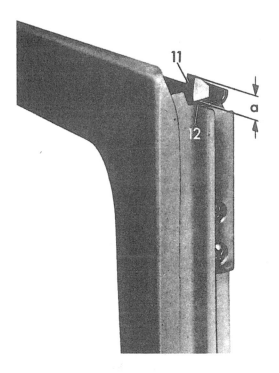

Fig. 12.38 Upper locking rod adjustment diagram (Sec 13)

11 Lug   12 Surface of rod guide

a   11.2 to 12.2 mm (0.44 to 0.48 in)

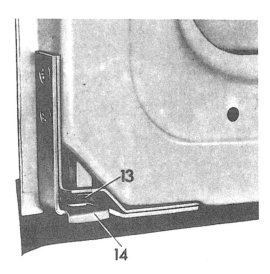

Fig. 12.39 Lower locking rod adjustment diagram (Sec 13)

13 Lug   14 Rod guide

Fig. 12.40 Van rear door rubber stop (16) and rod guide (17)
(Sec 13)

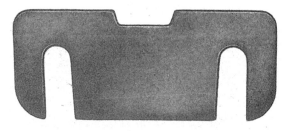

Fig. 12.41 Anti-rattle shim (Sec 13)

14 High roof panel van rear doors – removal, refitting and adjustment

The operations are similar to those described in Section 11.

15 High roof panel van rear door primary lock – removal, refitting and adjustment

1   Open the door and move the handle to the latched position.

Fig. 12.42 High roof panel van rear door lock handle (5) and sealing plugs (6) (Sec 15)

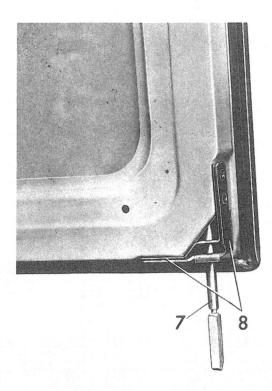

Fig. 12.44 Removing lower locking rod (Sec 15)

   7   *Locking rod*   8   *Rod guide*

2   Prise out the sealing plugs and extract the screws which are visible through the holes in the door panel to release the locking rods.
3   Pull the lower locking rod downwards out of the door.
4   Remove the upper locking rod guide and pull the locking rod out of the door.
5   Extract the securing screw from the interior handle and remove the handle.
6   Extract the three screws and withdraw the lock from the door cavity.
7   Refitting is a reversal of removal, but before tightening the locking rod clamp screws, carry out the following adjustment.
8   Make sure that the rods are in the unlocked position.
9   The correct setting of the upper rod is when the higher edge of the lug is level with the surface of the rod guide plate. Turn the lug in full turns only, and make sure that the crank of the rod is facing the correct way on completion.
10   The correct setting of the lower rod is when the higher edge of the lug is level with the surface of the guide cleat.

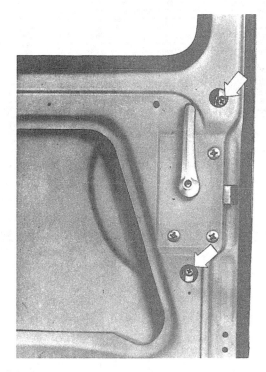

Fig. 12.43 Locking bar clamp screws visible through door panel holes (Sec 15)

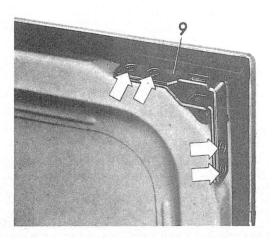

Fig. 12.45 Upper rod guide (9). Fixing screws arrowed (Sec 15)

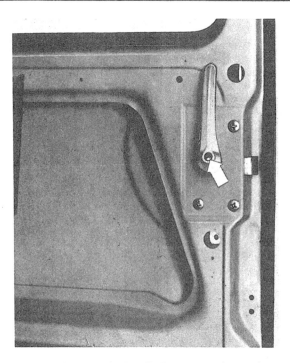

Fig. 12.46 Interior handle fixing screw (Sec 15)

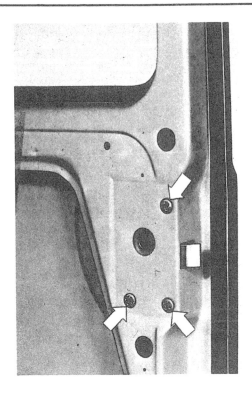

Fig. 12.47 Lock fixing screws (Sec 15)

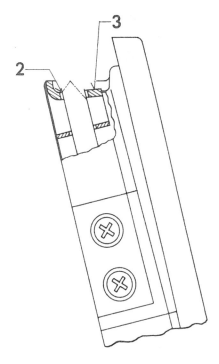

Fig. 12.48 Upper locking rod lug setting diagram (Sec 15)

2 Lug in unlocked position
3 Rod guide surface

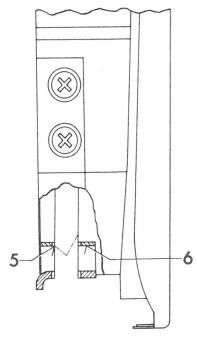

Fig. 12.49 Lower locking rod lug setting diagram (Sec 15)

5 Lug in unlocked position
6 Guide cleat

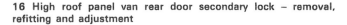

16 High roof panel van rear door secondary lock – removal, refitting and adjustment

1   Extract the securing screw and remove the interior handle.
2   Prise the clips from the lock pins to release the locking rods.
3   Pull the lower rod from the door.
4   Remove the upper guide and pull the upper rod from the door.

5   Extract the two screws and withdraw the lock from the door cavity.

6   Refitting is a reversal of removal. Adjust the locking rods as described in Section 15.

7   The door striker plates may be adjusted within the limits of their elongated bolt holes to provide smooth positive closure. If the door rattles when closed, refer to Section 13, paragraph 9.

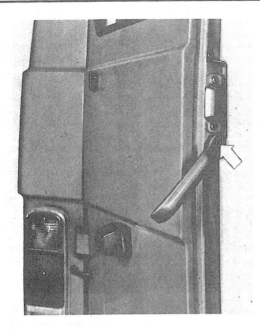

Fig. 12.50 Secondary lock handle fixing screw (Sec 16)

Fig. 12.51 High roof panel van rear door locking rods (Sec 16)

1	Clips	3	Locking rods
2	Locking pins	4	Lock

17 Van sliding door remote control lock – removal, refitting and adjustment

1   Prise out the small cap from the centre of the door interior handle.

2   Remove the handle fixing screw and the handle.

3   Extract the screw, which holds the exterior handle in position. Remove the handle.

4   Disconnect the remote control cable or rod from the lock by taking off the circlip and releasing the cable or rod adjustment by turning the adjuster nuts.

5   Unscrew the three securing screws and remove the lock.

6   Refitting is a reversal of removal, but adjust the remote control cable or rod by turning the adjuster nut to bring the end of the remote control lock pull rod up to its stop.

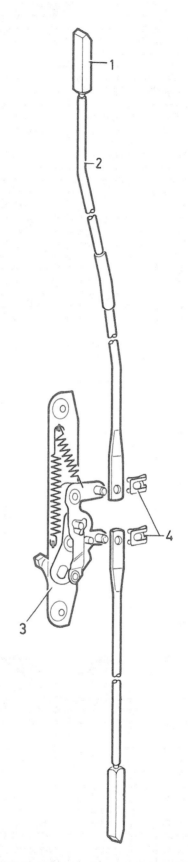

Fig. 12.52 High roof panel van rear door lock (Sec 16)

1	Lug	3	Lock
2	Locking rod	4	Clips

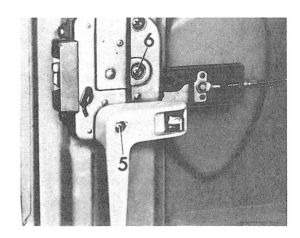

Fig. 12.54 Sliding door interior handle screw (5) and exterior handle screw (6) (Sec 17)

Fig. 12.53 High roof panel van lock securing screws (Sec 16)

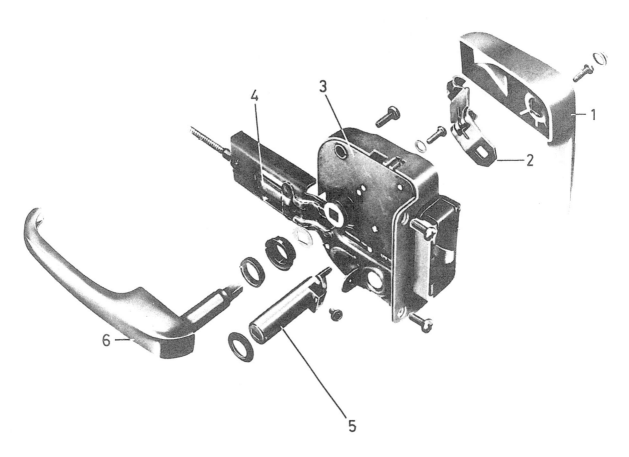

Fig. 12.55 Van sliding door remote control lock components (Sec 17)

1	Interior handle	3	Lock	5	Lock cylinder
2	Retainer	4	Cable/pullrod	6	Exterior handle

18 Van sliding door central lock – removal and refitting

1 Lock the central lock latch by pulling the interior handle.
2 Pull the tensioner spring rod towards the remote control lock, then unhook it from the door inner panel.
3 Release the remote control cable from its guide.
4 Extract the fixing screws and remove the central lock.
5 Refitting is a reversal of removal.

19 Van sliding door – removal, dismantling, reassembly, refitting and adjustment

1 Remove the sliding door guide rail cover as described in Section 2.
2 Slide the door open until the hinge guide piece can be lifted out of the opening in the guide rail.
3 Open the door fully and release the upper guide roller from its guide.
4 Pull the door slightly outwards and release the lower roller from

the opening in the guide rail.
5 The stop plate can be removed by unscrewing the two fixing bolts.
6 The hinge link can be unbolted and withdrawn from the door cavity for renewal of any worn components.
7 Reassembly is a reversal of dismantling. Grease the hinge link before installing.
8 Refitting is a reversal of removal, but adjustment must be carried out as necessary to achieve an even gap all round the edge of the door when it is closed, the door waistline must be level with the lines on the adjustment body panels and the door panel must be flush with the body panels.
9 If the upper edge of the door does not line up correctly, adjust the height of the upper roller to give the minimum possible running clearance between the roller and the guide rail. The position of the roller is also adjustable on its bracket.
10 If the front edge of the door does not line up correctly with the body panel, adjust the position of the lower roller guide. Do this by releasing the cross-head screw on the front edge of the door and the two bolts underneath. Vertical (height) adjustment can be altered by the inclusion of spacers as shown at point A (Fig. 12.65).
11 Adjustment at the rear end of the sliding door is made by altering

Fig. 12.56 Remote control cable/rod attachment (Sec 17)

7	Circlip	10	Cable adjuster
8	Pullrod	11	Fitting
9	Locknut	12	Cable

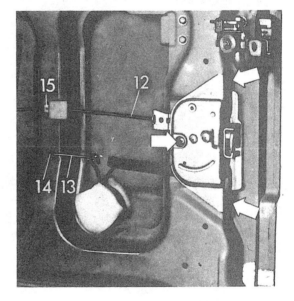

Fig. 12.57 Van sliding door central lock. Fixing screws arrowed (Sec 18)

12	Cable	14	Door inner panel
13	Tensioner spring pullrod	15	Guide

Fig. 12.58 Sliding door hinge guide piece (arrowed) (Sec 19)

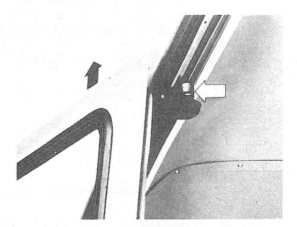

Fig. 12.59 Sliding door upper guide roller (arrowed) (Sec 19)

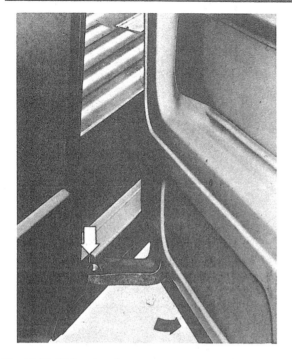

Fig. 12.60 Sliding door lower guide roller (arrowed) (Sec 19)

Fig. 12.61 Sliding door stop plate fixing screws (Sec 19)

Fig. 12.62 Sliding door hinge link components
(Sec 19)

1	Circlip	13	Spring washer
2	Pin	14	Operating cam
3	Locking lever	15	Nut
4	Spring	16	Spring washer
5	Spacer sleeve	17	Hinge attachment
6	Locking lever	18	Guide
7	Washer	19	Screw
8	Circlip	20	Hinge link
9	Housing	21	Roller
10	Spring	22	Spring washer
11	Spring anchor	23	Nut
12	Nut		

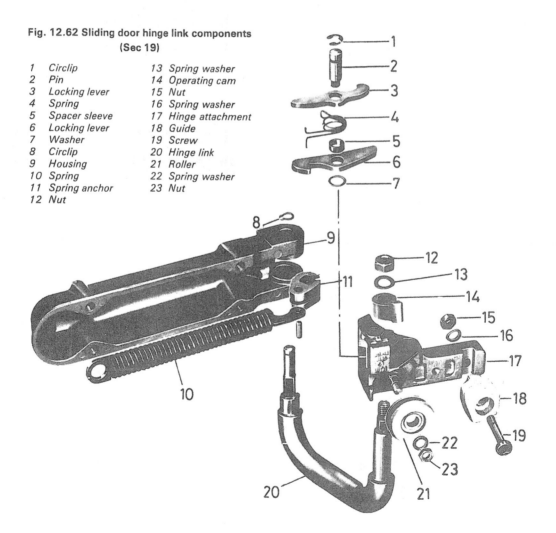

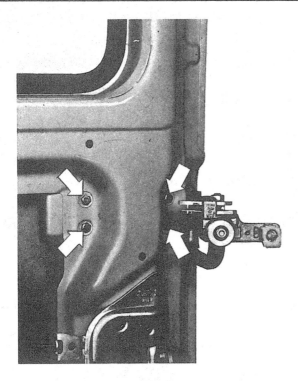

Fig. 12.63 Sliding door hinge link fixing screws (Sec 19)

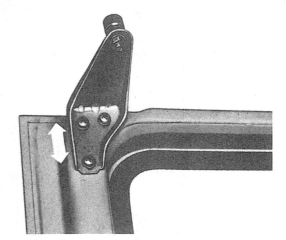

Fig. 12.64 Sliding door upper roller adjustment (Sec 19)

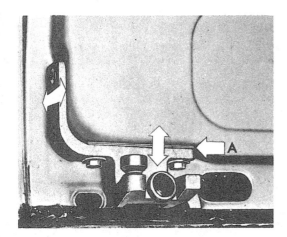

Fig. 12.65 Sliding door lower roller adjustment (Sec 19)

A Location of spacer

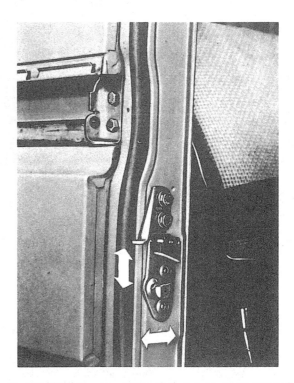

Fig. 12.66 Sliding door striker plate. Arrows show range of adjustment (Sec 19)

the position of the striker plate. If this is done, the hinge link must also be adjusted (see paragraph 13), also the latching lug at the interior handle.

12 The remote control striker plate should be released by unscrewing its fixing screws slightly. Close the sliding door from the inside, this will centralise the striker plate. Tighten the striker plate upper bolt before opening the door, open the door and tighten the second bolt. Where necessary, shims may be fitted beneath the striker plate to increase its projection.

13 To adjust the hinge link, close the door and working inside the vehicle, slacken the four hinge link fixing bolts.

14 Have an assistant pull the hinge link downwards until the roller and the guide piece contact the guide rail. Retighten the four bolts.

20 Safety belts and grab handle – general

1 At regular intervals examine the condition of the seat belts for fraying, cuts or other damage. Renew if evident.

2 Never be tempted to alter the belt or reel anchorage points, as the original ones are specially reinforced (photos).

3 When removing or refitting a belt, always maintain the original sequence of anchor plate, spacers and washers.

4 Cleaning of the belt fabric should be carried out using warm water and detergent only, never use a solvent of any kind.

5 The grab handle is secured to the windscreen A pillar. Slide off the end caps to reveal the fixing screws (photo).

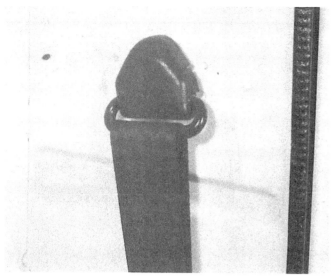

20.2a Safety belt upper anchorage

20.2b Safety belt lower anchorage

20.2c Safety belt inertia reel

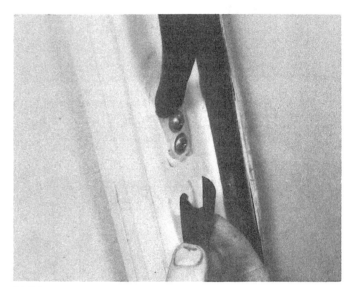

20.5 Exposing grab rail screws

Chapter 13 Supplement:
Revisions and information on later models

Contents

1 Introduction

This Supplement contains information which is additional to, or a revision of, material in the first twelve Chapters. Most of the material relates to the new 2.4 litre engine.

The Sections in the Supplement follow Chapter order. The Specifications are all grouped together for convenience, but they too follow Chapter order.

It is recommended that before any particular operation is undertaken, reference is made to the appropriate Section(s) in the Supplement. In this way, any changes to procedure or components can be noted before work commences.

For the purposes of maintenance, the LT Fleet model should be regarded as an LT 28 and the LT 50 bracketed with the LT 40 and 45 models.

2 Specifications

The values given here are additional to, or revisions of, the figures given in the first 12 Chapters.

Engine – 2.0 litre
Torque wrench setting

	Nm	lbf ft
Cylinder head multi-point socket-headed bolts	75 then a further 90°	55 then a further 90°

Engine – 2.4 litre
General

Type	Six-cylinder, in-line, overhead cam, water-cooled
Code	DL or HS (low compression)
Bore	76.5 mm (3.012 in)
Stroke	86.4 mm (3.402 in)
Cubic capacity	2396 cc
Compression ratio:	
DL	8.1:1
HS	7.1:1
Power output:	
DL	66 kW (90 bhp) at 4500 rpm
HS	61 kW (85 bhp) at 4500 rpm
Torque:	
DL	164 Nm (121 lbf ft) at 2800 rpm
HS	160 Nm (118 lbf ft) at 2800 rpm
Cylinder compression pressures:	
New:	
DL	8.5 to 12.0 bar (123 to 174 lbf/in²)
HS	8.0 to 11.5 bar (116 to 167 lbf/in²)
Minimum:	
DL	7.0 bar (102 lbf/in²)
HS	6.5 bar (94 lbf/in²)
Maximum variation between cylinders	3.0 bar (44 lbf/in²)

Crankshaft

Number of main bearings	7
Main journal diameter:	
Standard	58.00 mm (2.283 in)
Undersizes	0.25, 0.50 and 0.75 mm (0.010, 0.020 and 0.030 in)
Main bearing running clearance:	
New	0.016 to 0.075 mm (0.0006 to 0.0030 in)
Wear limit	0.15 mm (0.006 in)
Big-end bearing running clearance:	
New	0.015 to 0.062 mm (0.0006 to 0.0024 in)
Wear limit	0.12 mm (0.005 in)
Big-end bearing side clearance:	
New	0.05 to 0.31 mm (0.002 to 0.012 in)
Wear limit	0.40 mm (0.016 in)
Endfloat:	
New	0.07 to 0.18 mm (0.002 to 0.007 in)
Wear limit	0.25 mm (0.009 in)

Pistons and piston rings

Piston diameter:	
Standard	76.48 mm (3.011 in)
Oversizes	0.25, 0.50 and 1.00 mm (0.010, 0.020 and 0.039 in)
Corresponding bore diameter:	
Standard piston	76.51 mm (3.012 in)
Oversizes	0.25, 0.50 and 1.00 mm (0.010, 0.020 and 0.039 in)
Piston ring side clearance in groove:	
New	0.02 to 0.08 mm (0.001 to 0.003 in)
Wear limit:	
Upper compression ring	0.20 mm (0.008 in)
Lower compression ring	0.10 mm (0.004 in)
Oil scraper ring	0.10 mm (0.004 in)

Piston ring end gap:
    New .... 0.25 to 0.50 mm (0.010 to 0.020 in)
    Wear limit .... 1.0 mm (0.04 in)

Camshaft
Number of bearings 4
Endfloat (maximum) .... 0.15 mm (0.006 in)

Cylinder head
Block mating surface distortion limit .... 0.20 mm (0.008 in)
Minimum height after skimming .... 132.6 mm (5.22 in)

Valves and guides
Valve length (overall):
    Inlet .... 98.70 mm (3.886 in)
    Exhaust .... 98.50 mm (3.878 in)
Valve head diameter:
    Inlet .... 36.0 mm (1.42 in)
    Exhaust .... 31.0 mm (1.22 in)
Valve stem diameter:
    Inlet .... 7.97 mm (0.314 in)
    Exhaust .... 7.95 mm (0.313 in)
Valve face angle .... 45°
Valve stem-to-guide clearance (maximum):
    Inlet .... 1.0 mm (0.04 in)
    Exhaust .... 1.3 mm (0.05 in)

Valve timing (at 1.0 mm valve lift)
Inlet opens .... TDC
Inlet closes .... 29° ABDC
Exhaust opens .... 19° BBDC
Exhaust closes .... 3° ATDC

Valve clearances

	Cold	Hot
Inlet	0.15 to 0.25 mm (0.006 to 0.010 in)	0.20 to 0.30 mm (0.008 to 0.012 in)
Exhaust	0.35 to 0.45 mm (0.014 to 0.018 in)	0.40 to 0.50 mm (0.016 to 0.020 in)

Lubrication system
Oil capacity:
    With filter change .... 7.0 litres (12.3 Imp pints)
    Without filter change .... 6.0 litres (10.6 Imp pints)
Oil pressure (oil temperature 80°C, engine speed 2000 rpm) .... 2.0 bar (29 lbf/in²)
Oil pressure relief valve opens at .... 5.3 to 6.3 bar (76.9 to 91.4 lbf/in²)

Torque wrench settings

	Nm	lbf ft
Engine to gearbox:		
M12	80	59
M10	45	33
M8	25	18
Left-hand engine mounting	25	18
Right-hand engine mounting	45	33
Fan to vibration damper	25	18
Crankshaft bolt	460	339
Sump bolts	20	15
Oil drain plug	50	37
Valve cover nuts	10	7
Oil filter housing bolt	100	74
Vibration damper bolts	25	18
Crankshaft rear oil seal housing	10	7
Oil pump mounting bolts:		
Long	20	15
Short	10	7
Oil pick-up pipe bolts	10	7
Flywheel bolts	75 then a further 90°	55 then a further 90°
Main bearing cap bolts	65	48
Connecting rod cap bolts (new type)	30 then a further 180°	22 then a further 180°
Camshaft sprocket bolt	80	59
Cylinder head bolts:		
Stage 1	40	30
Stage 2	60	44
Stage 3	Turn an additional 180°	Turn an additional 180°
Camshaft bearing cap nuts	20	15

Cooling system – 2.4 litre models
General
System type .... Pressurised, with expansion tank. Pump driven from camshaft drive-belt with crankshaft-mounted cooling fan

Thermostat:
Opens .... 87° C (189°F)
Fully open .... 100° C (212°F)
Opening stroke .... 8 mm (0.315 in)
System capacity .... 13 litres (23 imp pints)
Expansion tank cap pressure .... 1.2 to 1.5 bar (17 to 22 lbf/in²)

Torque wrench settings

	Nm	lbf ft
Coolant pump bolts	25	18
Thermostat housing bolts	10	7
Temperature sensor	10	7
Housing at rear of cylinder head	20	15
Fan retaining bolts	25	18
Fan ring	20	15
Shroud bolts	10	7
Radiator support bracket	25	18
Radiator drain plug	10	7

Fuel system
Solex 1B1 carburettor
Type .... Single venturi, downdraught with manual choke

Jets and settings:
Venturi 28
Main jet .... x 135
Air correction jet with emulsion tube 140
Idling fuel/air jet 55/135
Auxiliary fuel/air jet 50/140
Float needle valve .... 2.0 mm
Accelerator pump injection volume .... 0.9 ± 0.10 cm³/stroke
Choke valve gap .... 4.3 ± 0.15 mm
Cold idling speed .... 4100 ± 100 rpm
Idling speed .... 900 ± 50 rpm
CO content at idle .... 1.5 ± 0.5%

Solex 2B6 carburettor
Type .... Twin progressive venturi, downdraught, automatic choke

	Stage 1	Stage 2
Jets and settings:		
Venturi	24	28
Main jet	x 115	x 112.5
Air correction jet with emulsion tube	110	100
Idle fuel/air jet	60/125	35/130
Auxiliary fuel/air jet	35/90	–
Idle air jet for progression reserve	–	180
Float needle valve	2.0 mm	2.0 mm
Enrichment valve	65	120
Pump injection tube	2 x 0.4	–
Accelerator pump injection volume	1.5 ± 0.2 cm³ stroke	
Choke valve gap	3.0 ± 0.15 mm	
Cold idle speed	1800 ± 50 rpm	
Idle speed	800 ± 50 rpm	
CO content at idle	1.0 ± 0.5%	

Torque wrench settings
2B6 and 1B1 carburettors

	Nm	lbf ft
Top cover screws	5	3.7
Bypass air cut-off valve	5	3.7
Air inlet elbow stud	5	3.7
Choke cover screws	5	3.7

Ignition system – 2.4 litre models
General
System type .... 12V, negative earth, coil, mechanical contact breaker and distributor.
Firing order .... 1–5–3–6–2–4 (number 1 cylinder at timing belt end of engine)

Distributor
Direction of rotation .... Clockwise (viewed from distributor cap end)
Dwell angle:
Setting .... 38 ± 3°
Wear limit .... 33 to 49°
Ignition timing (vacuum hose off) .... 7.5° ± 1 BTDC at 800 ± 50 rpm

Centrifugal advance:
Begins .... 1100 to 1400 rpm
Ends .... 4150 rpm
Vacuum advance:
Begins .... 100 to 180 m bar (75 to 135 mm Hg)
Ends .... 400 m bar (300 mm Hg)

Spark plugs
Type .... Bosch W7D
Beru 14-7D
Champion N7YC
Electrode gap .... 0.6 to 0.8 mm (0.024 to 0.031 in)

Ignition coil
Make .... Bosch
Primary resistance .... 1.7 to 2.1 ohms
Secondary resistance .... 7.0 to 12.0 k ohms

Torque wrench settings

	Nm	lbf ft
Distributor clamp nut	25	18
Spark plugs	20	15

Clutch – 1982 on
Clutch plate diameter .... 228 mm (8.98 in)

Gearbox
Application – 2.4 litre models
4-speed (015/1) .... LT 28, 31 and 35
5-speed (008/1) .... LT 40, 45 and 50 (may also be fitted to other models as an option)

Power Take Off gearbox
Initial shim thickness .... 0.75 mm (0.030 in)
Measured backlash .... 0.05 to 0.22 mm (0.002 to 0.009 in)
Shims available .... 0.15 mm (0.006 in), and 0.20 mm (0.008 in) to 0.60 mm (0.024 in) in
0.10 mm (0.004 in) increments

Torque wrench setting

	Nm	lbf ft
PTO housing bolts	50	37

Propeller shaft
Torque wrench setting

	Nm	lbf ft
Shaft-to-shaft retaining bolt	40	30

Electrical system – 1983 on
Battery
Type .... 12 volt, negative earth
Capacity .... 45 or 63 AH

Alternator
Make .... Bosch or Motorola
Rating .... 45, 65 or 90A
Brush length (Type A) .... 13 mm (0.51 in)
Wear limit .... 5 mm (0.20 in)
Brush length (Type B) .... 10 mm (0.39 in)
Wear limit .... 5 mm (0.20 in)

Torque wrench setting

	Nm	lbf ft
Alternator pulley nut	40	30

Suspension and steering – 2.4 litre models
Front suspension
Wheelbase .... 2500, 2950 or 3650 mm (98.4, 116.1 or 143.7 in) according to model
Track (at GVW) .... 1750 mm (68.9 in)

Front wheel alignment
Camber:
LT 40, 45 and 50 .... 0° 10' ± 35'
Castor:
LT 28, 31 and 35 .... 0° 40' ± 20'
LT 50 .... 3° 25' ± 20'
Toe-in:
LT 28, 31 and 35 .... 3.0 to 6.0 mm (0.118 to 0.236 in)
LT 40, 45 and 50 .... 0 to 2.2 mm (0 to 0.87 in)

Rear wheel alignment

Camber (non-adjustable) .... 0° ± 25'
Toe:
    LT 28, 31 and 35 .... 0° 20' ± 40'
    LT 40, 45 and 50 .... 0° ± 20'

Power-assisted steering

Ratio (overall) 19:1
Steering wheel turns (lock-to-lock) 4
Fluid:
    Type .... Dexron® ATF
    Capacity .... 1.5 litres (2.6 Imp pints)
    Pump delivery pressure .... 100 to 110 bar (1450 to 1595 lbf/in²)

Roadwheels and tyres

Wheel size:
    LT 50 .... 5J x 14
Tyre size:
    LT 50 .... 185 R14 C8PR or 195 R14 C8PR

Torque wrench settings

	Nm	lbf ft
Suspension		
Modified anti-roll bar:		
Mounting bolts	90	66
U-bolts	50	37
Power steering gear		
Reservoir bracket mounting bolts	10	7
Steering gear mounting bolts	200	148
Universal joint clamp bolt	25	18
Pump bracket mounting bolts	45	33
V-belt pulley bolts	20	15
Pump pivot bolt	33	24
Adjuster bracket bolt	20	15
Banjo unions on pump	40	30
Supply pipe union on gear	35	26
Return pipe union on gear	30	22
Pipe connection (in supply pipe)	40	30

3   Routine maintenance – 2.4 litre models

The weekly maintenance tasks are as described at the front of the book (photos).

On new or reconditioned exchange engines, or after major repairs carry out the following inspections when the vehicle has completed 600 miles (1000 km).

    Check the clutch and adjust if necessary
    Check and adjust valve clearances
    Check engine for leaks
    Renew engine oil and oil filter
    Check and adjust idling speed
    Tighten the cylinder head bolts

Every 10 000 miles (16 000 km) or 12 months, whichever comes first

    Check lights, indicators and horn
    Check and test windscreen wiper and washer systems, including headlamp wash where fitted
    Check and adjust clutch
    Check the battery acid level and top up with distilled water as necessary
    Engine – visually check for leaks
    Cooling system – check antifreeze strength and refill as necessary
    Check for condition and adjust tension of the alternator drive belt (and power steering pump belt if fitted)

Filling the engine with oil

Refit the cap on the valve cover

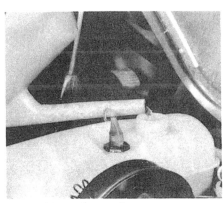
Filling the cooling system

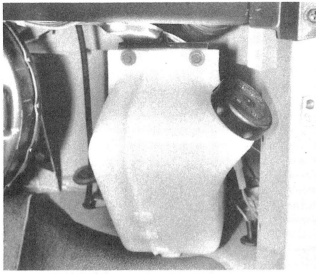

Windscreen wash reservoir (under dash)

Underbody view of LT 28 (2.4 litre)

1 Radius rods
2 Tie-rod
3 Steering knuckle
4 Idler arm
5 Lower balljoint
6 Lower shock absorber attachment
7 Lower wishbone
8 Brake caliper
9 Axle beam
10 Engine oil sump
11 Engine oil drain plug
12 Clutch housing
13 Gearbox
14 Gearbox drain plug

15 Gearbox filler plug (on side, hidden from view)
16 Gearshift rod
17 Speedometer cable
18 Exhaust mounting bracket
19 Starter motor
20 Handbrake linkage
21 Steering bell crank
22 Alternator
23 Crankshaft pulley
24 Cooling fan
25 Fan shroud
26 Accelerator cable
27 Drag link (from steering box)
28 Cooling system drain plug

View of LT 28 engine bay (2.4 litre)

1 Coolant expansion tank
2 Accelerator cable
3 Hot air collector pipe
4 Carburettor
5 Carburettor air intake duct
6 Valve cover breather pipe
7 Engine breather control valve
8 Engine oil filler cap
9 Valve cover
10 Inlet manifold
11 Exhaust manifold
12 Distributor
13 Coil (on underside)
14 Left-hand engine mounting
15 Engine oil dipstick
16 Spark plug HT leads
17 Thermostat housing (cooling system)
18 Engine air intake filter housing

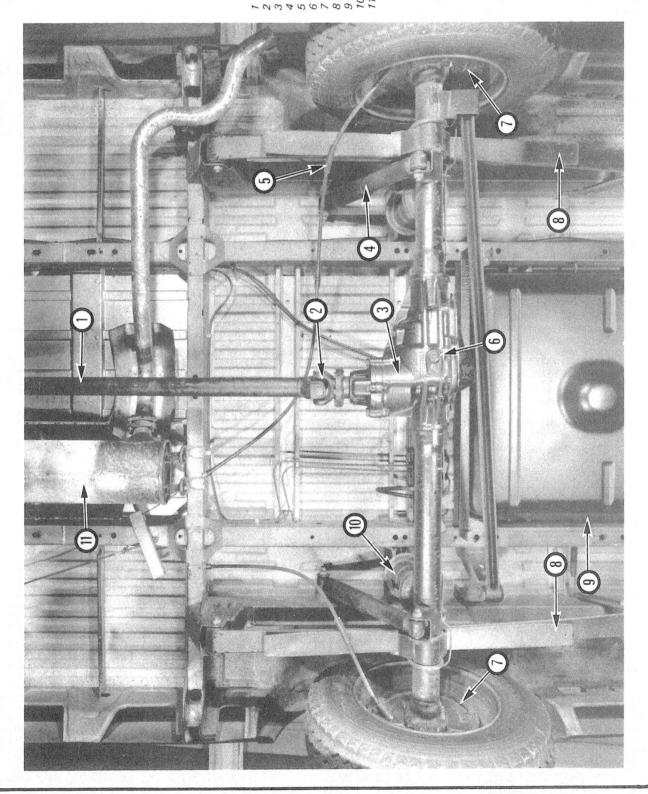

Underbody view of LT 28 rear axle (2.4 litre)

1 Propeller shaft
2 Universal joint
3 Differential/rear axle
4 Rear shock absorber
5 Handbrake cable
6 Rear axle oil drain plug
7 Rear brake drum
8 Leaf spring
9 Fuel tank
10 Auxiliary rubber spring
11 Exhaust silencer

Oil filter

Gearbox filler plug

Renew spark plugs
Check and adjust steering and wheel alignment. Check steering
gear for leaks
Clean and lubricate door locks, hinges and sliding door gear
Change the engine oil and filter (photo)
Check brake system for leaks and damage
Check exhaust system for damage and integrity
Check tie-rod ends for play and security. Check dust caps
Steering knuckles – check dust caps for damage
Check gearbox, final drive and joints for wear and damage. Check
rubber boots for integrity
Check the breather valve on rear axle tube for correct positioning
Grease front axle (LT 40, 45 and 50)
Check thickness of brake linings
Check tyres for tread depth, wear and damage
Check brake fluid level
Check wheel bolts/nuts for tightness
Renew contact breaker points
Check and adjust headlamp beam alignment
Carry out a road test and functionally check all systems
Check gearbox and final drive oil levels (photo)

**Every 20 000 miles (32 000 km) or 2 years, whichever comes
first**

In addition to those items listed under the 10 000 mile service
Renew the brake fluid
Check and adjust valve clearances, renew the valve cover gasket
Renew air filter element, clean out the filter housing
Renew the fuel filter
Visually inspect underbody sealant for damage, renew as
necessary
Check power-assisted steering system, check level in reservoir

4 Engine

Part A: 2.0 litre models

Cylinder head bolts
1   From September 1979 a multi-point socket-headed bolt is used in
place of the hexagon socket-headed bolt.
2   These multi-point bolts should be tightened to the specified torque
(see the Specifications in this Supplement) in stages and in the same
sequence as previously, and then tightened a further 90° (1/4 turn).
3   These new bolts do not need renewing at every repair, nor do they
need retightening after an initial length of service.

Part B: 2.4 litre models

General description
1   The 2.4 litre, 6-cylinder engine was introduced at the beginning of
1983, and replaced the 2 litre 4-cylinder unit, which is no longer fitted
to the LT range.
2   The 2.4 litre engine is a 4-stroke, 6-cylinder, in-line, water-cooled
unit, mounted longitudinally in the vehicle.
3   The overhead camshaft is driven by a toothed belt from a sprocket
on the crankshaft; the belt also driving the water pump of the cooling
system. The camshaft is mounted in 4 bearings machined directly in
the cylinder head.
4   The valve components are conventional, but the tappets are of
bucket type; valve clearance being set by adjusting discs of differing
thickness fitted in the bucket.
5   The pistons are conventional with two compression rings and one
oil control ring, the piston being held to the connecting rod by
gudgeon pin and circlip. The big-ends run in white metal bearing
shells.

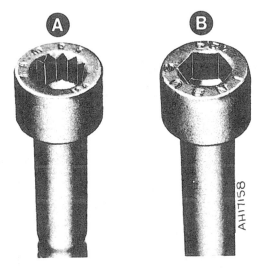

Fig. 13.1 Old and new type cylinder head bolts (Sec 4A)

A *New type*   B *Old type*

6 The crankshaft runs in seven main bearings which are also lined with white metal shoes; endfloat being controlled by thrust washers fitted either side of No 4 main bearing. A needle roller bearing is also fitted at the flywheel end of the crankshaft.

7 The flywheel is bolted to the crankshaft.

8 The oil pump is of the gear type, driven directly from the crankshaft.

9 The mechanical distributor is mounted at the rear of the cylinder head and is driven from the camshaft.

Major operations possible with engine in situ

10 The following operations are possible without removing the engine from the vehicle:

Removal and fitting of toothed timing belt
Removal and fitting of camshaft
Removal and refitting of cylinder head
Removal and refitting of sump
Removal and refitting of pistons/connecting rods
Removal and refitting of oil pump
Removal and refitting of flywheel
Removing and refitting of front and rear crankshaft oil seals
Removal and refitting of crankshaft needle bearing

11 The engine removal method is similar to that for the 2.0 litre engine; the engine being lifted upward and then out of the left-hand door, but it should be borne in mind that the 6 cylinder engine is obviously a much heavier unit and the lifting gear should be sufficiently robust.

12 Extensive use of socket-headed bolts is made throughout the construction of the engine and it is necessary that a good set of both hexagon and multi-point key wrenches are available during all overhaul operations.

13 Where procedures call for the removal of ancillary components, reference should be made to the relevant Sections either in this Supplement or in the main Chapters.

14 Remove the driver's and passenger's seats.

15 Disconnect the battery negative and positive leads, and, for complete safety, remove the battery.

16 Open the engine cover (photos), disconnect the earth lead from the valve cover (photo), and remove the cover.

17 From underneath the engine, remove the engine splash panels.

18 Drain the cooling system.

19 Drain the engine and gearbox oil.

20 Refer to Section 5 and remove the upper and lower air deflector plates, fan and fan cowl.

21 Disconnect the electrical leads and remove the starter motor (photos).

22 Make an alignment mark across the edges of the propshaft-to-rear axle flanges.

23 Remove the bolts securing the propshaft to the rear axle, then pull the propshaft rearward out of the gearbox.

24 Cover the propeller shaft entry hole in the gearbox to prevent the ingress of drt.

25 Disconnect the front and rear exhaust pipes from the exhaust manifold.

26 Disconnect the exhaust-to-gearbox support bracket (photo).

27 Disconnect the clutch cable from the clutch release lever, and remove the cap and return spring (photos).

28 Remove the clutch cable adjusting nut and detach the cable from the gearbox (photo).

29 Disconnect the gearbox earth lead.

30 Disconnect the gear lever bracket from the gearbox and disconnect the gearchange rod.

31 Remove the speedometer cable from the gearbox, and plug the resulting hole (photos).

32 Do not lose the nylon drivegear.

33 Disconnect the reversing light switch.

34 Remove all the engine-to-gearbox securing bolts (the gearbox will not fall until the rear mounting bolt is taken out).

35 Place a wooden block between the camshaft toothed belt cover and the bodywork to prevent the engine tilting once the gearbox is removed.

4B.16A Lifting the rubber carpet

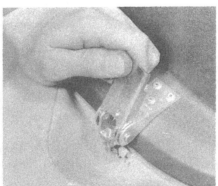
4B.16B Release the clips

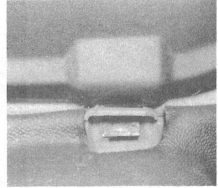
4B.16C Lift the cover from the hinge blocks

4B.16D Disconnecting the engine cover earth lead

4B.21A The starter motor and leads

4B.21B Starter motor mounting bolts

4B.26 Exhaust-to-gearbox support bracket

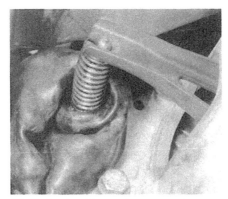

4B.27 Clutch cable attachment to release lever

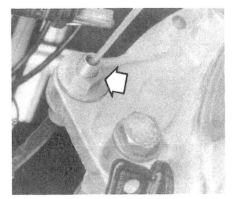

4B.28 Clutch cable adjusting nut (arrowed)

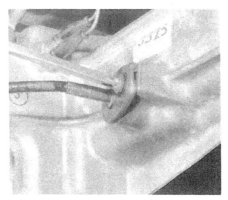

4B.31A Remove the screw ...

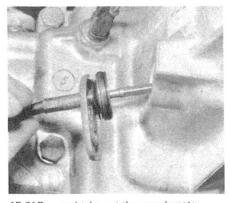

4B.31B ... and take out the speedometer cable

Fig. 13.2 Position a block of wood to prevent the engine tilting (Sec 4B)

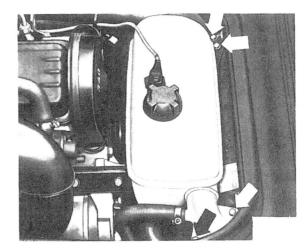

Fig. 13.3 Expansion tank bolts (arrowed) (Sec 4B)

36 With assistance, support the gearbox and remove the rear mounting bolt (photo) then slide the gearbox rearward, being careful not to put strain on the input shaft.

37 Once the shaft is clear, lower the gearbox.

38 Working at the top of the engine, disconnect the electrical lead from the coolant expansion tank.

39 Disconnect the coolant hoses which are attached to the expansion tank at their connection to the engine (photo). Be prepared for coolant spillage, as not all the coolant will have drained out.

40 Remove the bolts from the expansion tank mountings and remove the tank, complete with hoses.

41 Disconnect the oil pressure transmitter lead (photo).

42 Disconnect the coolant hose at the rear of the cylinder head.

43 Disconnect the electrical leads from the coil.

44 Disconnect the HT leads from the spark plugs and then remove the distributor (photo).

45 Disconnect the coolant temperature sender lead (photo).

46 Remove the air intake elbow securing nut from the carburettor, disconnect the vacuum hoses for the warm air control valve, and disconnect the other end of the duct at the air cleaner.

47 Disconnect the hose from the breather valve (photo), and lift away the air duct.

48 Remove the breather valve (photo) which is a push-fit in the housing.

49 Disconnect the coolant hose from the intake manifold.

4B.36 Gearbox rear mounting bolt (arrowed)

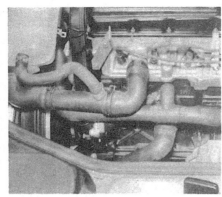

4B.39 Coolant hose connections on the engine

4B.41 Oil pressure transmitter

4B.44 Removing the distributor

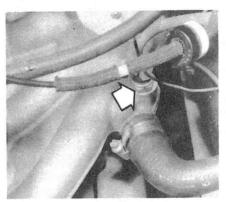

4B.45 Coolant temperature sender (arrowed)

4B.47 Breather valve hose connection (arrowed)

50 Disconnect the inlet and outlet fuel pipelines at the carburettor.
51 Disconnect the following electric leads:

Choke cover
Bypass air valve
Manifold heater
Thermoswitch for manifold heater

52 Remove the accelerator cable (photos).
53 Pull off the warm air intake hose from the exhaust collector plate.
54 Disconnect the brake servo vacuum pipe (photo).
55 Disconnect the remaining coolant hose from the inlet manifold.
56 Disconnect the earth lead from the valve cover (photo).
57 Fit suitable lifting equipment, using the lifting eyes fitted to the engine, and lightly tension it.
58 Disconnect the right-hand engine mounting (photo).
59 Disconnect the left-hand engine mounting (photo).
60 Lift the engine slowly, being particularly careful not to damage the bypass air valve and the ignition coil (photo).
61 Carefully remove the engine from the vehicle, and place it either on the ground or a suitable bench, supported on firm blocks of wood. **Note:** Because of the design of the sump, it is very difficult to support the engine in an upright position and it is advantageous to remove the sump and oil pick-up pipe prior to setting the engine down on blocks.
62 The remainder of the ancillary engine components may now be removed.
63 Remove the inlet manifold complete with carburettor.
64 Remove the exhaust manifold, taking note of the positions and fixings of the heat shields.
65 Remove the alternator.
66 Remove the oil filter, and the filter housing if desired.
67 Unbolt and remove the thermostat housing and thermostat.
68 Remove the clutch.
69 The remaining components will be dealt with in the following paragraphs on engine overhaul.

4B.48 Removing the breather valve

4B.52A Disconnecting the accelerator cable clamp ...

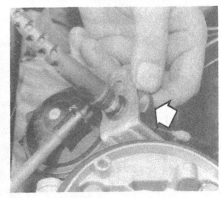

4B.52B ... and the clip on the support bracket (arrowed)

Sump – removal and refitting (engine in situ)

LT 28, 31 and 35 (wishbone axle)
70 Drain the engine oil.
71 Remove the sump retaining bolts. The two bolts at the rear of the sump are made more accessible by turning the flywheel until the two cut outs in the flywheel are in line with the two bolts.
72 Allow the sump to rest on the suspension cross-member.
73 Through the resultant gap, gain access to the oil pick-up pipe and undo its retaining bolts, allowing it to drop into the sump.
74 The sump can now be manoeuvred out rearward.

LT 40, 45 and 50 (beam axle)
75 The removal of the sump on these vehicles follows the same procedure as for the LT 28, 31 and 35, except that the engine should be hung on a suitable hoist or crane, which is then lightly tensioned.
76 The left-hand engine mounting should also be removed.

All models
77 Refitting is a reversal of removal, but use new gaskets on the oil pick-up pipe and sump, tighten all bolts to their specified torque. **Note:** the engine mounting (if removed) should not be torque loaded until the weight of the engine is back on the mounting.

Vibration damper – removal and refitting
78 Remove the crankshaft bolt (photo). The crankshaft bolt is extremely tight and a secure method of preventing the flywheel from turning is needed (photo).
79 Remove the four socket-headed bolts in the vibration damper (photo).

4B.54 Disconnecting the brake servo vacuum hose

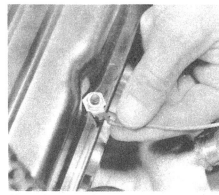
4B.56 Disconnecting the valve cover earth lead

4B.58 Right-hand engine mounting

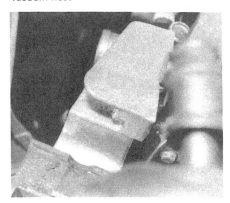
4B.59 Left-hand engine mounting

4B.60 Lifting the engine (note the positions of the lifting eyes)

4B.78A Removing the crankshaft bolt

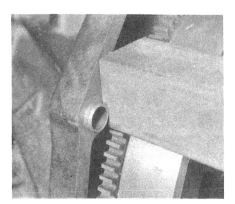
4B.78B One method of preventing the flywheel from turning

4B.79 Removing the bolts from the vibration damper

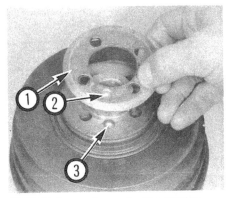
4B.80 Fit the washer under the vibration damper
1 Washer   3 Drilling
2 Dimple

80 When refitting the damper, do not forget the washer underneath, which is located by the dimple in the drilling (photo).
81 Tighten the crankshaft bolt to the specified torque.

Timing belt – removing, refitting and tensioning
82 The procedure is similar to that given in Chapter 1, Section 5, with the following differences.
83 The upper dust cover is held by quick release clips.
84 The crankshaft can be moved using the method described in paragraph 161.
85 Release the water pump bolts to relieve tension in the belt.

86 Remove the lower cover.
87 On models produced after May 1984, there is an additional idler roller, held by one bolt, screwed into the cylinder block.
88 If the same belt is to be refitted, mark its direction of travel, and be careful not to kink the belt while it is removed.
89 On refitting, tension the belt by turning the water pump to the left, before tightening the retaining bolts.
90 When correctly tensioned, the belt should just turn through 90°, using moderate force, when twisted.
91 When refitting the dust covers, make sure the seals are in place (photo).

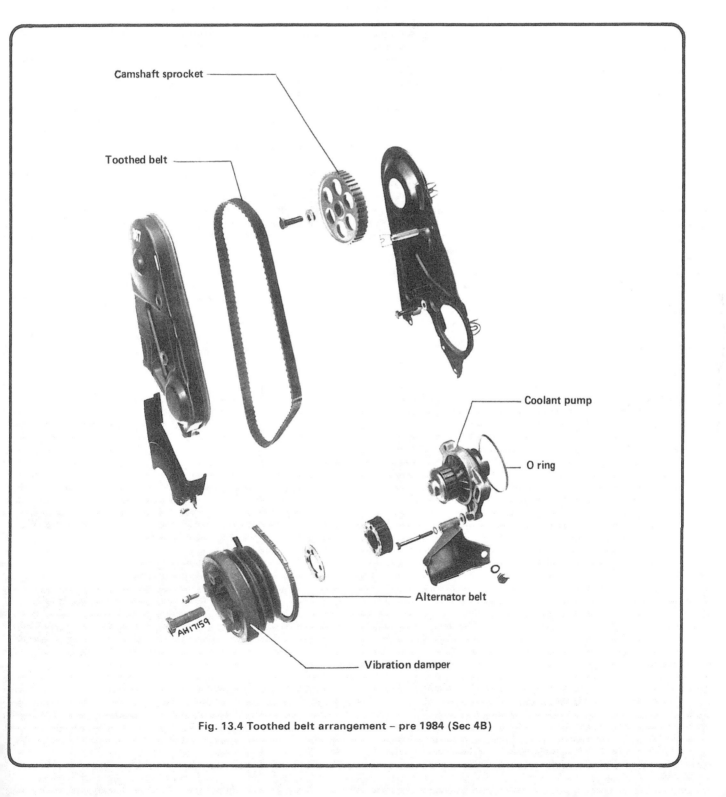

Fig. 13.4 Toothed belt arrangement – pre 1984 (Sec 4B)

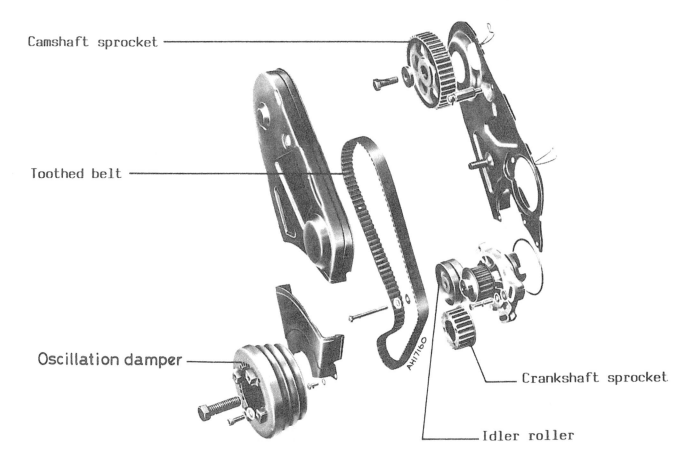

Camshaft sprocket

Toothed belt

Oscillation damper

Crankshaft sprocket

Idler roller

Fig. 13.5 Toothed belt arrangement – post 1984 (Sec 4B)

4B.91A Lower dust cover bolts being fitted

4B.91B Make sure the seals are fitted

Camshaft – removal and refitting
92 Remove the toothed timing belt as described earlier.
93 Remove the valve cover.
94 Remove the distributor; refer to Section 7 of this Supplement.
95 Remove the camshaft sprocket retaining bolt, preventing it from turning by using one of the methods described in Chapter 1, Section 6.

96 Remove numbers 1 and 4 caps first (the bearing caps are numbered 1 to 4 from the front of the engine). **Note**: Number 2 and 3 bearing caps are offset, so make alignment marks across the mating surfaces of the cap and bearing before removing them to ensure they are refitted correctly.
97 Remove number 2 and 3 caps a little at a time, working alternately

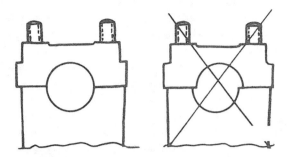

Fig. 13.7 Number 2 and 3 bearing are offset, so watch their fitting position (Sec 4B)

Fig. 13.6 Tensioning the toothed belt (Sec 4B)

100 Apply engine oil to the lip of a new oil seal and slide it on to the camshaft (photo).
101 Lower the camshaft carefully into the cylinder head bearings (photos).
102 Position the cam peaks for number one cylinder so that they are both facing upward uniformly (photo).
103 Fit number 2 and 3 caps; remembering the offset mentioned earlier.
104 Tighten the retaining nuts in the reverse manner of removal.
105 Fit and lightly tighten caps 1 and 4.
106 Ensure that the camshaft front oil seal is tapped fully home (photo), then tighten all bearing cap retaining nuts to the specified torque (photo).
107 Fit the toothed belt rear cover plate if this has been removed.
108 Fit the camshaft sprocket Woodruff key to the camshaft.
109 Fit the camshaft sprocket (photo).

and diagonally on the retaining nuts. (This is to relieve the pressure exerted by the valve springs on the camshaft).
98 Lift out the camshaft.
99 Oil the camshaft and bearing surfaces with clean engine oil before refitting.

4B.100 Fitting the camshaft front oil seal

4B.101A Lowering the camshaft into the front bearing ...

4B.101B ... and the rear

4B.102 Positioning both cam peaks facing upward

4B.106A Using a socket to tap the oil seal fully home ...

4B.106B ... before tightening the caps down

4B.109 Fitting the camshaft sprocket

4B.110 Fitting the camshaft sprocket bolt

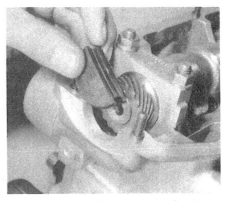
4B.111 Rubber sealing plug being fitted at the rear end of cylinder head

4B.112 Fit the seal over number one cap ...

4B.113 ... and the cork seal for the valve cover

110 Fit and tighten the sprocket retaining bolt to the specified torque (photo). **Note:** It is important when tightening the retaining bolt not to allow the camshaft to turn, or the valves may contact the pistons with resultant damage. Similarly, do not turn the engine unless the toothed belt is fitted and the valve timing set.
111 Fit the rubber sealing plug to the rear end of the cylinder head (photo).
112 Fit a new seal over number 1 bearing cap (photo).
113 Fit a new core seal for the valve cover (photo).
114 Fit the valve cover and tighten the nuts to the specified torque.
115 If a new camshaft has been fitted, the valve clearances will have to be checked and adjusted, once the toothed belt has been refitted.
116 Refit the toothed belt as described earlier.

Cylinder head – removal and refitting
117 The cylinder head can be removed with the camshaft installed, and with the inlet and exhaust manifolds, complete with carburettor, in place. These can be removed once the head is removed. If the engine is already out of the vehicle, it would be easier to remove them first. If an exchange cylinder head is to be fitted, which is complete with camshaft installed, there is no need to check valve clearances. Do not remove the protective plastic covers on the valves on a new head until just before the head is fitted.
118 If cracks are evident between the valve seats or between the valve seat rings and the first threads of the spark plug holes, consult your dealer – it does not necessarily render the cylinder unserviceable.
119 The procedure given here describes cylinder head removal with the engine removed from the vehicle.
120 Remove the toothed belt.
121 Remove the camshaft sprocket.
122 Remove the toothed belt cover backplate.
123 Remove the valve cover.
124 Remove the inlet and exhaust manifolds (the carburettor may be left on the inlet manifold).
125 **Note:** before removing any cylinder head bolts the cylinder head must be cold to avoid any risk of distortion. Remove the cylinder head

bolts, undoing each bolt a little at a time and in the reverse sequence of tightening, again to prevent risk of distortion (see Fig. 13.8).
126 Always use new cylinder head bolts and a new cylinder head gasket when refitting the cylinder head.
127 Clean both mating surfaces of the cylinder head and cylinder block, removing all traces of old gasket, and make sure both surfaces are dry.
128 VW produce special guide pins to ensure the cylinder head is correctly lined up on refitting.
129 These could be made from old cylinder head bolts by cutting off the heads of two bolts, then putting a saw cut across the exposed end, so that they can be removed once the head is fitted over them.
130 Insert the guide pins into holes 12 and 14 (see Fig. 13.8) on the cylinder block.
131 Ensure the rubber insert is fitted in a new cylinder head gasket (photo) then fit the gasket (dry) onto the cylinder block over the guide pins.
132 The word 'Oben' or the part number should face upward.
133 Ensure the TDC mark on the flywheel is lined up with the pointer on the bell housing.
134 Turn the camshaft to align the marks on the rear face of the camshaft sprocket and the cylinder head.

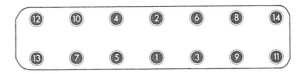
Fig. 13.8 Tightening sequence of cylinder head bolts (Sec 4B)

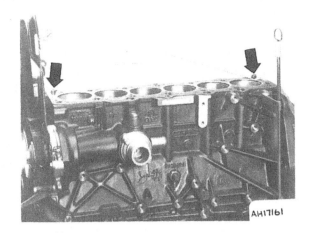

Fig. 13.9 Cylinder head guide pins (arrowed) (Sec 4B)

135 Lower the cylinder head carefully onto the cylnder block and over the guide pins (photo).
136 Always use new cylinder head bolts whenever the head is removed.
137 Fit the new bolts to all free holes and lightly tighten them, before removing the guide pins and fitting bolts in their place.
138 Now, using the sequence shown in Fig. 13.8, tighten the cylinder head bolts in the stages given in the Specifications. The additional 180° turn in Stage 3 should be done in one movement, although two 90° turns is acceptable (photo).
139 Refit the detached components.
140 After the engine has been started on completion of overhaul, and has reached operating temperature, stop the engine and tighten each bolt a further 90° ($^1/_4$ turn), in the same sequence as before and in one movement.
141 After the vehicle has completed its first 600 miles (1000km) in service, the bolts should be tightened a further 90° ($^1/_4$ turn), using the same sequence and in one movement.
142 This last operation may be completed with the engine hot or cold.

Cylinder head and valve gear – overhaul
143 Remove the cylinder head as described earlier.
144 Remove the camshaft.
145 Before removing any valve components, set up clean receptacles so that the valve gear can be kept together in sets. Do not mix any components.
146 Lift out the buckets (photo).
147 The procedure for removing the valves and for cylinder head overhaul is adequately described in Chapter 1, Section 18, but observe the following points, and the Specifications for the 2.4 litre engine in this Supplement.
148 Both the inlet and outlet valve springs sit in seats (photo).
149 The revised reworking dimensions are given in Figs. 13.10 and 13.11 (see also Chapter 1).
150 Exhaust valves cannot be reworked, and should only be ground in by hand.
151 Ensure the oil spray jets are positioned with their outlet holes on a left/right axis (photo).
152 After overhaul work on the cylinder head, check and adjust the valve clearances.

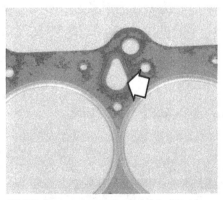

4B.131 Ensure the rubber insert is fitted in the cylinder head gasket (arrowed)

4B.135 Lower the head gently onto the block

4B.138 Tightening the cylinder head bolts

4B.146 Lifting out a tappet bucket

4B.148 Fitting a valve spring seat

4B.151 Oil spray jet axis

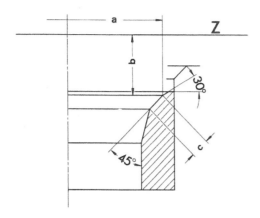

Fig. 13.10 Refacing inlet valve seats (Sec 4B)

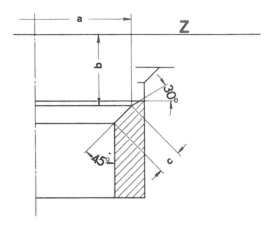

Fig. 13.11 Refacing exhaust valve seats (Sec 4B)

Valve clearances – checking and adjustment
153 Remove the valve cover.
154 The valve clearances are checked and adjusted with the engine at normal working temperature.
155 After overhaul work on the cylinder head, the clearances may be set cold, and should be checked and adjusted warm after the engine has completed 600 miles (1000 km) of service.
156 The clearances are set by inserting discs of the required thickness between the cam and the bucket.
157 The disc thickness is etched on the underside of the disc (photo).
158 Ensure the discs are fitted with this number facing the bucket.
159 Used discs may be refitted provided they are in good condition.
160 Oil the working surfaces of all parts before fitting.
161 To check the clearances, turn the engine using a screwdriver inserted in the bellhousing aperture and levering on the flywheel, until the cams on the cylinder to be adjusted both point upward uniformly.
162 Check the clearance between the cam and disc on that cylinder, using feeler gauges.
163 If the tolerances are exceeded, adjust as follows.
164 Turn the buckets so that the notches in their rim face inward toward the spark plug side.
165 There are special tools for depressing the buckets and removing the discs, but if these are not available, use two screwdrivers, pressing down on opposite sides of the bucket rim.
166 A helper will be required to remove and insert the disc.

4B.157 Valve clearance adjusting disc

3.80 = 3.80 mm

Fig. 13.12 Checking valve clearance with feeler gauge
(Sec 4B)

Fig. 13.13 Turn the buckets so their notches face inward
(arrows) (Sec 4B)

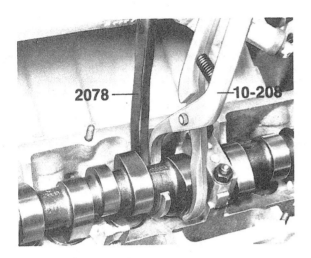

Fig. 13.14 The special tools for removing/fitting discs
(Sec 4B)

167 Discs are available in suitable sizes to allow the clearance to be brought into tolerance.
168 Check and adjust as necessary the remaining cylinders.
169 Refit the valve cover on completion.

Pistons/connecting rods – removal and refitting
170 Remove the cylinder head.
171 Remove the sump. **Note:** On LT 40, 45 and 50 vehicles where to remove the sump with the engine *in situ* the engine has to be raised slightly on a hoist, then the sump should be removed first and the engine then supported securely on an axle stand or jack before the cylinder head is removed, as the lifting eye is bolted to the cylinder head.
172 The procedure for removing and refitting the pistons is as described in Chapter 1, Section 9 with the following points.
173 Do not forget to mark each cap, conrod and piston with its cylinder number, and do not mix the bearing shells if they are to be used again.
174 Use the Specifications in this Supplement for all tolerances and torque loadings.
175 Care should be exercised when fitting the connecting rod bolts.
176 There are two types, identified in Fig. 13.15.
177 The new bolts are of the 'stretch' type and the torque loading procedure is different.
178 The correct procedure is to oil the contact surfaces of the bolt and nut (which are shouldered on the new type) and then fit and tighten the bolts to the specifed torque, then tighten by a further 180° (¹/₂ turn).
179 The new type bolts must be renewed wherever they are removed.

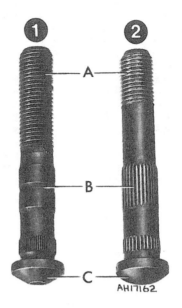

Fig. 13.15 The two types of connecting rod bolt (Sec 4B)

1   *New type*   A   *Revised thread length*
2   *Old type*   B   *Revised centre part*
                        C   *Revised head shape*

Oil pump and pick-up pipe assembly – removal and refitting
180 The oil pump is basically the same as fitted to the 4-cylinder engine, and can be removed after removing the alternator drivebelt, camshaft toothed belt, crankshaft pulley, camshaft pulley and the camshaft pulley backplate.
181 The inspection procedure given in Chapter 1 should be followed, using the Specifications in this Supplement for the 6-cylinder engine.
182 If the pump is worn beyond limits, it should be renewed as a complete unit.
183 Renewal of the oil seal is dealt with later in this Supplement.
184 The oil pick-up pipe is bolted to the crankcase by the flanged inlet to the pump (photo) and by a support bracket at its rear end (photo).
185 Some engines may also have an intermediate bracket halfway along the pick-up pipe.
186 When refitting the pick-up pipe, use a new gasket under the inlet flange.

Oil dipstick tube – removal and refitting
187 The oil dipstick and tube are bolted to the rear left-hand side of the engine.
188 To remove the tube, undo the bolt (photo) at the top of the tube.

4B.184A Oil pump pick-up pipe flanged inlet

4B.184B Pick-up pipe support bracket

4B.188 Oil dipstick tube top mounting

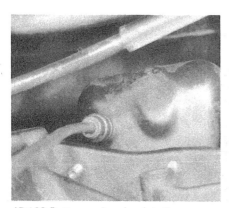

4B.189 Bottom end pushes into the sump housing

4B.194 Removing the flywheel bolts

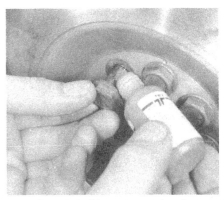

4B.196 Apply locking compound to the bolt threads when refitting

189 Then pull the bottom end out from the sump casting (photo).
190 Renew the rubber O-ring seals on the tube when refitting, push the tube back in the sump, and refit the top bolt.

Flywheel – removing and refitting
191 Access to the flywheel is gained after removal of the clutch assembly.
192 Mark the relationship of the flywheel to the crankshaft before removing it, so that it is replaced in exactly the same position.
193 Prevent the flywheel from turning (we used a piece of angle iron).
194 Undo the flywheel bolts, which will be tight (photo).
195 If the surface of the flywheel where the clutch pressure plate contacts it is worn, scored or badly burnt, or the teeth of the starter ring gear are worn or chipped, the flywheel should be renewed.
196 Refitting the flywheel is a reversal of removal but use new bolts, applying thread locking compound to the bolt threads (photo) and tightening them to their specfied torque.

Crankshaft oil seals – renewal
197 Both the front and rear crankshaft oil seals may be renewed with the engine *in situ*, and both seals should be renewed as a matter of course at major overhaul.
198 The procedure given here is with the engine installed; that with the engine removed being similar.
199 Access to the front oil seal is gained after removal of the vibration damper and the crankshaft toothed belt sprocket.
200 Prise out the old seal, being careful not to damage the oil pump housing.
201 Give the new seal a good covering of clean engine oil.
202 Wrap some protective tape around the crankshaft where there is a slight shoulder, which may be sharp and damage the lip of the seal.

203 The seal is fitted with its closed edge facing outward.
204 Place the seal squarely over the end of the crankshaft and begin to push it down the shaft.
205 If it is cocked in any way, it will probably be damaged and have to be discarded.
206 Once the seal is squarely on the shaft, make up a press using a suitable sized socket and the crankshaft pulley bolt, and press the seal firmly into the oil pump housing.
207 Refit those ancillary components removed for access.
208 The rear oil seal is contained within a housing which is bolted to the rear face of the cylinder block (photo).
209 Access to the oil seal can be gained by removal of the gearbox, clutch housing and clutch assembly, and the flywheel.
210 The seal can be renewed without removing the housing, but in the case of a major overhaul the housing has to be removed if the crankshaft is being taken out.
211 Prise out the oil seal, being careful not to damage the housing.
212 Coat the new seal liberally with clean engine oil.
213 The seal is fitted with its closed edge facing outward.
214 Carefully place the seal in position on the housing, then working around its circumference gently tap it into the housing with a suitable drift and hammer, working a little at a time to avoid undue distortion, until it is fully seated.
215 If the housing has been removed, fit a new gasket to the rear face of the cylinder block (photo).
216 Fit the housing over the crankshaft flange and the two location dowels (photo).
217 Fit and tighten the retaining bolts to their specified torque.
218 Refit the flywheel, clutch assembly and housing and gearbox.
Note: If the oil seal is being renewed *in situ* there is no need to remove the sump, but remember the two sump retaining bolts which fit into the oil seal housing.

4B.208 Crankshaft rear bearing oil seal and housing
1 Oil seal   2 Housing

4B.215 Fitting a new oil seal housing gasket to the cylinder block

4B.216 Positioning the housing over the crankshaft flange

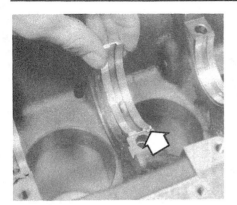

4B.222 Fitting a bearing shell into the recess in the cylinder block (arrowed)

4B.224A Fitting thrust washers to number 4 bearing ...

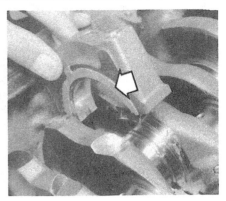

4B.224B ... and cap (thrust washer arrowed)

Crankshaft — removal, refitting and examination
219 Removal, refitting and examination of the crankshaft is covered in Chapter 1, Section 17, 19 and 21, with the following additional points.
220 There are seven main bearings, numbered 1 to 7 on the side which faces the oil filter.
221 No 1 bearing is at the vibration damper end of the engine and No 7 at the flywheel end.
222 When fitting the bearing shells, the retaining tabs must fit in the recess of the cylinder block (photo).
223 Crankshaft endfloat should be checked at Number 4 bearing.
224 There are also thrust washers fitted either side of Number 4 bearing (photos).
225 Radial play can be measured (with the engine *in situ* if required) using the Plastigage method as follows:
226 Remove the bearing cap.
227 Clean the bearing shell and crankshaft journal.
228 Place a strip of Plastigage across the journal.
229 Fit the bearing cap and shell and tighten to 65 Nm (48 lbf ft).
230 Remove the bearing cap.
231 Measure the width of the strip of Plastigage against the scale.

232 If the journals are worn beyond limits, then have the crankshaft reground (Refer to Chapter 1, Section 19).
233 Since February 1984 a longer needle roller bearing has been used in the crankshaft rear end (photo).
234 Because of this, the fitting depth is not so deep.
235 Lubricate the bearing with molybdenum disulphide.

Reassembly
236 Rebuild the engine following the general procedure given in Chapter 1, Section 21, and using a reverse of removal sequences given in this Supplement.

Refitting
237 Refitting the engine is basically a reversal of the removal procedures in this Supplement; following the general principles given in Chapter 1, Sections 22 and 23.

Low compression engine (code HS)
238 A 2.4 litre, low compression engine for use in countries using low octane fuel was introduced in April 1983.

4B.233 Crankshaft needle roller bearing

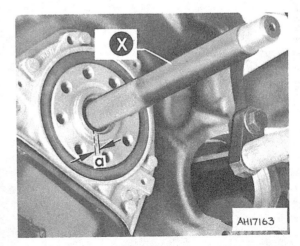

Fig. 13.16 Fitting the needle roller bearing in the flywheel (Sec 4B)

    a   *Fitted depth*
          Old type bearing = 5.5 mm (0.217 in)
          New type bearing = 1.5 mm (0.059 in)
    x   *Suitable driving tool*

239 This engine differs from the standard octane fuel version in various areas, including:

(a) *Output torque and compression pressures*
(b) *The pistons have recessed crowns*
(c) *A different distributor is fitted*

240 Refer to the Specifications for technical details.
241 Although no in-depth details were available at the time of going to press, it is believed that, apart from the Specification changes, the low compression engine is similar in all respects to the standard compression 2.4 litre engine.

5   Cooling system – 2.4 litre models

Draining, flushing and refilling
1   The general procedure given in Chapter 2 should be followed, using the following instructions as necessary.
2   The drain plug is fitted on the bottom of the radiator inlet pipe (photo).
3   After draining the radiator, the cylinder block should be drained by disconnecting either the hose at the rear of the cylinder head (photo) or the hoses on the thermostat housing.
4   When refilling the system, fill with coolant up to the 'max' mark in the expansion tank, then start the engine and keep topping-up the level as it falls when air is released.
5   Do this until the level remains constant.
6   Refit the expansion tank cap.
7   Run the engine until the thermostat opens and then check the level again.

Thermostat – removal, testing and refitting
8   Drain the system so that the coolant level is below that of the thermostat housing.

9   Disconnect the large inlet hose on the thermostat housing (photo). **Note:** the photo shows all other hoses removed as this sequence was done with the engine removed.
10 Remove the thermostat cover (photo).
11 Lift out the thermostat.
12 Test the thermostat as described in Chapter 2, but using the valves given in the Specifications in this Supplement.
13 When refitting the thermostat ensure the rubber sealing ring is in place round its periphery (photo) and that the air bleed valve is uppermost (photo).
14 Refit the cover and hose(s) and fill and bleed the system.

Radiator – removal, repair and refitting
15 Drain the cooling system.
16 Remove the lower air deflector (photo).
17 Disconnect the coolant level indicator wires in the coolant expansion tank (photo).
18 Remove the bolts from the expansion tank elbow connection, and separate the tank from the radiator (photo). **Note:** There is no need to remove the hoses from the tank, as shown in the photograph, if only the radiator is being removed.
19 Remove the upper air deflector plate.
20 Mark the fan blades in relation to the vibration damper, as the blades are offset, then remove the fan (photo).
21 Remove the fan ring (photo).
22 Remove the protective plate from underneath the radiator.
23 Disconnect the radiator bottom hose (photo).
24 Remove the bolts from the radiator support brackets (photo).
25 Lift the radiator up to clear the upper locating brackets, then down and out of the vehicle.
26 The repair procedures given in Chapter 2 should be followed.
27 Refit in the reverse order.
28 Ensure the radiator upper mountings are properly located, and that the rubber sealing ring between the fan ring and air deflector plates fits snugly (photo).
29 Fill and bleed the system on completion.

5.2 Radiator drain plug (arrowed)

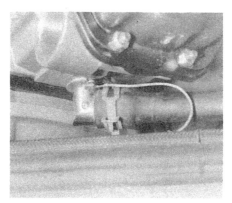

5.3 Coolant hose connection at the rear of the cylinder head

5.9 Connections at the thermostat housing

5.10 Removing the thermostat cover

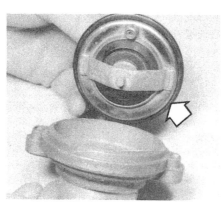

5.13A Rubber sealing ring (arrowed)

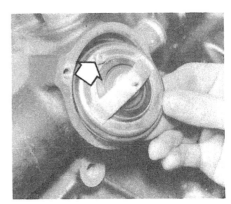

5.13B Air bleed valve in thermostat (arrowed)

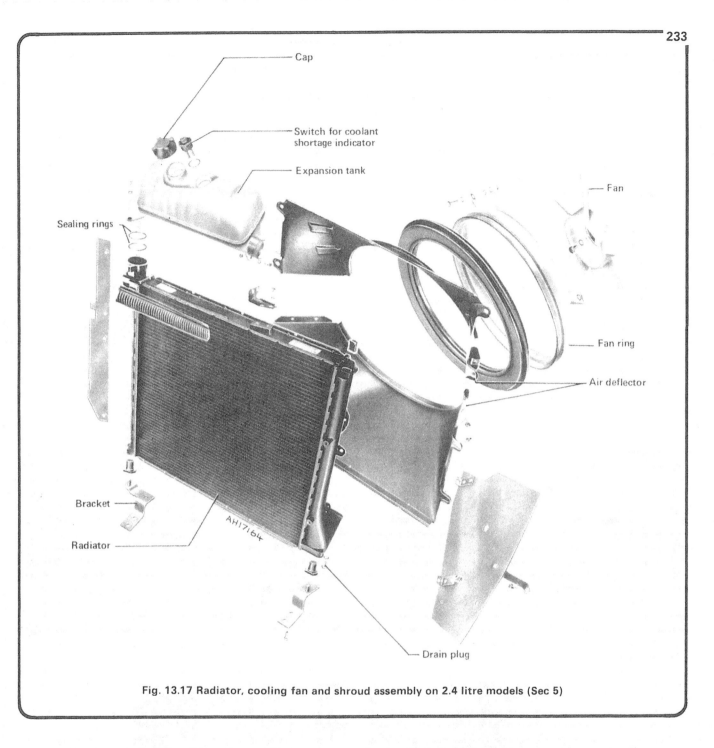

Fig. 13.17 Radiator, cooling fan and shroud assembly on 2.4 litre models (Sec 5)

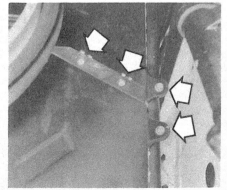

5.16 Air deflector plate retaining bolts (arrowed)

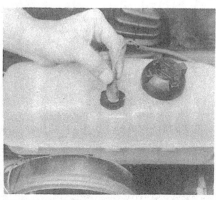

5.17 Disconnecting the level indicator

5.18 Removing the expansion tank from the radiator

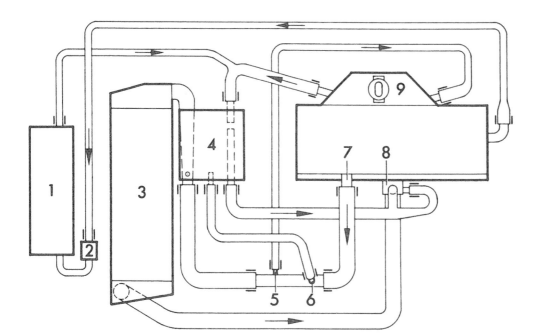

Fig. 13.18 Coolant flow diagram – 2.4 litre models (Sec 5)

1   Heater
2   Control valve
3   Radiator
4   Expansion tank
5   Restrictor
    (6.0 mm – 0.24 in)
6   Restrictor
    (4.5 mm – 0.18 in)
7   Cylinder head connection
8   Thermostat housing
9   Inlet manifold

5.20 The cooling fan retaining bolts (arrowed); two bolts are hidden

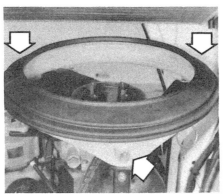

5.21 Fan ring retaining bolts (arrowed)

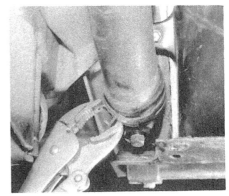

5.23 Disconnecting the bottom radiator hose

5.24 Radiator support bracket

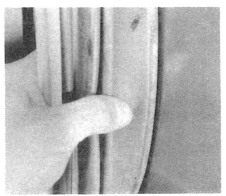

5.28 Fitting the sealing ring

Coolant pump – removal and refitting

Note: the coolant pump can be removed with the engine installed, but the fan and fan shroud, alternator drivebelt, and power steering pump belt, if fitted, must first be removed. The camshaft drivebelt must also be removed from the camshaft sprocket, and in order to remove the camshaft drivebelt rear cover, the camshaft sprocket must also be

removed. Refer to the engine removal section for details.
30 The coolant pump retaining bolts will have been loosened in order to remove the camshaft drivebelt, remove the bolts completely (photo).
31 Lift the camshaft drivebelt sufficiently for the pump to be lifted from its housing.

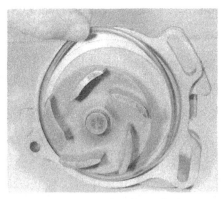

5.30 Coolant pump retaining bolts (arrowed)

A   Pivot bolt   B   Adjusting bolt

5.32 Coolant pump unit

5.33 Fitting a new seal to the coolant pump

32 The pump cannot be overhauled if it is faulty and should be renewed as a complete unit (photo).
33 Refit in the reverse order, using a new seal (photo).
34 If the operation has been carried out with the engine installed, fit those components removed to gain access, with reference to the relevant Sections.

Coolant temperature sensor
35 The coolant temperature sensor is screwed into the rear of the cylinder block, just above the coolant outlet elbow (photo).
36 When refitting a sensor, use a new sealing ring under its head.

Heater and heater controls
37 The same components are used in later models as in early models, but with the following differences:
38 The heater controls are different, being mounted vertically.
39 The control levers pull off (photo).
40 The control panel is screwed into the instrument panel (photos).
41 The water control valve is mounted under the cab floor behind the radiator grille (photo).
42 With these changes borne in mind, the removal, overhaul and refitting instructions in Chapter 2 should be followed.

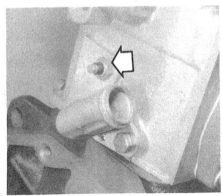

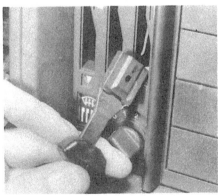

5.35 Coolant temperature sender (arrowed)

5.39 Pulling off a heater control lever

5.40A Removing a heater control panel screw

5.40B View of the heater control quadrant with instrument panel removed

5.41 Water control valve behind the radiator grille

6 Fuel, exhaust and emission control systems

Part A: 2.0 litre models, 1980 on

General description
1 From the beginning of 1980, all 2.0 litre LT models were fitted with the Solex 1B1 carburettor.
2 It is a single venturi, downdraught carburettor fitted with a manual choke.

3 The general comments on carburettor repair and overhaul in Chapter 3 should be followed, using the adjustment procedures and Specifications given in this Supplement.

Manual choke – removing, refitting and adjustment
4 Disconnect the choke cable at the carburettor attachment point.
5 Undo the choke cover screws and lift off the choke cover.
6 When refitting the cover, ensure the bi-metal spring fits over the operating lever spigot and that the opening lever is positioned to the left of the operating lever.

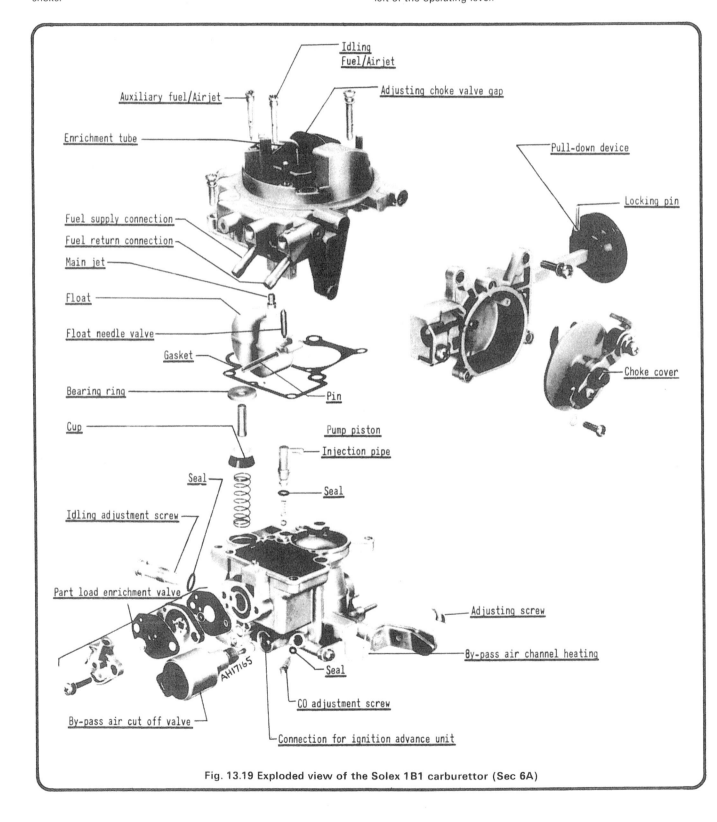

Fig. 13.19 Exploded view of the Solex 1B1 carburettor (Sec 6A)

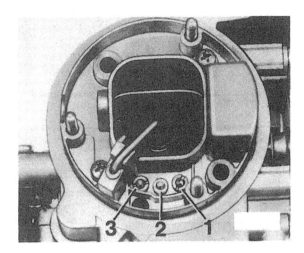

Fig. 13.20 Jet arrangement in top cover (1B1 carburettor)
(Sec 6A)

1   Idling fuel/air jet
2   Air correction jet with
    emulsion tube (cannot be
    removed)
3   Auxiliary fuel/air jet

7   Fit the choke cover screws, but before tightening them, adjust the
choke cover so that the index marks line up.
8   While tightening the screws, press the choke cover upward against
the guide lugs on the housing.

Idle and mixture adjustment
Idle speed
9   Connect a tachometer to the engine in accordance with the
manufacturer's instructions.
10 The engine should be at normal operating temperature.
11 All electrical consumers should be switched off (eg headlamps,
heater fan etc).
12 Pull off the crankcase breather hose and clamp it at the air filter
end.
13 Open the choke (ie push the control knob fully in).
14 If fitted, pull the afterburning system hose off at the intake pipe and
clamp the hose.
15 Ensure the ignition timing is correct.
16 Start the engine and allow it to idle for a few minutes.
17 Check the idling speed is as specified.
18 Adjust with the idling speed screw.

Fig. 13.21 Main jet arrowed (1B1 carburettor)
(Sec 6A)

Fig. 13.22 Fitting the choke cover (1B1 carburettor)
(Sec 6A)

1   Bi-metal spring   3   Opening lever
2   Operating lever

Fig. 13.23 Choke cover alignment marks – arrowed
(1B1 carburettor) (Sec 6A)

Fig. 13.24 Idling adjustment screw – arrowed
(1B1 carburettor) (Sec 6A)

Idling mixture
Note: The only accurate way to set the idling mixture is with a CO meter (exhaust gas analyser). If this equipment is not available then your local dealer should do the adjustment.
19 Set the idle speed as previously described.
20 Turn the adjustment screw (see Fig. 13.25) to give the specified CO content.
21 After adjustment, reset the idling speed if necessary.

General
22 On completion of both operations, reconnect all hoses, and remove test equipment.
23 See note on reconnection of hoses in the idle and mixture adjustment sub-section for 2.4 litre models.

Cold idle speed – adjustment
24 The engine should be at normal operating temperature.
25 The idle speed should be correctly set.
26 Pull the choke fully out.
27 The choke valve operating lever must be contacting the stop (see Fig. 13.26).
28 The index mark on the curved disc should be pointing to the centre of the adjusting screw, or to slightly above it.
29 Adjust the index mark (arrowed in Fig. 13.26) to this position by loosening the screw.
30 Start the engine without touching the accelerator pedal, and allow it to run at a fast idle.
31 The cold idle speed should be as specified.
32 Adjust if necessary with the screw (2).

Pull-down device – checking
33 Checking the choke pull-down device is as described for the 2B6 carburettor (see part B of this Section, paragraphs 26 to 34).

Choke valve gap – adjustment
34 Check the operation of the choke pull-down device.
35 Pull the choke cable out, so that the curved washer is turned to its highest point.
36 Loosen the choke cover securing screws and rotate the cover anti-clockwise until the choke valve is closed, then tighten the screws.
37 Push the choke operating rod onto its stop by using a screwdriver placed on the adjusting screw (1) and pushed in direction of arrow (Fig. 13.27) – towards the pull-down device.
38 Check the gap between the choke flap and the carburettor wall using a drill of suitable dimensions.
39 Adjust the gap to that specified by turning the screw (1).
40 Now turn the choke cover back again until the marks on the cover and carburettor are aligned.

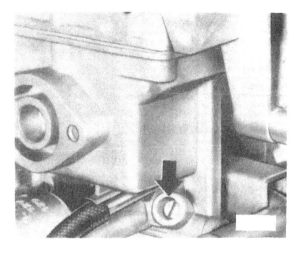

Fig. 13.25 CO adjustment screw – arrowed
(1B1 carburettor) (Sec 6A)

Fig. 13.26 Cold idle speed adjustment (1B1 carburettor)
(Sec 6A)

1   Operating lever   3   Operating lever screw
2   Adjustment screw

Fig. 13.27 Choke valve gap adjusting screw – 1
(1B1 carburettor) (Sec 6A)

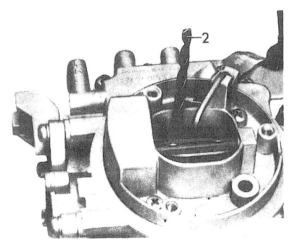

Fig. 13.28 Checking the choke flap gap (1B1 carburettor)
(Sec 6A)

2   Twist drill gap

Accelerator pump – adjustment

41 The procedure for checking the amount of fuel injected by the accelerator pump is as described in Chapter 3, Section 11, with the following differences.

42 Hold the choke valve operating lever in the fully open position.

43 Operate the throttle valve lever fully over its complete stroke 10 times, taking at least 3 seconds per stroke.

44 Divide the injected quantity by 10, the result should be as specified.

45 Adjust the quantity injected by loosening screw a, and turning the quadrant b to the right to increase and to the left to decrease (refer to Fig. 13.29).

Manual choke cable – operation and adjustment
Operation

46 Although the choke is manual, in that it is opened or closed by hand via the choke cable, there is still an element of automatic control.

47 This is achieved by a bi-metal spring fitted on the choke valve shaft, which reacts to ambient temperature.

48 The bi-metal spring will control the point at which the choke valve closes.

49 At low temperature the tension of the spring is increased and at high temperature, reduced.

50 At temperatures above 17°C (63°F) the choke valve will remain slightly open even though the choke cable is pulled firmly out.

Adjustment

51 Thread the choke cable into the clamp on the choke lever.

52 The choke control knob should be pushed fully home and the lever on its stop.

53 Pull the choke cable out fully, then carefully push it back again, until all play in the cable is eliminated.

54 From this position, push the cable through the clamping roller a maximum of 0.5 mm (0.020 in), then tighten the clamp screw.

55 Check that there is a clearance between the curved washer and plastic rubbing pad when the choke cable is pushed fully in.

56 The choke lever should be touching the stop or have a maximum play of 0.5 mm (0.020 in), with the choke pulled fully out.

57 Check that the choke warning light illuminates with the choke out and goes off when it is pushed in.

Throttle damper and delay valve – checking and adjustment

58 The engine should be at normal operating temperature.

59 Start the engine and allow it to idle.

60 Press the operating lever onto the stop in the direction of the vacuum unit.

61 Check the engine speed which should be 1400 ± 50 rpm.

62 Adjust on the screw on the end of the vacuum unit.

63 Increase engine speed to approximately 3000 rpm, so that the lever of the throttle damper is operated.

64 Release the throttle, and the operating lever should return to the idling position after 3 to 5 seconds.

Modified enrichment valve (from November 1980)

65 From the above date, a modified enrichment valve has been used on 1B1 carburettors.

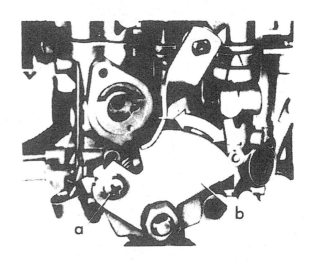

Fig. 13.29 Accelerator pump injection capacity adjustment (1B1 carburettor) (Sec 6A)

a   Screw   b   Quadrant

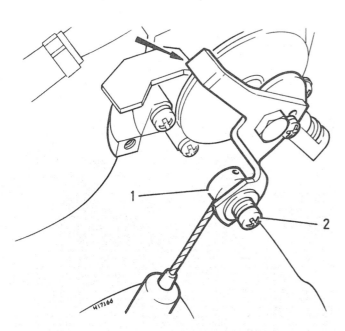

Fig. 13.30 Manual choke cable connection at carburettor (1B1 carburettor) (Sec 6A)

1   Clamp   2   Clamp screw

Arrow indicates stop

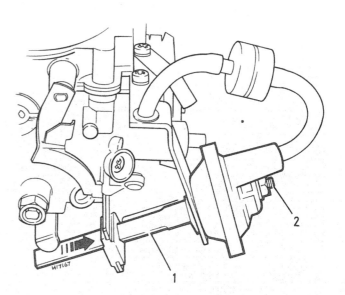

Fig. 13.31 Throttle damper and delay valve (1B1 carburettor) (Sec 6A)

1   Operating lever   2   Adjustment screw

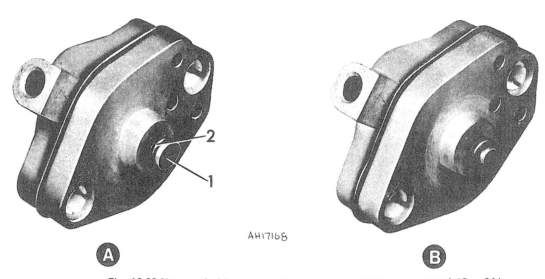

AH17168

Fig. 13.32 New and old type enrichment valves (1B1 carburettor) (Sec 6A)

A   Modified valve with shoulder (1) – sealing ring colour
    brown (2)

B   Old type valve with no shoulder – sealing ring colour black

66 Fig. 14.43 shows the two valves, the new one having a shoulder on the diaphragm rod, to prevent the sealing ring slipping off.
67 On the older valves, where the seal has become swollen and slipped off the rod, uneven idle and high fuel consumption could be the result.
68 Where this is proved to be the case, or during overhaul of the carburettor, a modified valve should always be fitted.

Part B: 2.4 litre models

General description
1   The fuel system fitted to the 2.4 litre, 6-cylinder LT models introduced in 1983 is basically the same as that fitted to earlier 2.0 litre, 4-cylinder models.
2   Fuel pump, fuel tank and accelerator and choke controls are all similar, and the servicing procedures given in Chapter 3 should be used

for these components, noting any relevant changes given in this Supplement.
3   The 2.4 litre models are fitted with a Solex 2B6 carburettor, which is described in the following paragraphs.

Air intake filter – removing and refitting
4   Remove the cover from behind the driver's seat (photo).
5   Release the clips on the filter housing cover and remove the top cover (photo).
6   Undo the nut securing the filter and lift the filter out (photos).
7   Refit in the reverse order.

Idle and mixture adjustment
8   Both the idle speed and mixture (CO content) adjustments are adequately covered in Chapter 3, Section 8, with the following observations:
9   Turn the adjustment screws (photo and Fig. 13.33) in to increase and out to decrease.

6B.4 Removing the cover from behind the driver's seat

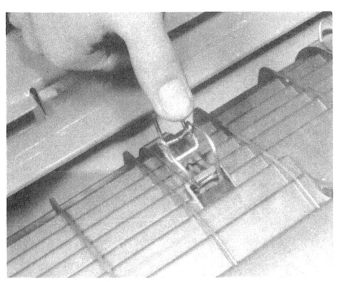

6B.5 Releasing the clips on the filter housing lever

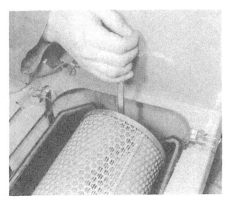

6B.6A Undoing the filter securing nut

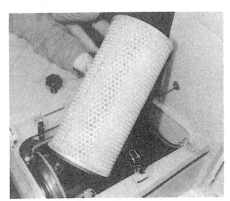

6B.6B Lifting out the filter

6B.9 Idle speed adjusting screw (arrowed) – 2B6 carburettor

Fig. 13.33 CO adjustment screw (2B6 carburettor) (Sec 6B)

10 The following engine conditions must be observed during the adjustments:

> Engine at normal operating temperature
> Throttle lever to be contacting the limiting screw (choke fully open)
> All electrical consumers switched off (headlamps, blower motor etc)
> Hose for the crankcase breather pulled off from intake hose and connection sealed
> Ignition timing should be adjusted correctly
> Vacuum hose disconnected

11 On completion reconnect all hoses and remove test equipment. **Note:** When the disconnected hoses are refitted, the engine speed will increase by approximately 150 rpm. If the CO content also increases it does not mean the settings are incorrect, but is due to oil dilution by fuel from the crankcase. A long, fast drive will reduce the amount of fuel in the oil and the CO content will return to normal.

Carburettor – removal and refitting
12 Follow the procedure in Chapter 3, Section 9.
13 Remember there are two fuel lines, inlet and outlet (photo). Mark the hoses accordingly before removal.
14 The carburettor is held to the inlet manifold by four bolts (photo).

6B.13 Inlet and outlet fuel hoses on 2B6 carburettor (arrowed)

6B.14 Carburettor-to-manifold securing bolts (arrowed)

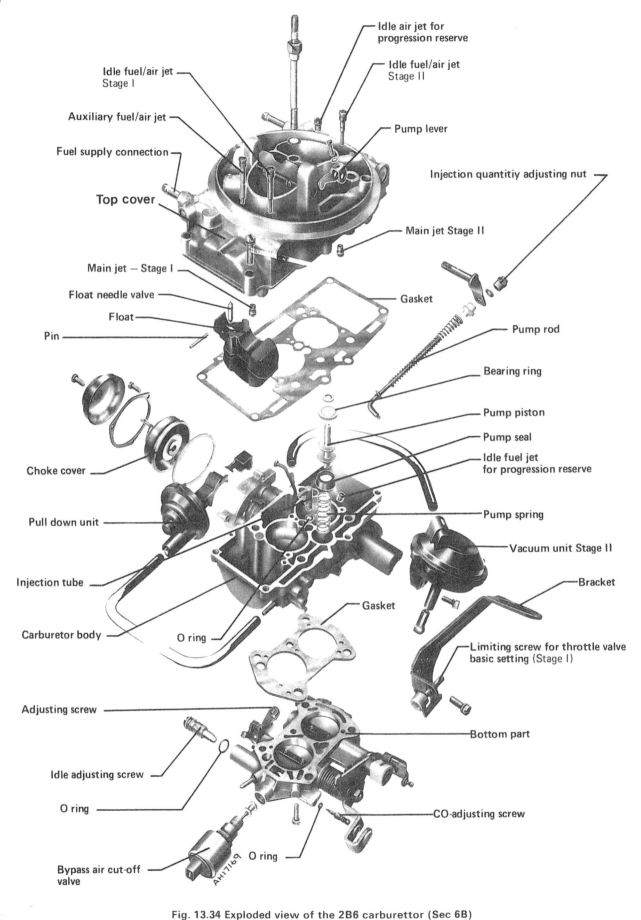

Idle air jet for progression reserve

Idle fuel/air jet Stage II

Idle fuel/air jet Stage I

Auxiliary fuel/air jet

Pump lever

Fuel supply connection

Injection quantitiy adjusting nut

Top cover

Main jet Stage II

Main jet — Stage I

Gasket

Float needle valve

Float

Pin

Pump rod

Bearing ring

Pump piston

Pump seal

Idle fuel jet for progression reserve

Choke cover

Pull down unit

Pump spring

Vacuum unit Stage II

Injection tube

Bracket

Carburetor body

O ring

Gasket

Limiting screw for throttle valve basic setting (Stage I)

Adjusting screw

Bottom part

Idle adjusting screw

O ring

CO-adjusting screw

Bypass air cut-off valve

O ring

AH17169

Fig. 13.34 Exploded view of the 2B6 carburettor (Sec 6B)

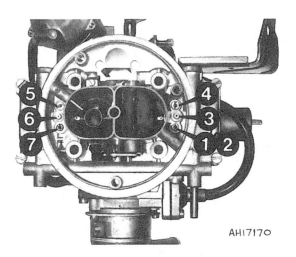

AHI7170

Fig. 13.35 Jet positions in top cover of 2B6 carburettor (Sec 6B)

1 and 2 Auxiliary fuel/air jet
3 Air correction jet with emulsion tube – Stage 1 (cannot be removed)
4 Idle fuel/air jet – Stage 1
5 Idle fuel/air jet – Stage 2
6 Air correction jet with emulsion tube – Stage 2 (cannot be removed)
7 Idle air jet for progression reserve

Carburettor – overhaul

15 An exploded view of the carburettor is shown in Fig. 13.34.
16 The carburettor can be separated into its three component parts following the general procedure in Chapter 3, Section 10.

Cold idle speed – checking and adjustment

17 The engine should be at normal operating temperature.
18 The ignition timing and normal engine idle speed should be adjusted correctly.
19 Open the throttle and turn the cam of the choke so that the adjusting screw is resting on the highest cam.
20 The choke must be fully open.
21 Without using the accelerator, start the engine.
22 Adjust the cold idle speed to that specified by turning the adjuster screw in or out as necessary (photo).
23 The adjusting screw should be sealed with sealing paint on completion.

Carburettor – setting and adjustment

Note: The following procedures require the use of a hand-held vacuum pump and gauge. If this equipment is not available, the tests will have to be conducted by your VW dealer.

Automatic choke
24 The index marks on the choke cover and housing should be in alignment (photo).
25 Adjust by loosening the cover screws, turning the cover as desired and tightening the screws on completion.

Choke pull-down device
26 Remove the air intake elbow.
27 Set the cold idle speed adjusting screw on the highest step of the cam.
28 Without using the accelerator, start the engine.
29 Close the choke by hand.
30 If the pull-down device is working correctly, the choke should close easily to a gap of approximately 4 mm (0.16 in) and then resistance should be felt, which will increase as the choke is closed completely.

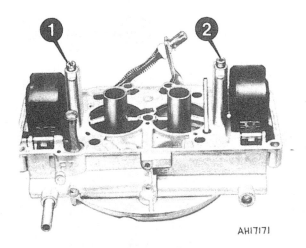

AHI7I7I

Fig. 13.36 Main jets in bottom part of 2B6 carburettor (Sec 6B)

1 Main jet – Stage 1 *2 Main jet – Stage 2*

Fig. 13.37 Idle fuel jet for progression reserve – arrowed (2B6 carburettor) (Sec 6B)

6B.22 Cold idle speed adjusting screw (arrowed) – 2B6 carburettor

6B.24 Automatic choke alignment marks (arrowed) – 2B6 carburettor

Fig. 13.38 Checking and adjusting the choke valve gap (2B6 carburettor) (Sec 6B)

A Cam   B Adjusting screw

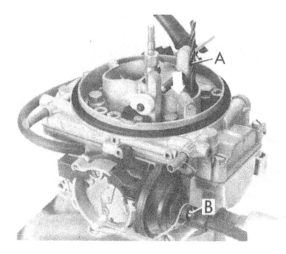

Fig. 13.39 Checking the choke flap gap (2B6 carburettor) (Sec 6B)

A   Twist drill   B   Adjusting screw

31 If there is no resistance, the pull-down device should be checked as follows:
32 Pull the vacuum hose of the pull-down device from the carburettor connection, and attach a vacuum pump to it.
33 Create a vacuum, and check that it is held within the pull-down device for 2 minutes.
34 If not, renew the pull-down device.

Choke valve gap
35 Remove the air intake elbow.
36 Remove the choke cover.
37 Operate the accelerator and position the adjusting screw (B) on the highest step of the cam (A) (Fig. 13.38).
38 Use a rubber band to lightly tension the choke operating lever.
39 Connect the vacuum pump to the choke pull-down device.
40 Create a vacuum within the pull-down device.
41 Press down lightly on the choke flap spindle with a finger to eliminate bearing play and check the gap between the choke valve flap and carburettor venturi wall is as specified.
42 Adjust the gap by turning screw (B) (Fig. 13.39).

Vacuum unit – leak test
43 Pull the vacuum unit hose from the carburettor and connect the vacuum pump to the hose.
44 Create a vacuum in the unit, and check that it remains for at least 2 minutes.
45 If the vacuum falls during this time, renew the unit.

Vacuum unit – operation
46 If the leak test proves satisfactory, recreate a vacuum in the unit.
47 Fully open the Stage 1 throttle valve flap. The Stage 2 throttle valve flap should be opened by the vacuum unit operating rod (photo).
48 Renew the vacuum unit if it does not open the Stage 2 throttle.

Accelerator pump
49 The amount of fuel injected by the pump can only be accurately set by using a special tester, and should be left to your dealer.

Stage 2 throttle valve flap
50 The limiting screw is set during manufacture and should not normally be altered. However, if during overhaul the throttle valve is dismantled, or the gasket renewed, the Stage 2 throttle can be set as follows:
51 The adjusting screw is fitted with a tamperproof cap, which should be renewed on completion.
52 Remove the tamperproof cap. Fully open the choke flap and close the throttle.
53 Turn the screw (A) out until there is a gap between it and the throttle lever (Fig. 13.40).

6B.47 Vacuum unit operating rod – 2B6 carburettor (arrow shows direction of rod movement during test)

Fig. 13.40 Adjusting the Stage 2 throttle (2B6 carburettor) (Sec 6B)

A   Adjusting screw

54 Open and close the throttle quickly.
55 Turn the screw in until it just contacts the throttle lever. This can be determined by holding a thin piece of paper between the screw and throttle lever and moving it in and out until the screw nips it against the lever.
56 From this point, turn the screw in a further $1/4$ turn.
57 Adjust idle speed and CO content.

Inlet manifold preheating
General description
58 A heater is mounted on the underside of the inlet manifold (photo).
59 It consists of an electrically-heated element which is controlled by a thermo-switch, also mounted in the inlet manifold (photo), but in contact with the coolant which passes through the manifold.
60 With the engine cold, the thermo-switch will switch on the heater element.
61 As the coolant temperature increases, the thermo-switch will sense the rise in temperature and, at a predetermined level, switch the heater off.
62 With the engine cold, and a voltage supply of at least 11.5 volts (battery in good condition), measure the resistance between the connecting cable and a good earth, which should be between 0.25 and 0.50 ohms.
63 A similar check can be carried out on the thermo-switch, which should be removed from the inlet manifold and warmed in hot water.
64 The resistance between the two terminals should be zero ohms below approximately 55°C (131°F) and infinity above 65°C (149°F).

Warm air intake system
65 In addition to the inlet manifold heater, there is a warm air intake system designed to admit warm air to the air intake system at temperatures below 17°C (63°F).
66 The control valve is situated in the warm air intake ducting mounted along the right-hand side of the engine bay, which collects warmed air from around the exhaust manifold, and feeds it into the main air intake system (photo).
67 The control valve is operated by a temperature-sensitive vacuum unit, the vacuum been drawn from within the carburettor air intake elbow.
68 To check the control valve for correct operation it must be withdrawn from the ducting system by undoing its retaining clips, and disconnecting the vacuum hoses.
69 Refer to Fig. 13.28 and, with the temperature above 17°C (63°F), observe that the warm air intake valve is closed.

6B.58 Inlet manifold preheater

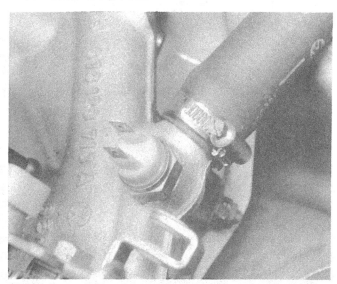

6B.59 Inlet manifold preheater thermo-switch

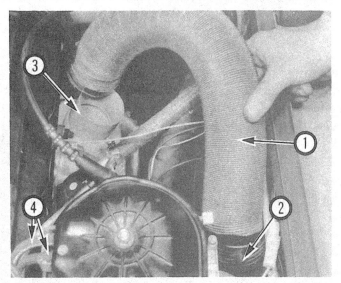

6B.66 Warm air intake system

1   Intake duct   3   Collector plate
2   Control valve   4   Vacuum connections

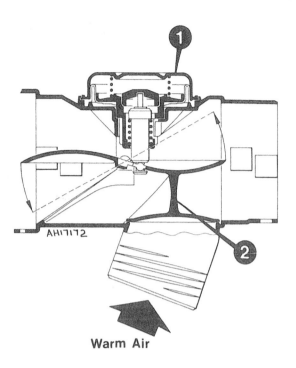

Warm Air

Fig. 13.41 Sectional view of warm air control valve (Sec 6B)

    1 *Warm air control valve*
    2 *Control flap (in closed position)*

70 Place the unit in a refrigerator and check that, with the temperature below 17°C (63°F), the warm air intake valve is open by at least $^4/_5$ of its travel.

71 Check that the valve operates freely without binding by sucking on the vacuum hose.

72 To check the temperature regulator, the control valve must be operating correctly.

73 The engine should be no more than warm.

74 The vacuum hoses should be connected.

75 Start the engine and let it idle.

76 Under these conditions the control flap must be open (disconnect the warm air collector hose to view).

77 Pull the vacuum hoses of the carburettor temperature regulator off at the carburettor (photo).

78 After 20 seconds, the control valve flap should have closed.

Manifolds and exhaust system
Manifolds

79 The inlet manifold is a one-piece casting bolted to the cylinder head using socket-headed bolts.

80 The carburettor is bolted to the inlet manifold.

81 The manifold is warmed by the coolant of the cooling system, and if the inlet manifold is being removed, the cooling system should be drained.

82 The exhaust manifold consists of two castings (photo) and is secured to the cylinder head using studs and nuts.

83 The general removal and fitting instructions in Chapter 3, Section 14, should be followed.

84 When refitting the manifolds always use new gaskets (photos).

85 Ensure that the heat shields and collector ducts are fitted correctly (photos).

Exhaust system

86 The exhaust system is very similar to that used on earlier models, but different flange connections are employed (photos).

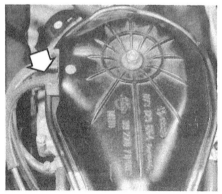

6B.77 Air inlet elbow and vacuum connections for warm air control valve (arrowed)

6B.82 Front section of exhaust manifold being fitted

6B.84A Fitting new gaskets to the exhaust manifold ...

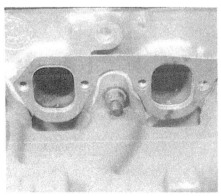

6B.84B ... and the inlet manifold

6B.85A The lower front (arrowed) ...

6B.85B ... and upper collector shields ...

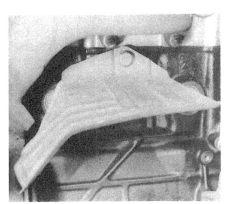

6B.85C ... and rear heat shield

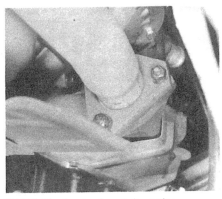

6B.86A The front exhaust downpipe flange ...

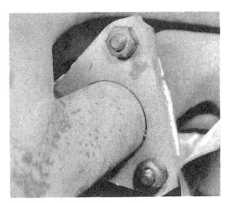

6B.86B ... and the rear

6B.86C Typical exhaust system flange

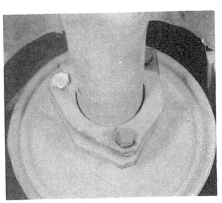

6B.86D The silencer flange

Fuel filler pipe – installation

87 Due to the varying types of body fitted to the chassis, there are different types of fuel filler pipe.

88 When refitting a fuel filler pipe at a fuel tank, ensure the correct type of filler pipe is used.

89 To ensure correct sealing, the end of the filler pipe in the seal area has been reworked.

90 The correct filler pipe is installed if the distance from the beginning of this rework up to the seal in the fuel tank is as given in Fig. 13.42.

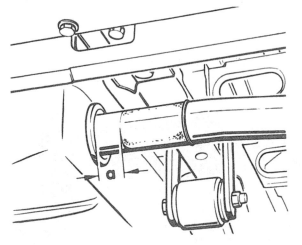

Fig. 13.42 Reworked area on fuel filler pipes (2.4 litre models) (Sec 6B)

a = 30 mm (1.18 in) approximately

7   Ignition system – 2.4 litre models

General description

1   The general description of the ignition system given in Chapter 4 applies equally to the system fitted to the 6-cylinder engine, except for the obvious difference of supplying 6 cylinders and not 4.

2   Equally, the servicing and renewal procedures are similar, but where major differences occur, they will be dealt with in the following paragraphs.

3   When carrying out any servicing or repair work, the Specifications in this Supplement should be used.

Contact breaker points – renewal

4   Release the clips securing the distributor cap to the distributor, remove the cap and put it to one side. (There is no need to remove the HT leads).

5   Pull off the rotor arm (photo).

6   Remove the moisture proof cover (photo).

7   Disconnect the LT lead.

8   Remove the screw securing the points to the distributor baseplate (photo).

9   Fit a new set of points in the reverse order of removal, but do the securing screw up only lightly.

10 Turn the engine until the heel of the moving contact is on the highest point of one of the lobes on the distributor cam.

11 Using feeler blades, and moving the fixed part of the points with a screwdriver inserted in the slot in the fixed contact and levered against one of the two rivet heads, set the gap to the nominal value given in the Specifications (Chapter 4).

12 This is a temporary setting only, and the ignition timing should be set by the dwell angle method at the earliest opportunity.

13 Tighten the points retaining screw fully.

14 Refit the moisture proof cover.

15 Refit the rotor.

16 Refit the distributor cap (photo).

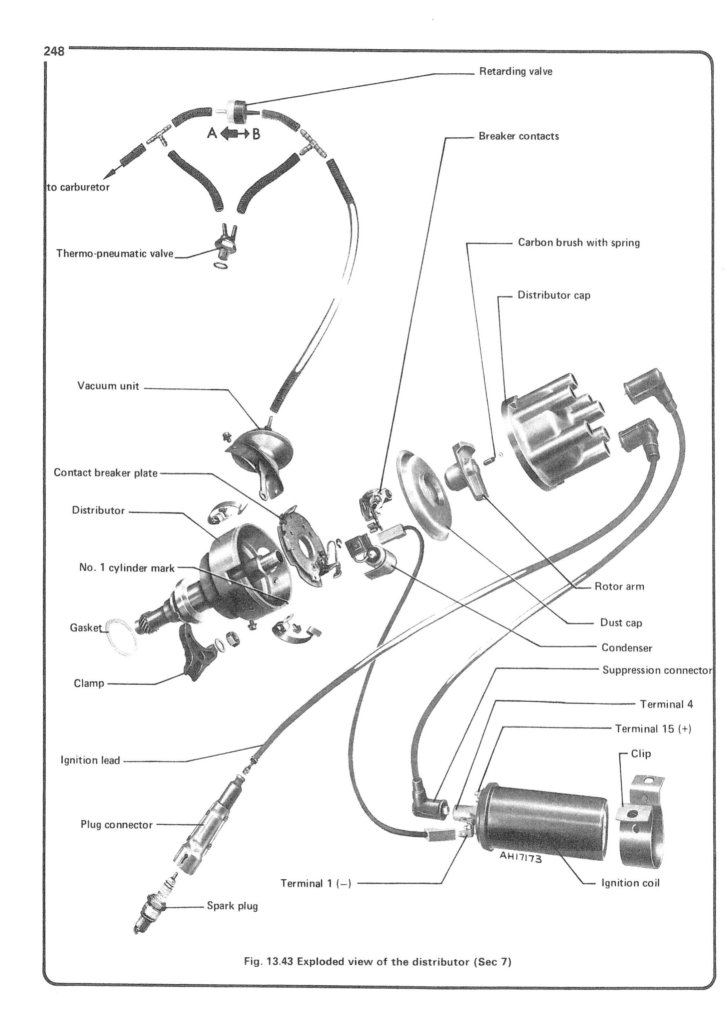

Retarding valve

A ← → B

to carburetor

Thermo-pneumatic valve

Breaker contacts

Carbon brush with spring

Distributor cap

Vacuum unit

Contact breaker plate

Distributor

No. 1 cylinder mark

Gasket

Clamp

Rotor arm

Dust cap

Condenser

Suppression connector

Terminal 4

Terminal 15 (+)

Clip

Ignition lead

Plug connector

AH17173

Terminal 1 (−)

Ignition coil

Spark plug

Fig. 13.43 Exploded view of the distributor (Sec 7)

7.5 Distributor rotor arm (arrowed)

7.6 Removing the moisture proof cover

7.8 Distributor contact points

1   LT lead connection   2   Securing screw

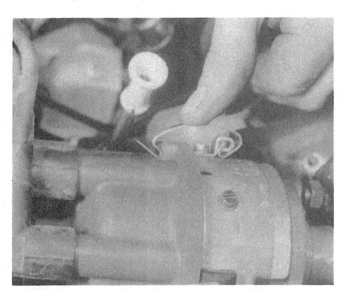

7.16 Distributor cap retaining clips

Distributor – removal and refitting

17 Before removing the distributor, mark its relationship to the distributor housing on the cylinder head as an additional aid to refitting.

18 Remove the distributor cap.

19 If it is desired to remove the HT leads from the cap, mark each one with its cylinder number, and similarly mark its connection in the cap (a waterproof felt tip pen will suffice).

20 This will save a lot of time trying to work out which lead goes where later.

21 Disconnect the LT lead (photo).

22 Disconnect the vacuum hose (photo).

23 Remove the nut from the clamp securing the distributor to the cylinder head (photo).

24 Withdraw the distributor.

25 Before refitting the distributor, set the engine to the TDC mark on the flywheel (photo). (Refer to Chapter 4, Section 6, for other methods of setting the engine timing).

26 Use a new gasket under the distributor.

27 Temporarily fit the rotor to the distributor and turn the rotor so that it is in line with the No 1 cylinder mark on the edge of the distributor (photo). (In this position, the vacuum unit should be facing

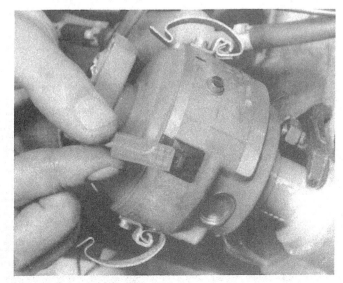

7.21 Disconnecting the LT lead

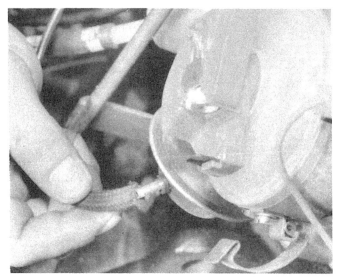

7.22 Disconnecting the vacuum hose

7.23 Distributor clamp nut (arrowed)

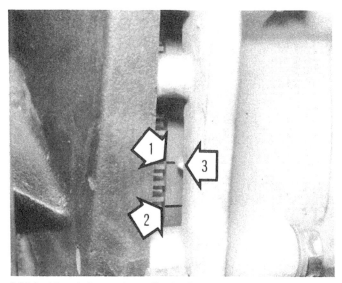

7.25 Ignition timing marks on flywheel

1   TDC mark   3   Pointer on bellhousing
2   7.5° BTDC mark

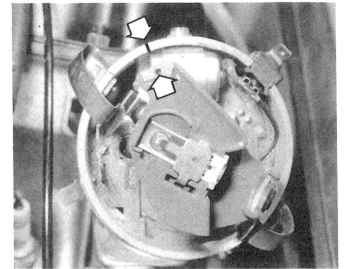

7.27 Lining up the rotor arm and distributor index mark (arrows)

downward when the distributor alignment marks made before removal are lined up).

28 Now turn the distributor rotor arm a few degrees clockwise, so that when the distributor is inserted in the housing and washed with its driving gear, the rotor arm, which will move anti-clockwise as the meshing takes place, is still in alignment with the mark on the distributor when the distributor is pushed fully home.

29 Fit the clamp and tighten the nut.

30 Check and adjust the points as described earlier.

31 Remove the rotor arm, refit the moisture proof cover, then the rotor arm.

32 Refit the distributor cap.

33 Check the ignition timing.

Engine speed limiter

34 Some ignition systems may be fitted with a distributor which incorporates a speed limiter device in the rotor arm (photo).

35 This is simply a spring-loaded contact in the rotor arm, which, due to centrifugal force, will open should the engine speed exceed 5500 to 5900 rpm, cutting off the current flow to the spark plugs.

7.34 Rotor arm with speed limiter

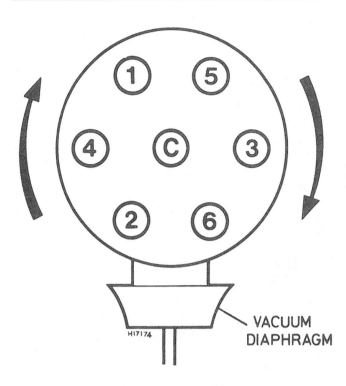

Fig. 13.44 HT lead connections in the distributor cap (Sec 7)

Spark plugs and HT leads
36 The remarks given in Chapter 4, Section 9, are relevant with the following additions.
37 Before disconnecting the plug leads, mark each one with the cylinder it supplies (photo).
38 Before removing a plug from the cylinder, brush out the plug housing with a dry paint brush or similar to prevent any dirt falling into the cylinder.
39 A plug can be tested for serviceability by removing it from the cylinder and, using insulated pliers, hold the plug against a good earth and crank the engine on the starter (photo).

40 The spark should be visible between the electrodes.
41 Before refitting, make sure the sealing washers are in good condition, and lightly grease the threads with high melting-point grease.
42 Start the plugs in their threads by hand, ensuring they are not cross-threaded, before tightening to the Specified torque – take care as alloy cylinder heads are easily damaged.

Distributor vacuum advance system
43 The vacuum advance system is shown in Fig. 13.45.
44 Incorporated in the vacuum line are a retarding valve and a thermo-pneumatic valve.
45 The retarding valve allows a free flow from the distributor to the carburettor, but provides a restriction in the reverse direction.
46 Note the fitting with the dark side toward the distributor.
47 The thermo-pneumatic valve is screwed into the intake manifold.
48 Above 46°C (115°F) it is open, allowing a free flow, below 30°C (86°F) it is closed, thus above this temperature it provides a bypass to the retarding valve.
49 This provides improved engine running on starting and when the engine is cold.
50 Both units are best tested by substitution with a known serviceable item.

8 Clutch

Clutch release bearing – modification
1 From 1982 a larger clutch has been used (see Specifications).
2 This larger clutch has a smaller diameter thrust area on the spring fingers of the diaphragm.
3 A release bearing of smaller diameter was also employed.
4 Release bearings with the smaller diameter should not be fitted to clutches with the larger diameter spring fingers.
5 However, the release bearings produced for the earlier, larger diaphragms may be fitted to the smaller diameter diaphragms, because the thrust area is much larger.

Clutch release linkage – modification
6 The clutch release arm locates on a ball-stud in the bellhousing (photos).
7 On models produced after 1982 the ball-stud was replaced by a stud with a cap.
8 This cap should be renewed whenever the clutch is renewed, or repairs to the release linkage carried out.
9 Neither the cap, nor the release arm bearing point, should be greased on reassembly.

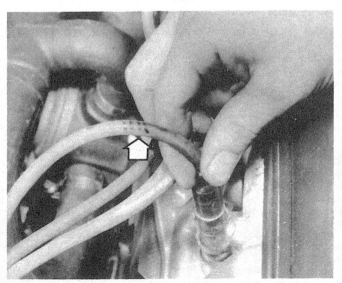

7.37 Disconnecting a spark plug HT lead. Note felt pen marks (arrowed) indicating cylinder

7.39 Testing a spark plug

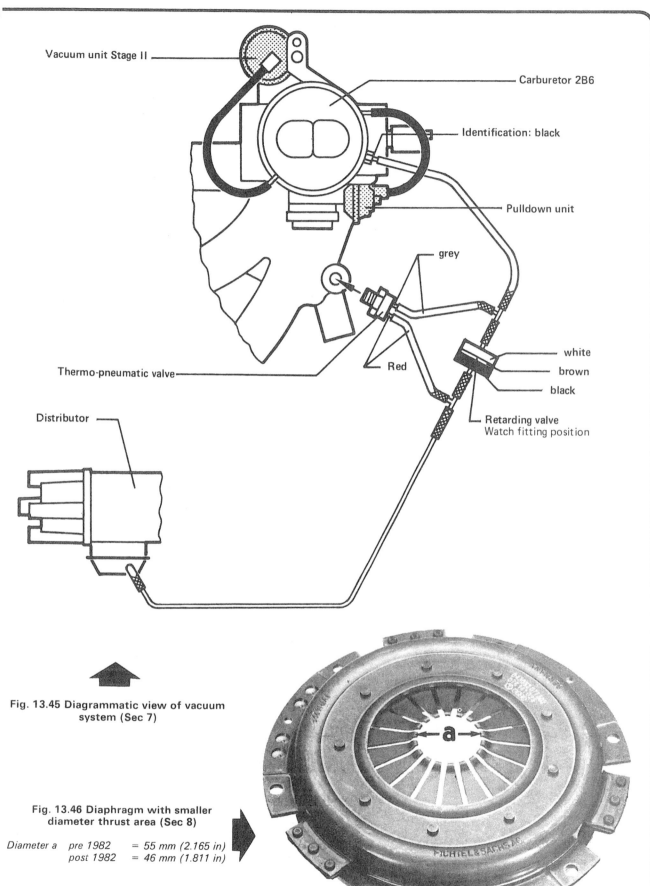

Vacuum unit Stage II

Carburetor 2B6

Identification: black

Pulldown unit

grey

Thermo-pneumatic valve

Red

white
brown
black

Distributor

Retarding valve
Watch fitting position

**Fig. 13.45 Diagrammatic view of vacuum
system (Sec 7)**

**Fig. 13.46 Diaphragm with smaller
diameter thrust area (Sec 8)**

Diameter a   *pre 1982   = 55 mm (2.165 in)*
               post 1982   = 46 mm (1.811 in)

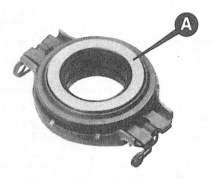

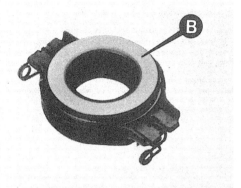

Fig. 13.47 Release bearing thrust areas (Sec 8)

A   Narrow thrust area for 46 mm (1.811 in) diaphragms only

B   Wide thrust area for both diaphragms

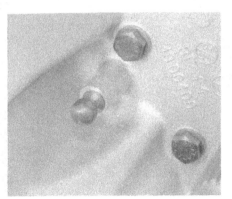

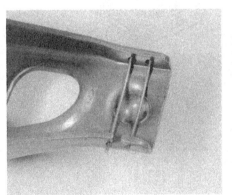

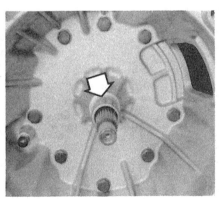

8.6A Release arm ball-stud

8.6B Release arm ball-stud retaining spring

8.10 Release bearing guide sleeve (arrowed)

Clutch release bearing guide sleeve

10   The clutch release bearing is centralised on the clutch shaft by a guide sleeve (photo).

11   On early models this sleeve was made of sheet metal and on later models of plastic. If the sleeve becomes worn it can be renewed as follows:

12   Remove the release lever and make a saw cut through the sleeve along its entire length.

13   Using pliers or similar, tear the sleeve from the shaft.

14   Be extremely careful during both operations not to damage the shaft.

15   The new guide sleeve is knocked back onto the shaft with a suitable mandrel and mallet.

16   Ensure the sleeve engages with the groove in the shaft.

17   Sheet metal sleeves should be greased with molybdenum disulphide grease, plastic sleeves are not greased.

18   Check the bearing slides freely along the sleeve.

Pressure plate bolts – modification

19   The clutch pressure plate securing bolts were originally fitted with spring washers under the head of the bolts (M8 x 20).

20   Since December 1979, the washers have been deleted and shorter bolts used (M8 x 15).

21   If the longer bolts are used on any clutch plate, they must be fitted with washers.

22   Do not mix long and short bolts on the same clutch.

23   Tighten the bolts in diagonal sequence to the specified torque.

24   Clutch cables from 1983 on have an adjuster at the pedal end of the cable.

25   The free play in the clutch pedal is adjusted to the same value as given in the Specifications in Chapter 5, but using the adjuster at the pedal.

9   Gearbox

General description (2.4 litre models)

1   The 4-speed gearbox, designated code 015/1 and the 5-speed gearbox, designated code 008/1, have been developed from their predecessors and used in LT models since 1983.

2   The 015/1 4-speed is used in LT 28, 31 and 35 models.

3   The 008/1 5-speed is used in LT 40, 45 and 50 models, and can be fitted to LT 28, 31 and 35 models as an option.

4   The main difference between these later gearboxes and the earlier ones is in the use of taper roller bearings instead of caged roller bearings.

5   During removal and fitting of the gearboxes minor differences may be evident in the gearbox-to-bellhousing area, due to different gearbox casings and bellhousings being used in different models.

6   This will be clearly evident on LT 40, 45 and 50 models produced after 1983, which have an engine inclination angle greater than that on 28, 31, and 35 models, and the bellhousing-to-gearbox flange is altered to suit.

7 The gearchange linkage is different and is dealt with in the following paragraphs.

Gearchange linkage – adjustment (2.4 litre models)
8 An exploded view of the gearchange linkage appears in Fig. 13.48.
9 Adjust the gearchange linkage as follows:
10 Remove the gearchange lever knob and the rubber boot.
11 Select neutral.

12 Adjust the clamping stop vertically on the gearchange lever to the dimension shown in Fig. 13.50.
13 Do not press down on the gearchange lever.
14 The clamping stop lug must be square on to the stop bracket (photo).
15 Slacken off the clamp (photo).
16 Check that the shift rod moves freely.
17 Align the shift rod and the inner lever shift finger with one another.

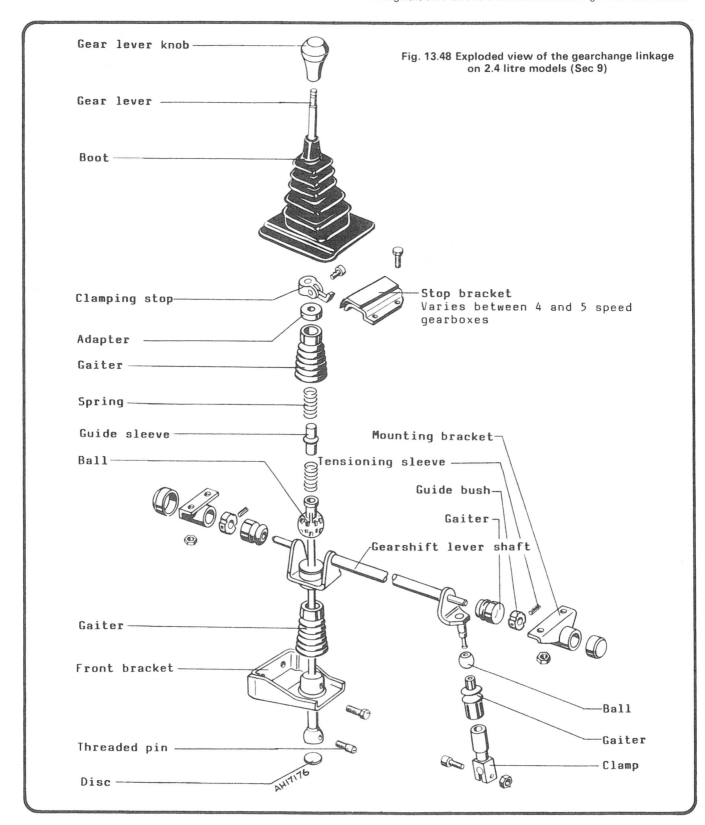

Gear lever knob

Gear lever

Boot

Clamping stop

Adapter

Gaiter

Spring

Guide sleeve

Ball

Gaiter

Front bracket

Threaded pin

Disc

Stop bracket
Varies between 4 and 5 speed gearboxes

Mounting bracket

Tensioning sleeve

Guide bush

Gaiter

Gearshift lever shaft

Ball

Gaiter

Clamp

Fig. 13.48 Exploded view of the gearchange linkage on 2.4 litre models (Sec 9)

AH17176

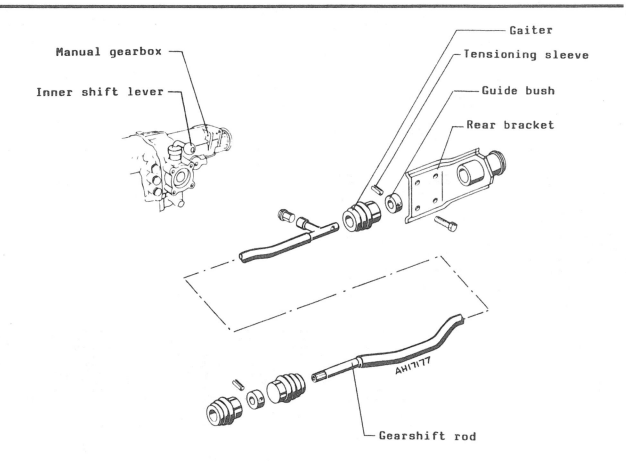

Manual gearbox

Inner shift lever

Gaiter

Tensioning sleeve

Guide bush

Rear bracket

AH17177

Gearshift rod

Fig. 13.49 Exploded view of the gearchange rod and rear bearing/bracket on 2.4 litre models (Sec 9)

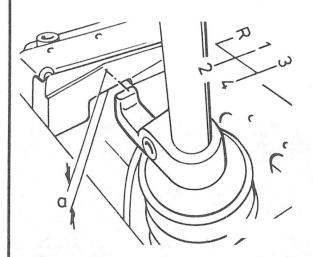

Fig. 13.50 Adjusting the clamping stop (Sec 9)

a   = 6 mm (0.236 in)

9.14 View of lower end of gearchange lever

1   Clamping stop   3   Gearchange rod
2   Stop bracket

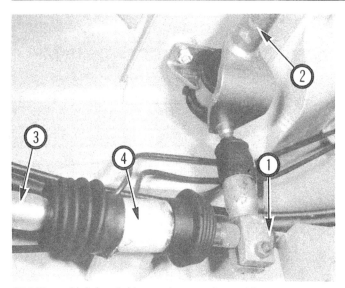

9.15 View of left-hand side gearchange rod assembly

1 Clamp bolt	3 Gearchange rod
2 Crosstube	4 Support bearing/bracket

18 Pull the inner shift lever downward into the 3rd/4th gear position.
19 Now set the gearchange lever to the clearance dimensions shown in Fig. 13.51.
20 Tighten the clamp bolt, refit the rubber boot and lever knob, and check that all gears can be engaged easily without sticking using normal gearchange techniques.

Removal and refitting (2.4 litre models)
21 **Note:** On models fitted with 6-cylinder engines and the 015/1 or 008/1 gearboxes, the gearbox must be removed from the engine before engine removal. The removal/refitting procedure is similar to that given in Chapter 6 for the earlier gearboxes, with the following qualifications.
22 Mark the propshaft-to-rear axle drive flange before removal (photo).
23 Disconnect the earth strap and reversing light leads (photo).
24 Refitting is a reversal of removal.
25 Do not forget to refit the plastic oil guard to the gearbox before refitting the propeller shaft (photo).
26 Make sure the plastic speedometer gear is in place before refitting the drive cable (photo).

Overhaul (2.4 litre models)
27 It is felt that, due to the need for special pullers to dismantle these gearboxes which are unlikely to be available to the home mechanic, that overhauling be left to your VW dealer.

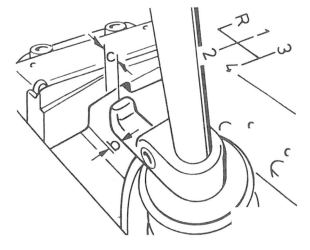

Fig. 13.51 Adjusting the gearchange lever (Sec 9)

b = 4 mm (0.157 in) c = 21 mm (0.827 in)

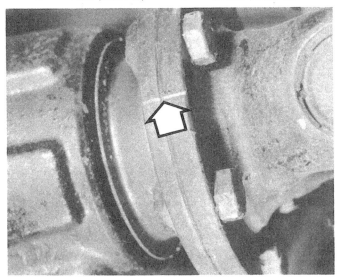

9.22 Make alignment marks on the flanges (arrowed)

Lubrication
28 Originally the gearboxes were fitted with Dexron ATF.
29 Later in production, normal gear oil was re-introduced as the gearbox lubricant.
30 The specified gear oil should be used in all LT gearboxes.

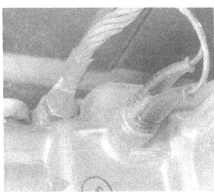

9.23 Earth strap and reversing light leads

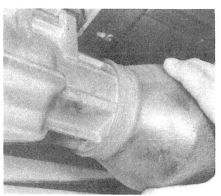

9.25 Don't forget the plastic oil guard

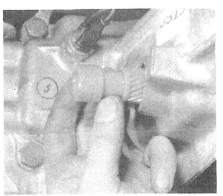

9.26 Speedometer drive gear being fitted

31 Dexron ATF and gear oil can be safely mixed, so there is no need to drain gearboxes filled with ATF, just top up with gear oil in the normal manner.

Power Take Off (PTO) – general

32 A Power Take Off gearbox for driving auxiliaries from the vehicle gearbox is available. It is normally supplied as an optional extra on new vehicles, but may be fitted subsequently by your VW dealer.

33 The Power Steering Take Off gearbox is bolted to the side of the main gearbox, and is driven from the gearbox through the clutch via a sliding intermediate gear.

34 The PTO is engaged or disengaged by a cable-operated lever in the cab. There is also a hand throttle in the cab to control PTO speed.

35 The PTO should be operated in accordance with the owner's handbook.

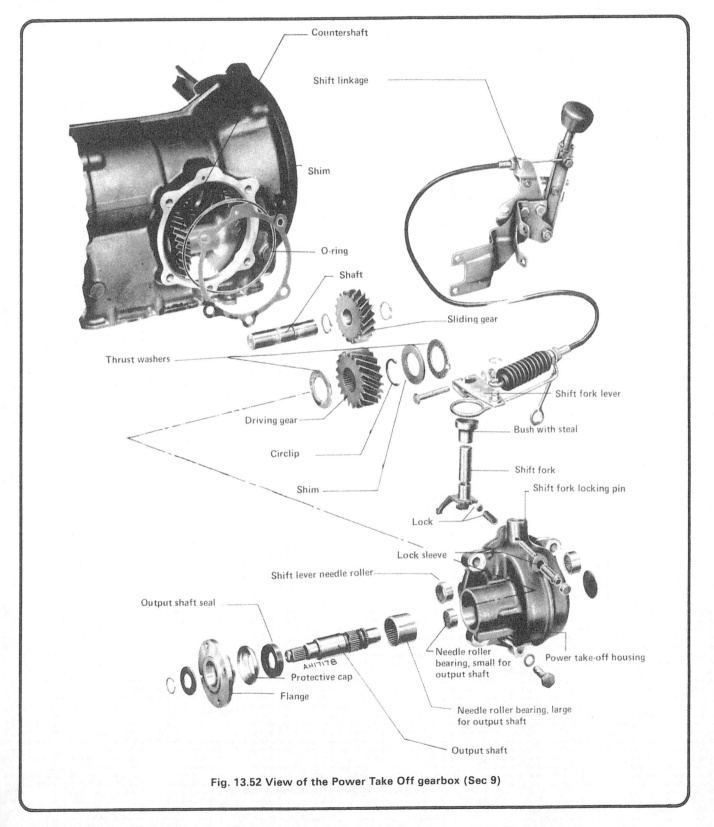

Fig. 13.52 View of the Power Take Off gearbox (Sec 9)

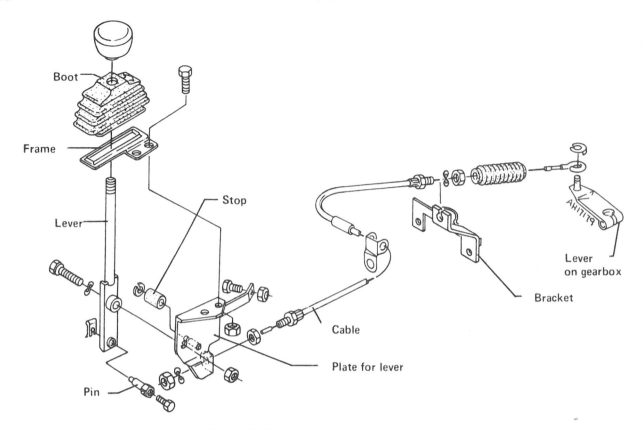

Fig. 13.53 PTO control cable assembly (Sec 9)

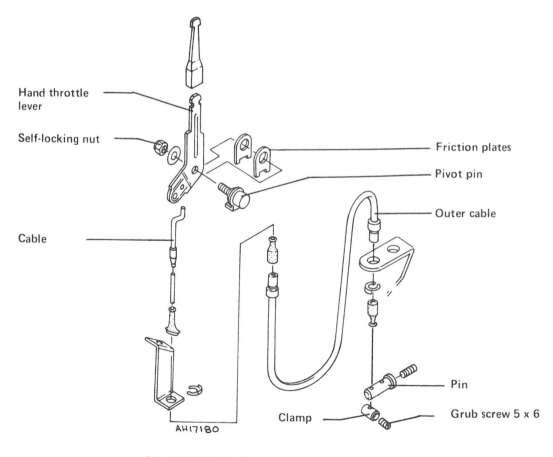

Fig. 13.54 PTO hand throttle cable assembly (Sec 9)

PTO control cable – adjustment

36 Loosen the clamp screw in the pin at the lower end of the control lever.
37 Press the lever forward against its stop.
38 At the same time, push the lever on the PTO gearbox outward (anti-clockwise) into its detent.
39 With both levers held in this position, tighten the clamp screw.

PTO hand throttle – adjustment

40 To adjust the PTO hand throttle cable, loosen the grub screw in the clamp, press the hand throttle lever against its stop and tighten the grub screw in the clamp.

PTO gearbox – removal and refitting

41 The PTO gearbox is bolted to the side of the main gearbox. It is best renewed as a complete unit should it fail during service. To remove the PTO gearbox, proceed as follows.
42 Drain the gearbox oil.
43 Disconnect the control cable at the operating lever on the PTO gearbox.
44 Remove the bolts securing the PTO gearbox to the main gearbox casing, and lift the PTO gearbox away.
45 Refitting is a reversal of removal, but the backlash between 3rd gear on the countershaft and the driven gear in the PTO must be set to the Specified value.
46 This is achieved by fitting shims between the PTO housing and the main gearbox casing.
47 Fit the PTO with shims to the initial value shown in the Specifications.
48 Remove the main gearbox sump.
49 Measure the backlash in the gears using a dial test indicator as shown in Fig. 13.55.
50 Adjust the backlash in the gears to the value specified by putting a thicker or thinner shim between the housings.

---

10 Propeller shaft – 2.4 litre models

Removal and refitting

1 The propeller shaft fitted with the 008/1 5-speed gearbox is in two sections with a central universal joint and bearing.
2 The comments in Chapter 7 apply equally to this propeller shaft, in that none of the universal joints are repairable in service, neither should the front or rear sections be changed individually as both halves of the propeller shaft are balanced together.
3 The centre bearing can be changed by removing the propeller shaft,

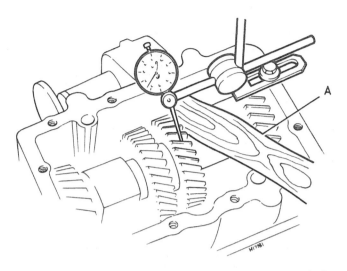

Fig. 13.55 Measuring backlash on 3rd gear on countershaft (Sec 9)

A   Holding sliding gear with a hammer handle

as described in Chapter 7, with the addition of removing the bolts from the centre bearing.
4 Separate the two halves of the propeller shaft by releasing the locking tab and undoing the bolt.
5 Knock the central bearing off the front shaft.
6 Refit in the reverse order, using a new U-plate and locking tab, noting that the two halves of the shaft should be refitted in exactly the same relationship to each other as before, in order to maintain balance.

---

11 Rear axle

Breather valve – modification

1 From November 1984 the breather valve on the rear axle is secured by a plastic cable clip.
2 Vehicles produced before this date had no clip, and the breather valve should be checked for security at the earliest opportunity.
3 If the breather valve is found to be missing, the oil in the rear axle should be renewed.
4 All breather valves should be secured with a cable clip, to prevent them being dislodged.

Fig. 13.56 Centre joint of two section propeller shaft (Sec 10)

1   Bolt   3   U-plate
2   Locking tab

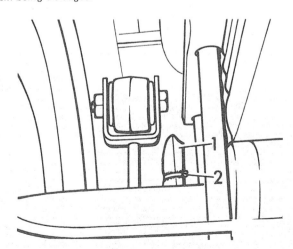

Fig. 13.57 Breather valve on rear axle (Sec 11)

1   Breather valve   2   Cable clip

12 Braking system

Pressure regulating valve – adjustment

1   The adjustment data given in Chapter 9, Section 10, has been amended as follows for the LT 28 and 31 vehicle.
2   If doubt exists as to which springs are fitted, consult your VW dealer, who can determine the spring 'weight' from the part number, as several different types of spring are in use.

Distance a (mm)	Pressures (bar) Normal springs	Strong springs
440 to 435	12 to 14	15 to 19
435 to 430	14 to 16	19 to 23
430 to 425	16 to 18	23 to 27
425 to 420	18 to 21	27 to 31
420 to 415	21 to 24	31 to 36
415 to 410	24 to 28	36 to 42
410 to 405	28 to 32	42 to 48
405 to 400	32 to 36	48 to 54
400 to 395	36 to 41	54 to 61
395 to 390	41 to 46	60 to 68
390 to 385	46 to 51	–
385 to 380	51 to 56	–
380 to 375	56 to 61	–
375 to 370	61 to 66	–
370 to 365	66 to 71	–

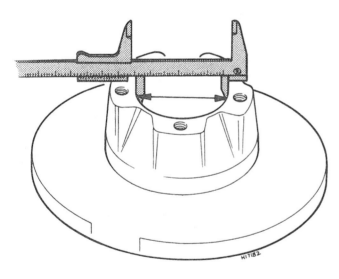

Fig. 13.58 Brake disc inner diameter (Sec 12)

Old   = 87 mm (3.43 in)   *New = 91 mm (3.58 in)*

Discs and wheel hubs – modification

3   From April 1981 on the diameter of the inner part of the brake disc has been increased. Similarly, the hub seat diameter has been increased.
4   Three new diameter components may not be mixed with older components, as the brake disc will not centralise.
5   See also Section 14 on hubs.

Brake fluid reservoir

6   All models from 1983 are equipped with a brake hydraulic reservoir incorporating a low level warning device in the filler cap (photo).

Handbrake 'on' warning

7   All models produced since 1983 are equipped with a handbrake 'on' warning light.
8   This is operated from a micro switch situated directly under the handbrake lever.
9   To remove the micro switch, undo the securing screw (photo), disconnect the electrical lead and lift out the switch.
10 Refit in the reverse order.

13 Electrical system

Part A: General
Battery – 1982 on

1   The battery is still housed behind the co-driver's seat, although the covers are different (photo).
2   The electrolyte cell plugs can only be removed by inserting a screwdriver into the slot on the side of the plug, piercing the plastic skin, and turning the plug anti-clockwise to its stop. The plugs are screwed back in the normal manner.

Alternator – removal and refitting (2.4 litre models)

3   Disconnect the battery earth lead.
4   Remove the wire clip and disconnect the air intake duct.
5   Disconnect the electrical leads.
6   Remove the transport bar (photo).
7   This will reveal the top mounting bolt, the nut of which should be removed (use a new nut on re-assembly).
8   Remove the nut and bolt from the alternator adjustment strap (photo).
9   Slip the drivebelt off the alternator pulley (on vehicles equipped with power steering, the power steering pump drivebolt will have to be removed to remove the alternator belt completely).
10 Remove the bolt from the upper mounting and lift out the alternator.
11 To remove the alternator mounting bracket from the cylinder block, undo the two bolts (photo).
12 Refit in the reverse order.
13 Tension the drivebelt(s) as described in Chapter 2 for the alternator belt – remember the coolant pump is now driven by the toothed belt – and in Section 14 of this Supplement for the power steering pump drivebelt.

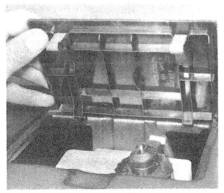

12.6 Brake fluid hydraulic reservoir with low level indicator

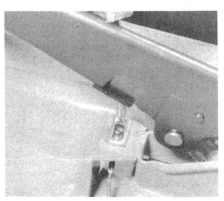

12.9 Removing the handbrake 'on' warning light microswitch

13A.1 Battery, housing and cover

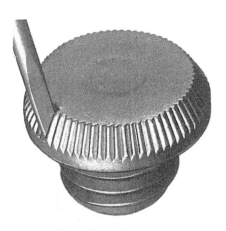

Fig. 13.59 Piercing a battery plug for removal (Sec 13A)

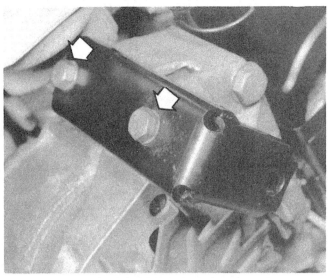

13A.6 Transport bar retaining bolts (arrowed)

13A.8 Alternator adjustment strap nut and bolt (arrowed)

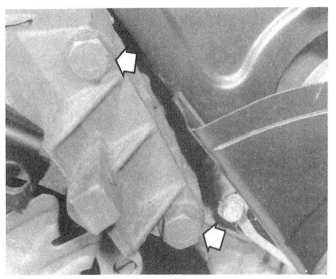

13A.11 Alternator mounting bracket bolts (arrowed)

Alternator – overhaul (2.4 litre models)

14 The overhaul procedure is similar to that described in Chapter 10.
15 There are two types of brush holder (see Figs. 13.60 and 13.61).
16 Type A brush holders require the brush leads to be soldered as described in Chapter 10 for renewal, but type B brush holders require no soldering and can be renewed with the alternator *in situ.*
17 Brush lengths and wear limits are shown in the Specifications.

Alternator – alternative terminal connections

18 The alternator connector on the back of the alternator may be held in place by one of two methods shown in Figs. 13.62 and 13.63.
19 If a connector with a flat clip is removed and replaced by one intended for a wire clip, then the central web on the connector must be filed flat so the flat clip fits properly.

Switches – removing and refitting

20 The facia-mounted switches are held in position by spring arms (photo).
21 Gain access to the rear of the switch, depress the two springs and push the switch forward out of its housing.
22 Disconnect the electrical wires and remove the switch.
23 Refit by reconnecting the wires and pushing the switch home.

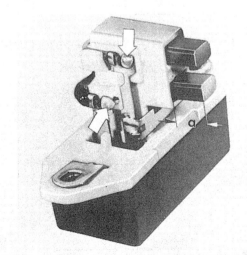

Fig. 13.60 Checking voltage regulator brush length – Type A
(Sec 13A)

a = Brush length
Arrows indicate soldered connections

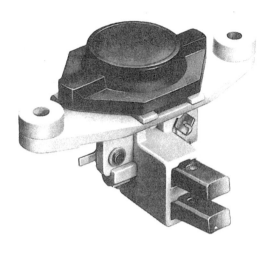

Fig. 13.61 Checking voltage regulator brush length – Type B
(Sec 13A)

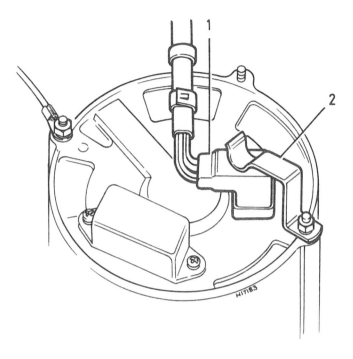

Fig. 13.62 Alternator connector with flat clip (Sec 13A)

1   Connector   2   Flat clip

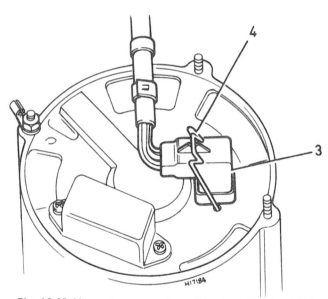

Fig. 13.63 Alternator connector with wire clip (Sec 13A)

3   Connector   4   Wire clip

Rear exterior lamps – bulb renewal (2.4 litre models)

24 The rear lamp cluster bulbs can be renewed after removal of the
lens cover (photos).
25 Access to the rear of the lamp unit is by prising the plastic cover
(photo).
26 The rear fog lamp is held in the bumper assembly by a clamp and
nut (photo).
27 The number plate light is mounted on the rear door (panel van).
28 Remove the cover (photo) to renew the bulb (photo).

Instrument panel – removal and refitting (2.4 litre models)

29 The procedure for removal of the instrument panel is very similar to
that described in Chapter 10.
30 The instrument panel is slightly different in design and due to an
increase in instrumentation there will be more connections (photos).
31 There is also an air distribution duct attached to the instrument
panel (photo).

13A.20 Facia mounted switch – spring arm
arrowed

13A.24A Removing the rear lamp cluster
lens ...

13A.24B ... and bulb

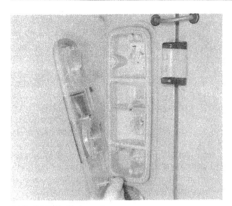

13A.24C Refitting the cover

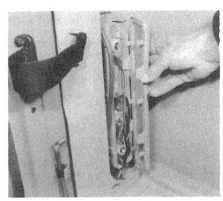

13A.25 Plastic cover for rear of lamp unit

13A.26 Rear foglamp securing clamp

13A.28A Removing the panel van number plate light cover ...

13A.28B ... and bulb

13A.30A Typical multi-block connections on the instrument panel ...

13A.30B ... and facia panel

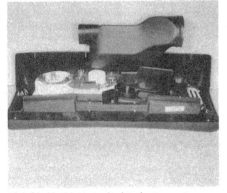

13A.31 Air duct attached to instrument panel

32 There are no clips on the ducting, which simply pulls off.
33 When refitting the panel, ensure the ducting is lined up and fitting over its mating duct.
34 The fitting and removal of the heater controls are dealt with in Section 5.

Oil pressure warning system – 1986 on

General description
35 All models produced after 1986 are fitted with an optical and acoustic oil pressure warning system.
36 Basically, the system consists of an additional 1.8 bar oil pressure switch (white insulation) along with the existing 0.3 bar oil pressure switch (brown insulation) fitted to the engine.
37 The pressure warning control unit is mounted in the rev counter housing.
38 Once the ignition switch is turned, the oil pressure warning lamp

will light up, until an oil pressure in excess of 0.3 bar is generated, when it will go out.
39 Should oil pressure in the lower speed ranges drop below 0.3 bar, the warning light will come on again.
40 If at engine speeds in excess of 2200 rpm, if the oil pressure should drop below 1.8 bar, then the warning light will come on and the warning buzzer sound.
41 Apart from changing the oil pressure switches, if the system malfunctions you should consult your VW dealer, who has the necessary test and fault diagnosis equipment.

Radio – removal and refitting (2.4 litre models)
42 Disconnect the battery.
43 Remove the bracket at the rear of the radio.
44 Push a wire loop of suitable size through the holes at each side of the radio front panel to disengage the side retainers.

45 Pull the radio forward out of the facia and disconnect the aerial and connector.
46 Refit by reconnecting the wiring and pushing the radio home.

Aerial – removal and refitting (2.4 litre models)

47 Disconnect the battery.
48 Detach the relay plate.
49 Remove the self-tapping screw from the support bracket.
50 From outside the vehicle, undo and remove the aerial cap nut.
51 Disconnect the aerial from the radio.
52 Remove the aerial from under the facia.
53 Refit in the reverse order.

Part B: Radios and tape players – general information

Loudspeakers

Speakers should be matched to the output stage of the equipment, particularly as regards the recommended impedance. Power transistors used for driving speakers are sensitive to the loading placed on them.

Before choosing a mounting position for speakers, check whether the vehicle manufacturer has provided a location for them. Generally door-mounted speakers give good stereophonic reproduction, but not all doors are able to accept them. The next best position is the rear parcel shelf, and in this case speaker apertures can be cut into the shelf, or pod units may be mounted.

For door mounting, first remove the trim, which is often held on by 'poppers' or press studs, and then select a suitable gap in the inside door assembly. Check that the speaker would not obstruct glass or winder mechanism by winding the window up and down. A template is often provided for marking out the trim panel hole, and then the four fixing holes must be drilled through. Mark out with chalk and cut cleanly with a sharp knife or keyhole saw. Speaker leads are then threaded through the door and door pillar, if necessary drilling 10 mm diameter holes. Fit grommets in the holes and connect to the radio or tape unit correctly. Do not omit a waterproofing cover, usually supplied with door speakers. If the speaker has to be fixed into the metal of the door itself, use self-tapping screws, and if the fixing is to the door trim use self-tapping screws and flat spire nuts.

Rear shelf mounting is somewhat simpler but it is necessary to find gaps in the metalwork underneath the parcel shelf. However, remember that the speakers should be as far apart as possible to give a good stereo effect. Pod-mounted speakers can be screwed into position through the parcel shelf material, but it is worth testing for the best position. Sometimes good results are found by reflecting sound off the rear window.

Unit installation

Many vehicles have a dash panel aperture to take a radio/audio unit, a recognised international standard being 189.5 mm x 60 mm. Alternatively a console may be a feature of the car interior design and this, mounted below the dashboard, gives more room. If neither facility is available a unit may be mounted on the underside of the parcel shelf; these are frequently non-metallic and an earth wire from the case to a good earth point is necessary. A three-sided cover in the form of a cradle is obtainable from car radio dealers and this gives a professional appearance to the installation; in this case choose a position where the controls can be reached by a driver with his seat belt on.

Installation of the radio/audio unit is basically the same in all cases, and consists of offering it into the aperture after removal of the knobs (not push buttons) and the trim plate. In some cases a special mounting plate is required to which the unit is attached. It is worthwhile supporting the rear end in cases where sag or strain may occur, and it is usually possible to use a length of perforated metal strip attached between the unit and a good support point nearby. In general it is recommended that tape equipment should be installed at or nearly horizontal.

Connections to the aerial socket are simply by the standard plug terminating the aerial download or its extension cable. Speakers for a stereo system must be matched and correctly connected, as outlined previously.

Note: *While all work is carried out on the power side, it is wise to disconnect the battery earth lead.* Before connection is made to the vehicle electrical system, check that the polarity of the unit is correct. Most vehicles use a negative earth system, but radio/audio units often

have a reversible plug to convert the set to either + or – earth. *Incorrect connection may cause serious damage.*

The power lead is often permanently connected inside the unit and terminates with one half of an in-line fuse carrier. The other half is fitted with a suitable fuse (3 or 5 amperes) and a wire which should go to a power point in the electrical system. This may be the accessory terminal on the ignition switch, giving the advantage of power feed with ignition or with the ignition key at the 'accessory' position. Power to the unit stops when the ignition key is removed. Alternatively, the lead may be taken to a live point at the fusebox with the consequence of having to remember to switch off at the unit before leaving the vehicle.

Before switching on for initial test, be sure that the speaker connections have been made, for running without load can damage the output transistors. Switch on next and tune through the bands to ensure that all sections are working, and check the tape unit if applicable. The aerial trimmer should be adjusted to give the strongest reception on a weak signal in the medium wave band, at say 200 metres.

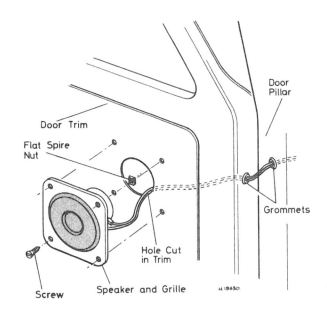

Fig. 13.64 Door mounted speaker installation (Sec 13B)

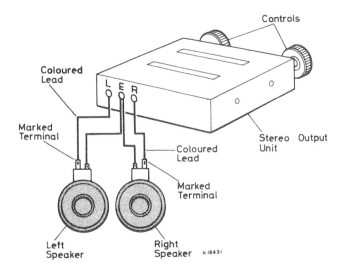

Fig. 13.65 Speaker connections must be correctly made as shown (Sec 13B)

Interference

In general, when electric current changes abruptly, unwanted electrical noise is produced. The motor vehicle is filled with electrical devices which change electric current rapidly, the most obvious being the contact breaker.

When the spark plugs operate, the sudden pulse of spark current causes the associated wiring to radiate. Since early radio transmitters used sparks as a basis of operation, it is not surprising that the car radio will pick up ignition spark noise unless steps are taken to reduce it to acceptable levels.

Interference reaches the car radio in two ways:

(a) by conduction through the wiring.
(b) by radiation to the receiving aerial.

Initial checks presuppose that the bonnet is down and fastened, the radio unit has a good earth connection *(not* through the aerial downlead outer), no fluorescent tubes are working near the car, the aerial trimmer has been adjusted, and the vehicle is in a position to receive radio signals, ie not in a metal-clad building.

Switch on the radio and tune it to the middle of the medium wave (MW) band off-station with the volume (gain) control set fairly high. Switch on the ignition (but do not start the engine) and wait to see if irregular clicks or hash noise occurs. Tapping the facia panel may also produce the effects. If so, this will be due to the voltage stabiliser, which is an on-off thermal switch to control instrument voltage. It is located usually on the back of the instrument panel, often attached to the speedometer. Correction is by attachment of a capacitor and, if still troublesome, chokes in the supply wires.

Switch on the engine and listen for interference on the MW band. Depending on the type of interference, the indications are as follows.

A harsh crackle that drops out abruptly at low engine speed or when the headlights are switched on is probably due to a voltage regulator.

A whine varying with engine speed is due to the dynamo or alternator. Try temporarily taking off the fan belt – if the noise goes this is confirmation.

Regular ticking or crackle that varies in rate with the engine speed is due to the ignition system. With this trouble in particular and others in general, check to see if the noise is entering the receiver from the wiring or by radiation. To do this, pull out the aerial plug, (preferably shorting out the input socket or connecting a 62 pF capacitor across it). If the noise disappears it is coming in through the aerial and is *radiation noise.* If the noise persists it is reaching the receiver through the wiring and is said to be *line-borne.*

Interference from wipers, washers, heater blowers, turn-indicators, stop lamps, etc is usually taken to the receiver by wiring, and simple treatment using capacitors and possibly chokes will solve the problem. Switch on each one in turn (wet the screen first for running wipers!) and listen for possible interference with the aerial plug in place and again when removed.

Electric petrol pumps are now finding application again and give rise to an irregular clicking, often giving a burst of clicks when the

ignition is on but the engine has not yet been started. It is also possible to receive whining or crackling from the pump.

Note that if most of the vehicle accessories are found to be creating interference all together, the probability is that poor aerial earthing is to blame.

Component terminal markings

Throughout the following sub-sections reference will be found to various terminal markings. These will vary depending on the manufacturer of the relevant component. If terminal markings differ from those mentioned, reference should be made to the following table, where the most commonly encountered variations are listed.

Alternator	Alternator terminal (thick lead)	Exciting winding terminal
DIN/Bosch	B+	DF
Delco Remy	+	EXC
Ducellier	+	EXC
Ford (US)	+	DF
Lucas	+	F
Marelli	+B	F

Ignition coil	Ignition switch terminal	Contact breaker terminal
DIN/Bosch	15	1
Delco Remy	+	–
Ducellier	BAT	RUP
Ford (US)	B/+	CB/–
Lucas	SW/+	–
Marelli	BAT/+B	D

Voltage regulator	Voltage input terminal	Exciting winding terminal
DIN/Bosch	B+/D+	DF
Delco Remy	BAT/+	EXC
Ducellier	BOB/BAT	EXC
Ford (US)	BAT	DF
Lucas	+/A	F
Marelli	F	F

Suppression methods – ignition

Suppressed HT cables are supplied as original equipment by manufacturers and will meet regulations as far as interference to neighbouring equipment is concerned. It is illegal to remove such suppression unless an alternative is provided, and this may take the form of resistive spark plug caps in conjunction with plain copper HT cable. For VHF purposes, these and 'in-line' resistors may not be effective, and resistive HT cable is preferred. Check that suppressed cables are actually fitted by observing cable identity lettering, or measuring with an ohmmeter – the value of each plug lead should be 5000 to 10 000 ohms.

A 1 microfarad capacitor connected from the LT supply side of the ignition coil to a good nearby earth point will complete basic ignition interference treatment. *NEVER fit a capacitor to the coil terminal to the contact breaker – the result would be burnt out points in a short time.*

If ignition noise persists despite the treatment above, the following sequence should be followed:

(a) Check the earthing of the ignition coil; remove paint from fixing clamp.
(b) If this does not work, lift the bonnet. Should there be no change in interference level, this may indicate that the bonnet is not electrically connected to the car body. Use a proprietary braided strap across a bonnet hinge ensuring a first class electrical connection. If, however, lifting the bonnet increases the interference, then fit resistive HT cables of a higher ohms-per-metre value.
(c) If all these measures fail, it is probable that re-radiation from metallic components is taking place. Using a braided strap between metallic points, go round the vehicle systematically – try the following: engine to body, exhaust system to body, front suspension to engine and to body, steering column to body (especially French and Italian cars), gear lever to engine and to body (again especially French and Italian cars), Bowden cable to body, metal parcel shelf to body. When an offending component is located it should be bonded with the strap permanently.

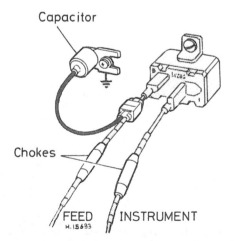

Fig. 13.66 Voltage stabiliser interference suppression (Sec 13B)

(d) As a next step, the fitting of distributor suppressors to each
     lead at the distributor end may help.
(e) Beyond this point is involved the possible screening of the
     distributor and fitting resistive spark plugs, but such advanced
     treatment is not usually required for vehicles with enter-
     tainment equipment.

Electronic ignition systems have built-in suppression components,
but this does not relieve the need for using suppressed HT leads. In
some cases it is permitted to connect a capacitor on the low tension
supply side of the ignition coil, but not in every case. Makers'
instructions should be followed carefully, otherwise damage to the
ignition semiconductors may result.

Suppression methods – generators

For older vehicles with dynamos a 1 microfarad capacitor from the
D (larger) terminal to earth will usually cure dynamo whine.
Alternators should be fitted with a 3 microfarad capacitor from the B +
main output terminal (thick cable) to earth. Additional suppression
may be obtained by the use of a filter in the supply line to the radio
receiver.

It is most important that:

(a) *Capacitors are never connected to the field terminals of either
     a dynamo or alternator.*
(b) *Alternators must not be run without connection to the
     battery.*

Suppression methods – voltage regulators

Voltage regulators used with DC dynamos should be suppressed by
connecting a 1 microfarad capacitor from the control box D terminal to
earth.

Alternator regulators come in three types:

(a) *Vibrating contact regulators separate from the alternator.
     Used extensively on continental vehicles.*
(b) *Electronic regulators separate from the alternator.*
(c) *Electronic regulators built-in to the alternator.*

In case (a) interference may be generated on the AM and FM
(VHF) bands. For some cars a replacement suppressed regulator is
available. Filter boxes may be used with non-suppressed regulators.
But if not available, then for AM equipment a 2 microfarad or 3
microfarad capacitor may be mounted at the voltage terminal marked
D+ or B+ of the regulator. FM bands may be treated by a
feed-through capacitor of 2 or 3 microfarad.

Electronic voltage regulators are not always troublesome, but
where necessary, a 1 microfarad capacitor from the regulator +
terminal will help.

Integral electronic voltage regulators do not normally generate
much interference, but when encountered this is in combination with
alternator noise. A 1 microfarad or 2 microfarad capacitor from the
warning lamp (IND) terminal to earth for Lucas ACR alternators and
Femsa, Delco and Bosch equivalents should cure the problem.

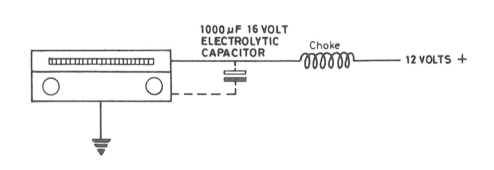

Fig. 13.67 Line-borne interference suppression (Sec 13B)

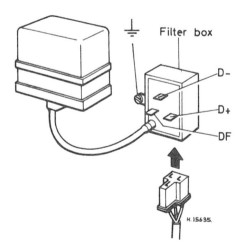

Fig. 13.68 Typical filter box for vibrating contact voltage
regulator (alternator equipment) (Sec 13B)

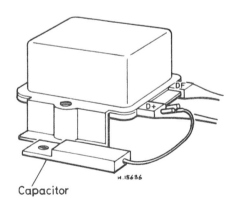

Fig. 13.69 Suppression of AM interference by vibrating
contact voltage regulator (alternator equipment)
(Sec 13B)

Suppression methods – other equipment

Wiper motors – Connect the wiper body to earth with a bonding strap. For all motors use a 7 ampere choke assembly inserted in the leads to the motor.

Heater motors – Fit 7 ampere line chokes in both leads, assisted if necessary by a 1 microfarad capacitor to earth from both leads.

Electronic tachometer – The tachometer is a possible source of ignition noise – check by disconnecting at the ignition coil CB terminal. It usually feeds from ignition coil LT pulses at the contact breaker terminal. A 3 ampere line choke should be fitted in the tachometer lead at the coil CB terminal.

Horn – A capacitor and choke combination is effective if the horn is directly connected to the 12 volt supply. The use of a relay is an alternative remedy, as this will reduce the length of the interference-carrying leads.

Electrostatic noise – Characteristics are erratic crackling at the receiver, with disappearance of symptoms in wet weather. Often shocks may be given when touching bodywork. Part of the problem is the build-up of static electricity in non-driven wheels and the acquisition of charge on the body shell. It is possible to fit spring-loaded contacts at the wheels to give good conduction between the rotary wheel parts and the vehicle frame. Changing a tyre sometimes helps – because of tyres' varying resistances. In difficult cases a trailing flex which touches the ground will cure the problem. If this is not acceptable it is worth trying conductive paint on the tyre walls.

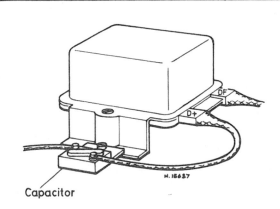

Fig. 13.70 Suppression of FM interference by vibrating contact voltage regulator (alternator equipment) (Sec 13B)

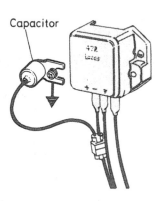

Fig. 13.71 Electronic voltage regulator suppression (Sec 13B)

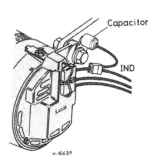

Fig. 13.72 Suppression of interference from electronic voltage regulator when integral with alternator (Sec 13B)

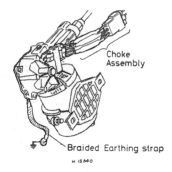

Fig. 13.73 Wiper motor suppression (Sec 13B)

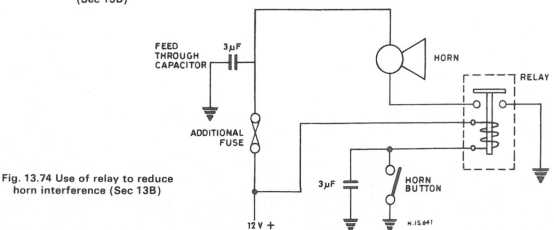

Fig. 13.74 Use of relay to reduce horn interference (Sec 13B)

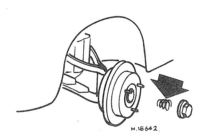

Fig. 13.75 Use of spring contacts at wheels (Sec 13B)

Fuel pump – Suppression requires a 1 microfarad capacitor between the supply wire to the pump and a nearby earth point. If this is insufficient a 7 ampere line choke connected in the supply wire near the pump is required.

Fluorescent tubes – Vehicles used for camping/caravanning frequently have fluorescent tube lighting. These tubes require a relatively high voltage for operation and this is provided by an inverter (a form of oscillator) which steps up the vehicle supply voltage. This can give rise to serious interference to radio reception, and the tubes themselves can contribute to this interference by the pulsating nature of the lamp discharge. In such situations it is important to mount the aerial as far away from a fluorescent tube as possible. The interference problem may be alleviated by screening the tube with fine wire turns spaced an inch (25 mm) apart and earthed to the chassis. Suitable chokes should be fitted in both supply wires close to the inverter.

Radio/cassette case breakthrough

Magnetic radiation from dashboard wiring may be sufficiently intense to break through the metal case of the radio/cassette player. Often this is due to a particular cable routed too close and shows up as ignition interference on AM and cassette play and/or alternator whine on cassette play.

The first point to check is that the clips and/or screws are fixing all parts of the radio/cassette case together properly. Assuming good earthing of the case, see if it is possible to re-route the offending cable – the chances of this are not good, however, in most cars.

Next release the radio/cassette player and locate it in different positions with temporary leads. If a point of low interference is found, then if possible fix the equipment in that area. This also confirms that local radiation is causing the trouble. If re-location is not feasible, fit the radio/cassette player back in the original position.

Alternator interference on cassette play is now caused by radiation from the main charging cable which goes from the battery to the output terminal of the alternator, usually via the + terminal of the starter motor relay. In some vehicles this cable is routed under the dashboard, so the solution is to provide a direct cable route. Detach the

original cable from the alternator output terminal and make up a new cable of at least 6 mm² cross-sectional area to go from alternator to battery with the shortest possible route. *Remember – do not run the engine with the alternator disconnected from the battery.*

Ignition breakthrough on AM and/or cassette play can be a difficult problem. It is worth wrapping earthed foil round the offending cable run near the equipment, or making up a deflector plate well screwed down to a good earth. Another possibility is the use of a suitable relay to switch on the ignition coil. The relay should be mounted close to the ignition coil; with this arrangement the ignition coil primary current is not taken into the dashboard area and does not flow through the ignition switch. A suitable diode should be used since it is possible that at ignition switch-off the output from the warning lamp alternator terminal could hold the relay on.

Connectors for suppression components

Capacitors are usually supplied with tags on the end of the lead, while the capacitor body has a flange with a slot or hole to fit under a nut or screw with washer.

Connections to feed wires are best achieved by self-stripping connectors. These connectors employ a blade which, when squeezed down by pliers, cuts through cable insulation and makes connection to the copper conductors beneath.

Chokes sometimes come with bullet snap-in connectors fitted to the wires, and also with just bare copper wire. With connectors, suitable female cable connectors may be purchased from an auto-accessory shop together with any extra connectors required for the cable ends after being cut for the choke insertion. For chokes with bare wires, similar connectors may be employed together with insulation sleeving as required.

VHF/FM broadcasts

Reception of VHF/FM in an automobile is more prone to problems than the medium and long wavebands. Medium/long wave transmitters are capable of covering considerable distances, but VHF transmitters are restricted to line of sight, meaning ranges of 10 to 50 miles, depending upon the terrain, the effects of buildings and the transmitter power.

Because of the limited range it is necessary to retune on a long journey, and it may be better for those habitually travelling long distances or living in areas of poor provision of transmitters to use an AM radio working on medium/long wavebands.

When conditions are poor, interference can arise, and some of the suppression devices described previously fall off in performance at very high frequencies unless specifically designed for the VHF band. Available suppression devices include reactive HT cable, resistive distributor caps, screened plug caps, screened leads and resistive spark plugs.

For VHF/FM receiver installation the following points should be particularly noted:

(a) Earthing of the receiver chassis and the aerial mounting is important. Use a separate earthing wire at the radio, and scrape paint away at the aerial mounting.

(b) If possible, use a good quality roof aerial to obtain maximum height and distance from interference generating devices on the vehicle.

(c) Use of a high quality aerial downlead is important, since losses in cheap cable can be significant.

(d) The polarisation of FM transmissions may be horizontal, vertical, circular or slanted. Because of this the optimum mounting angle is at 45° to the vehicle roof.

Citizens' Band radio (CB)

In the UK, CB transmitter/receivers work within the 27 MHz and 934 MHz bands, using the FM mode. At present interest is concentrated on 27 MHz where the design and manufacture of equipment is less difficult. Maximum transmitted power is 4 watts, and 40 channels spaced 10 kHz apart within the range 27.60125 to 27.99125 MHz are available.

Aerials are the key to effective transmission and reception. Regulations limit the aerial length to 1.65 metres including the loading coil and any associated circuitry, so tuning the aerial is necessary to obtain optimum results. The choice of a CB aerial is dependent on whether it is to be permanently installed or removable, and the performance will hinge on correct tuning and the location point on the vehicle. Common practice is to clip the aerial to the roof gutter or to

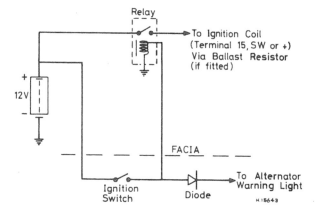

Fig. 13.76 Use of ignition coil relay to suppress case breakthrough (Sec 13B)

employ wing mounting where the aerial can be rapidly unscrewed. An alternative is to use the boot rim to render the aerial theftproof, but a popular solution is to use the 'magmount' – a type of mounting having a strong magnetic base clamping to the vehicle at any point, usually the roof.

Aerial location determines the signal distribution for both transmission and reception, but it is wise to choose a point away from the engine compartment to minimise interference from vehicle electrical equipment.

The aerial is subject to considerable wind and acceleration forces. Cheaper units will whip backwards and forwards and in so doing will alter the relationship with the metal surface of the vehicle with which it forms a ground plane aerial system. The radiation pattern will change correspondingly, giving rise to break-up of both incoming and outgoing signals.

Interference problems on the vehicle carrying CB equipment fall into two categories:

(a) Interference to nearby TV and radio receivers when transmitting.
(b) Interference to CB set reception due to electrical equipment on the vehicle.

Problems of break-through to TV and radio are not frequent, but can be difficult to solve. Mostly trouble is not detected or reported because the vehicle is moving and the symptoms rapidly disappear at the TV/radio receiver, but when the CB set is used as a base station any trouble with nearby receivers will soon result in a complaint.

It must not be assumed by the CB operator that his equipment is faultless, for much depends upon the design. Harmonics (that is, multiples) of 27 MHz may be transmitted unknowingly and these can fall into other user's bands. Where trouble of this nature occurs, low pass filters in the aerial or supply leads can help, and should be fitted in base station aerials as a matter of course. In stubborn cases it may be necessary to call for assistance from the licensing authority, or, if possible, to have the equipment checked by the manufacturers.

Interference received on the CB set from the vehicle equipment is, fortunately, not usually a severe problem. The precautions outlined previously for radio/cassette units apply, but there are some extra points worth noting.

It is common practice to use a slide-mount on CB equipment enabling the set to be easily removed for use as a base station, for example. Care must be taken that the slide mount fittings are properly earthed and that first class connection occurs between the set and slide-mount.

Vehicle manufacturers in the UK are required to provide suppression of electrical equipment to cover 40 to 250 MHz to protect TV and VHF radio bands. Such suppression appears to be adequately effective at 27 MHz, but suppression of individual items such as alternators/dynamos, clocks, stabilisers, flashers, wiper motors, etc, may still be necessary. The suppression capacitors and chokes available from auto-electrical suppliers for entertainment receivers will usually give the required results with CB equipment.

Other vehicle radio transmitters

Besides CB radio already mentioned, a considerable increase in the use of transceivers (ie combined transmitter and receiver units) has taken place in the last decade. Previously this type of equipment was fitted mainly to military, fire, ambulance and police vehicles, but a large business radio and radio telephone usage has developed.

Generally the suppression techniques described previously will suffice, with only a few difficult cases arising. Suppression is carried out to satisfy the 'receive mode', but care must be taken to use heavy duty chokes in the equipment supply cables since the loading on 'transmit' is relatively high.

Wiring diagrams commence overleaf

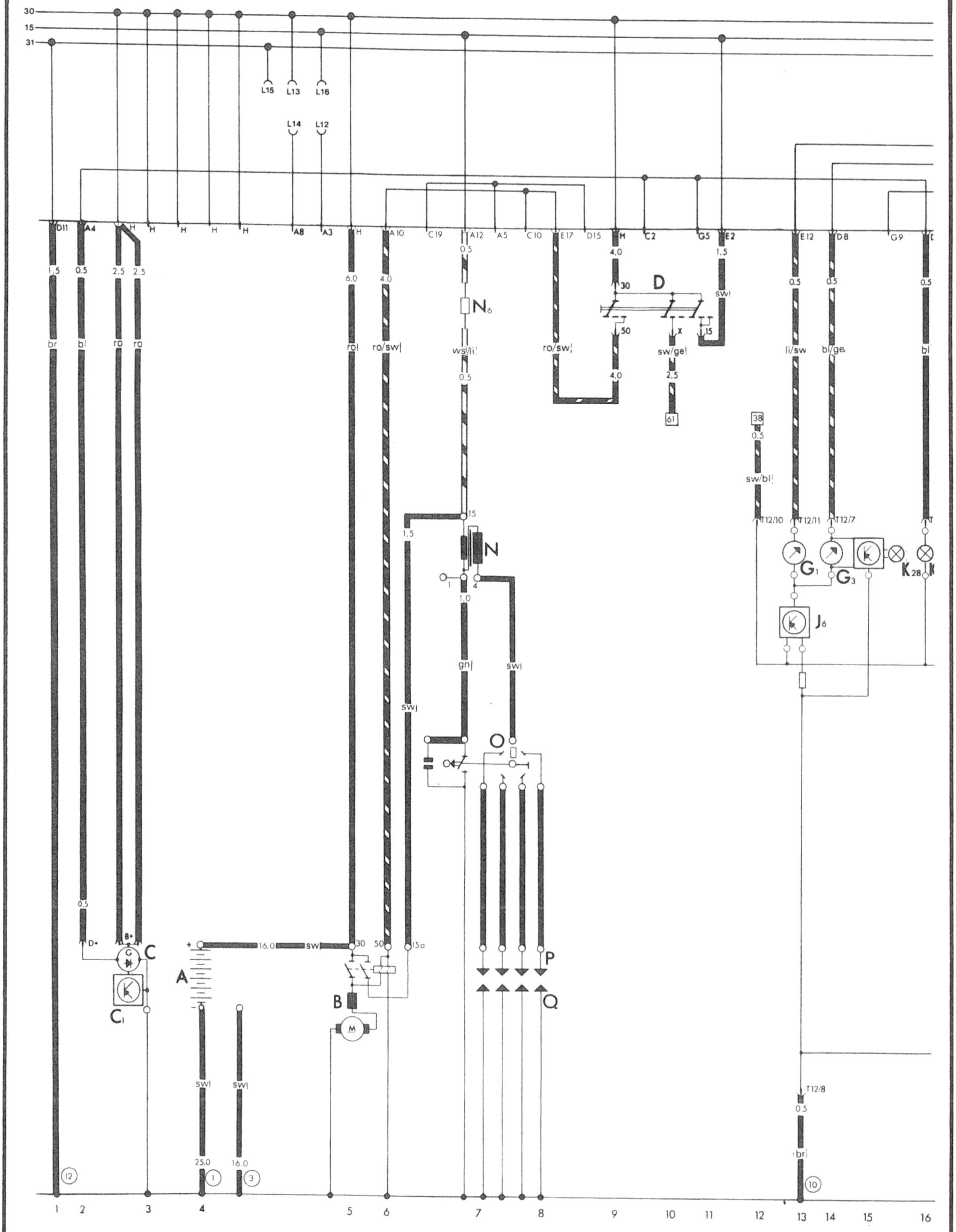

Fig. 13.77 Main current flow diagram – 4-cylinder models, 1981 on

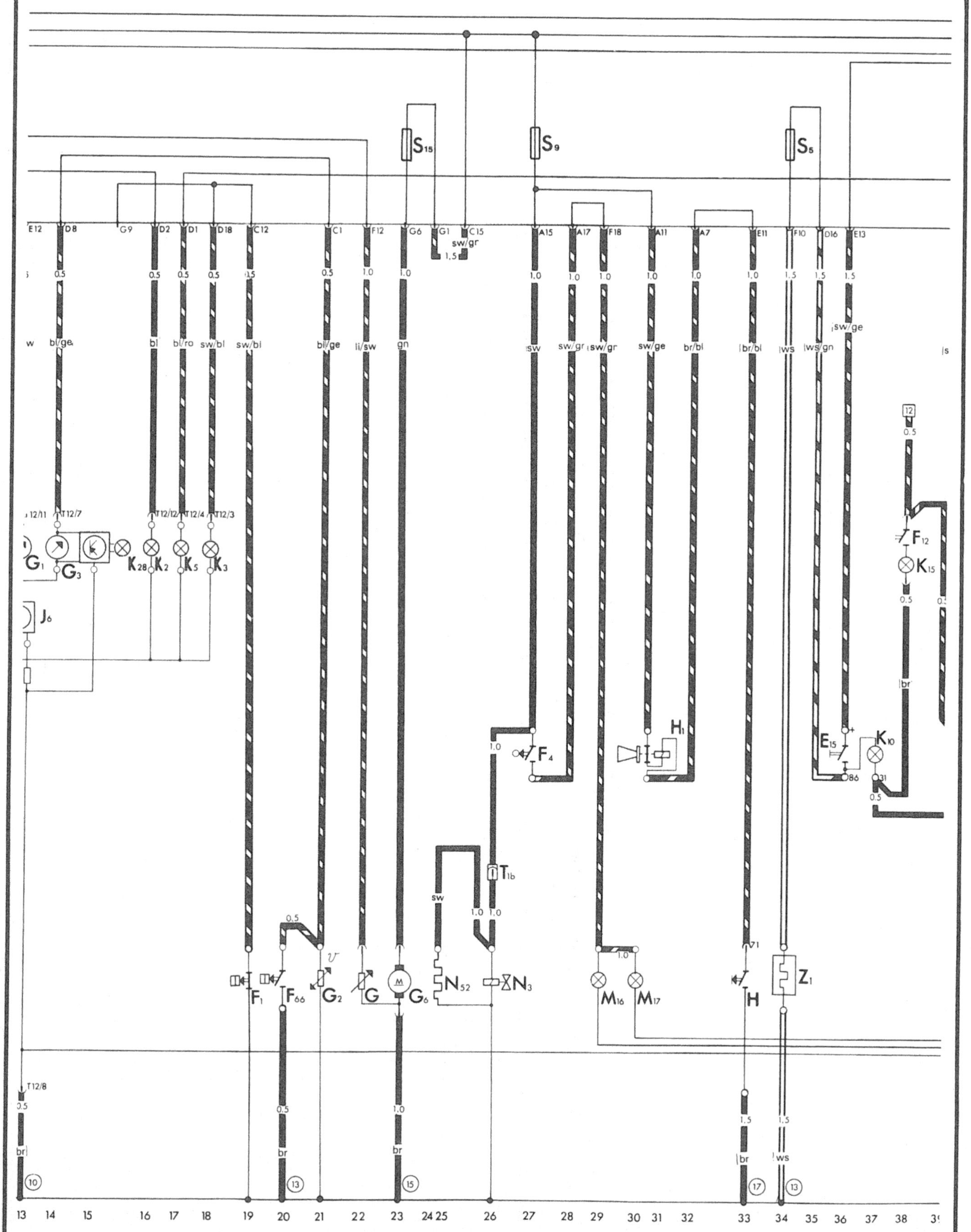

Fig. 13.77 Main current flow diagram – 4-cylinder models, 1981 on (continued)

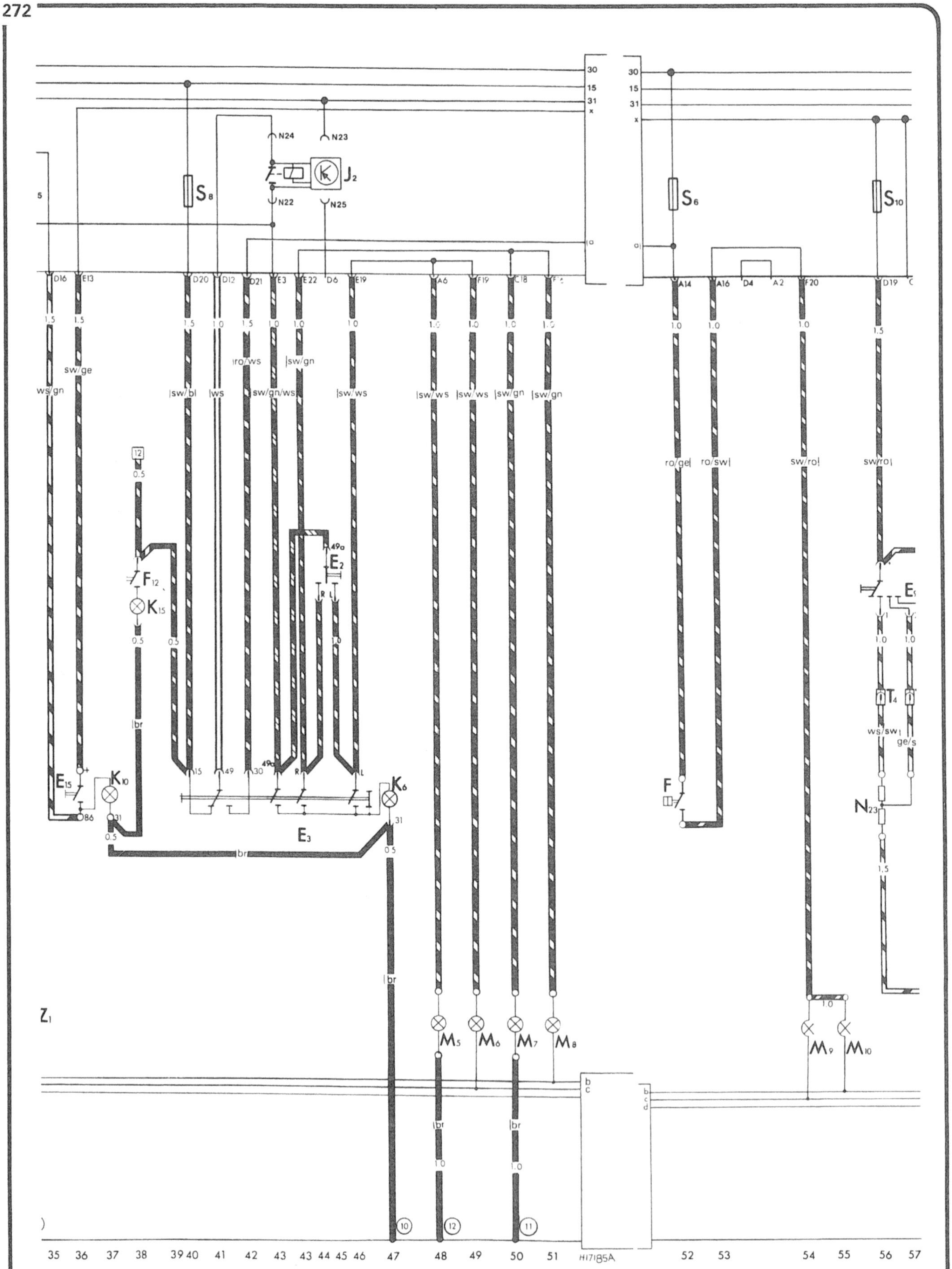

Fig. 13.77 Main current flow diagram – 4-cylinder models, 1981 on (continued)

Fig. 13.77 Main current flow diagram – 4-cylinder models, 1981 on (continued)

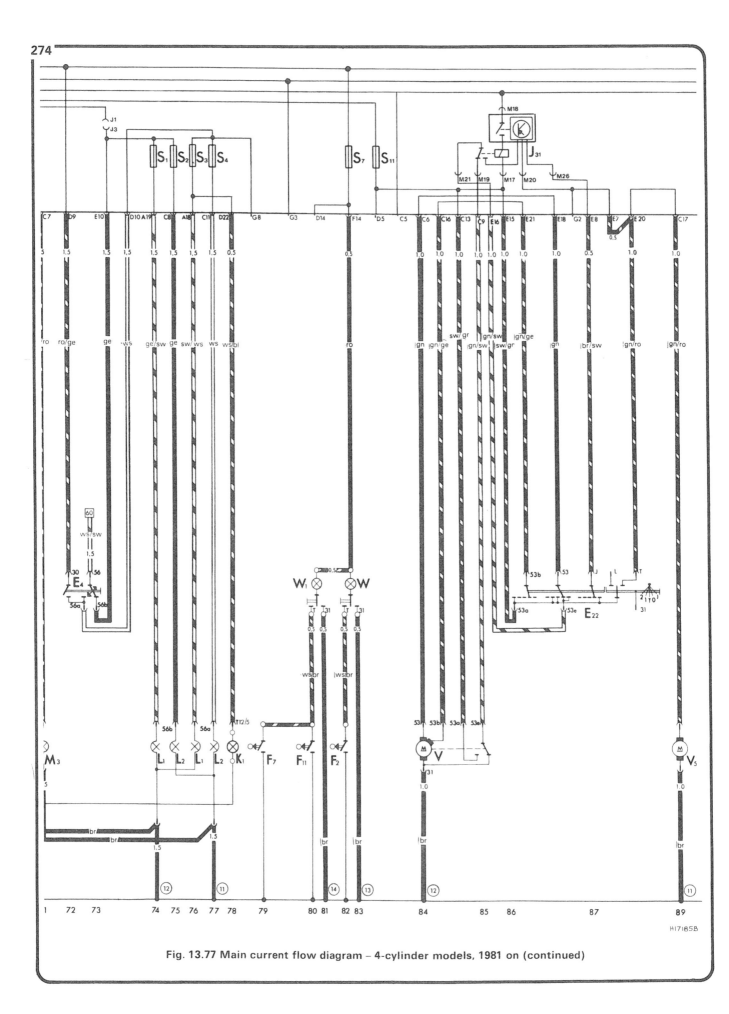

Fig. 13.77 Main current flow diagram – 4-cylinder models, 1981 on (continued)

H17185B

Key to Fig. 13.77

Designation		in current track
A	Battery	4
B	Starter	5, 6
C	Generator	2, 3
C1	Voltage regulator	2, 3
D	Igniton/starter switch	9 to 11
E1	Lighting switch	61 to 63
E2	Turn signal switch	44 to 46
E3	Emergency light switch	40 to 47
E4	Headlight dimmer/flasher switch	72, 73
E9	Fresh air blower switch	56 to 58
E15	Heated rear window switch	36 to 37
E22	Windscreen wiper switch for intermittent operation	86, 87
F	Brake light switch	52
F1	Oil pressure switch	19
F2	Door contact switch, front left	82
F4	Reversing light switch	27
F7	Door contact switch, rear (sliding door)	79
F11	Door contact switch, rear right	80
F12	Starting aid warning lamp contact	38
F66	Low level coolant indicator switch	20
G	Fuel gauge sender	22
G1	Fuel gauge	13
G2	Coolant temperature sender	21
G3	Coolant temperature gauge	14
G6	Electric fuel pump	23
H	Horn control	33
H1	Horn	30
J2	Emergency light relay	43, 44
J6	Voltage stabilizer	13
J31	Wash/wipe intermittent relay	85 to 87
J59	Relief relay (for X contact)	59, 60
K1	Main beam warning lamp	78
K2	Generator warning lamp	16
K3	Oil pressure warning lamp	18
K5	Turn signal warning lamp	17
K6	Emergency light warning lamp	47
K10	Heated rear window warning lamp	37
K15	Choke warning lamp	38
K28	Coolant temperature warning lamp	15
L1	Headlight, left	74, 76
L2	Headlight, right	75, 77
L9	Lighting switch light	59
L10	Instrument panel insert light	62 to 64
L16	Fresh air controls light	65
M1	Sidelight, left	69
M2	Tail light, right	70
M3	Sidelight, right	71
M4	Tail light, left	68
M5	Turn signal front left	48
M6	Turn signal rear left	49

Designation		in current track	
M7	Turn signal front right	50	
M8	Turn signal rear right	51	
M9	Brake light, left	54	
M10	Brake light, right	55	
M16	Reversing light, left	29	
M17	Reversing light, right	30	
N	Coil	7	
N3	Bypass air cut-off valve	26	
N6	Series resistance wire	7	
N23	Fresh air blower series resistance	56	
N52	Heat resistance (part throttle channel heating – carburettor)	25	
O	Distributor	7, 8	
P	Spark plug connector	7, 8	
Q	Spark plugs	7, 8	
S1 to S15	Fuses in fusebox		
T1b	Connector, single, in engine compartment near carburettor		
T1c	Connector, single, in rear door		
T1d	Connector, single, in rear door		
T1e	Connector, single, behind instrument panel		
T4	Connector, 4 point, behind instrument panel		
T12/	Connector, 12 point, on instrument panel	84, 85	
V	Windscreen wiper motor	58	
V2	Fresh air blower	89	
V5	Windscreen washer pump	82	
W	Interior light, front	80	
W1	Interior light, rear/load compartment (delivery van only)	66, 67	
X	Number plate light (delivery van only)	66, 67	
Z1	Heated rear window	34	
1	Battery earthing strap		
3	Earthing strap from engine to body		
10	Earthing point, instrument panel insert		
11	Earthing point, behind instrument panel		
12	Earthing point, underneath instrument panel near fusebox		
13	Earthing point, on rear door		
14	Earthing point, on roof crossmember, passenger compartment		
15	Earthing point, at rear of longitudinal member		
17	Earthing point, on steering gear		
18	Earthing point, on longitudinal member, rear left		
19	Earthing point, on longitudinal member, rear right		

Refer to Figs. 10.32 and 10.33 for guidance

Colour code

sw	–	Black	gr	–	Grey
ws	–	White	ge	–	Yellow
be	–	Blue	pi	–	Pink
ro	–	Red	br	–	Brown
gn	–	Green	li	–	Lilac

Key to Fig. 13.78

Designation		in current track
A	Battery	4
B	Starter	5, 6
C	Alternator	2, 3
C1	Voltage regulator	2, 3
D	Ignition/starter switch	8 to 10
E1	Lighting switch	60 to 63
E2	Turn signal switch	44
E3	Emergency light switch	40 to 47
E4	Headlight dimmer/flasher switch	69, 70
E9	Fresh air blower switch	56, 57
E15	Heated rear window switch	37 to 39
E20	Instrument/instrument panel lighting control	64
E22	Intermittent wiper switch	84 to 87
F	Brake light switch	53, 54
F1	Oil pressure switch	21
F2	Door contact switch, front left	79
F4	Reversing light switch	27
F7	Door contact switch rear sliding door	76
F9	Handbrake warning system switch	20
F11	Rear door contact switch	77
F34	Brake fluid level warning contact	17
F35	Thermoswitch for manifold preheating	31
F66	Coolant shortage indicator switch	23
G	Fuel gauge sender	25
G1	Fuel gauge	11
G2	Coolant temperature gauge sender	24
G3	Coolant temperature gauge	12
G6	Electric fuel pump	26
H	Horn control	35
H1	Horn	34
J2	Emergency light relay	42, 43
J6	Voltage stabiliser	11
J31	Intermittent wash/wipe relay	82, 86
J59	Relief relay (for X contact)	58, 59
J81	Intake manifold preheating relay	30, 31
J120	Switch unit for coolant low level indicator	22, 23
K1	Main beam warning lamp	75
K2	Generator warning lamp	15
K3	Oil pressure warning lamp	17
K5	Turn signal warning lamp	16
K6	Emergency light system warning lamp	46
K7	Dual circuit brake and handbrake warning lamp	19
K10	Heated rear window warning lamp	39
K28	Coolant temperature warning lamp (red)	14
L1	Twin filament headlight bulb, left	71, 73
L2	Twin filament headlight bulb, right	72, 74
L9	Lighting switch light bulb	58
L10	Instrument panel insert light bulb	61, 62
L16	Fresh air controls light bulb	47
L39	Heated rear window switch bulb	38
M1	Sidelight bulb, left	66
M2	Tail light bulb, right	67
M3	Sidelight bulb, right	68
M4	Tail light bulb, left	65

Designation		in current track
M5	Front left turn signal bulb	49
M6	Rear left turn signal bulb	50
M7	Front right turn signal bulb	51
M8	Rear right turn signal bulb	52
M9	Brake light bulb, left	55
M10	Brake light bulb, right	54
M16	Reversing light bulb, left	27
M17	Reversing light bulb, right	29
N	Ignition coil	7
N1	Automatic choke, left	32
N3	Bypass air cut-off valve	33
N6	Resistance wire	7
N23	Series resistor for fresh air blower	56
N51	Heater element for intake manifold preheating	30
O	Ignition distributor	7 to 9
P	Plug connector	7 to 9
Q	Spark plugs	7 to 9
S1 to S15	Fuses in fusebox	
T1	Connector, single, in engine compartment	
T1a	Connector, single, behind dash	
T1c	Connector, single behind dash	
T2	Connector, two point, behind dash	
T2a	Connector, two point, behind dash	
T2b	Connector, two point, behind dash	
T2c	Connector, two point, behind dash	
T4	Connector, four point, behind dash	
T14/	Connector, fourteen point on dash insert	
V	Windscreen wiper motor	81, 82
V2	Fresh air blower	57
V5	Windscreen washer pump	88
W	Interior light, front	80
W1	Interior light, rear	78
X	Number plate light	64
Z1	Heated rear window	36

1	Earth strap from battery to body
3	Earth strap from engine to body
10	Earth point, dash insert
11	Earth point, behind dash
12	Earth point behind dash near fusebox
13	Earth point on roof cross member left
14	Earth point on roof cross member passenger compartment
15	Earth point, sidemember, rear
17	Earth wire, steering box
18	Earth point on sidemember, rear left
19	Earth point on sidemember, rear right
20	Earth point, in engine compartment

Refer to Figs. 10.32 and 10.33 for guidance

Colour code

sw	–	Black	gr	–	Grey
ws	–	White	ge	–	Yellow
be	–	Blue	pi	–	Pink
ro	–	Red	br	–	Brown
gn	–	Green	li	–	Lilac

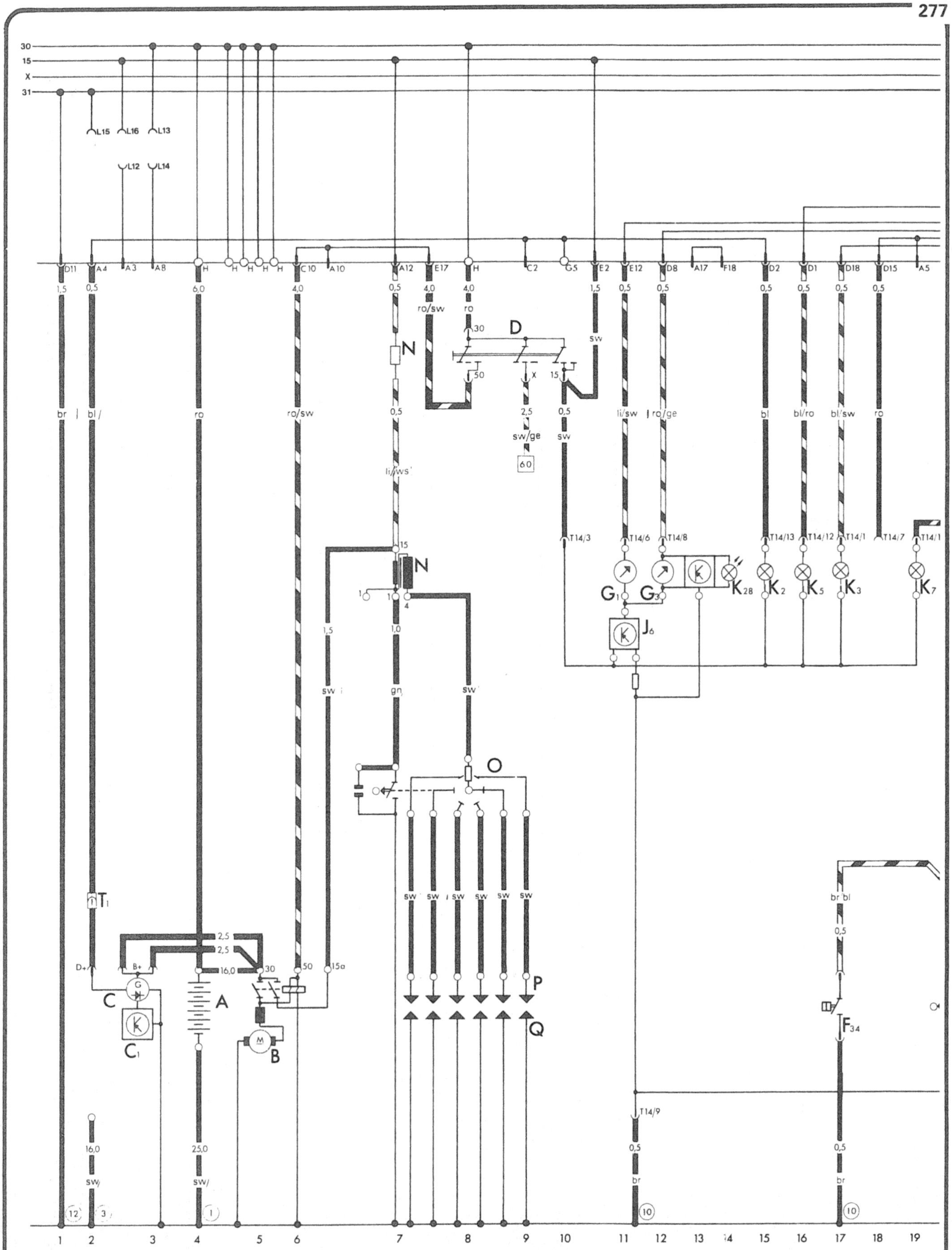

Fig. 13.78 Main current flow diagram – 6-cylinder models, 1983 to 1985

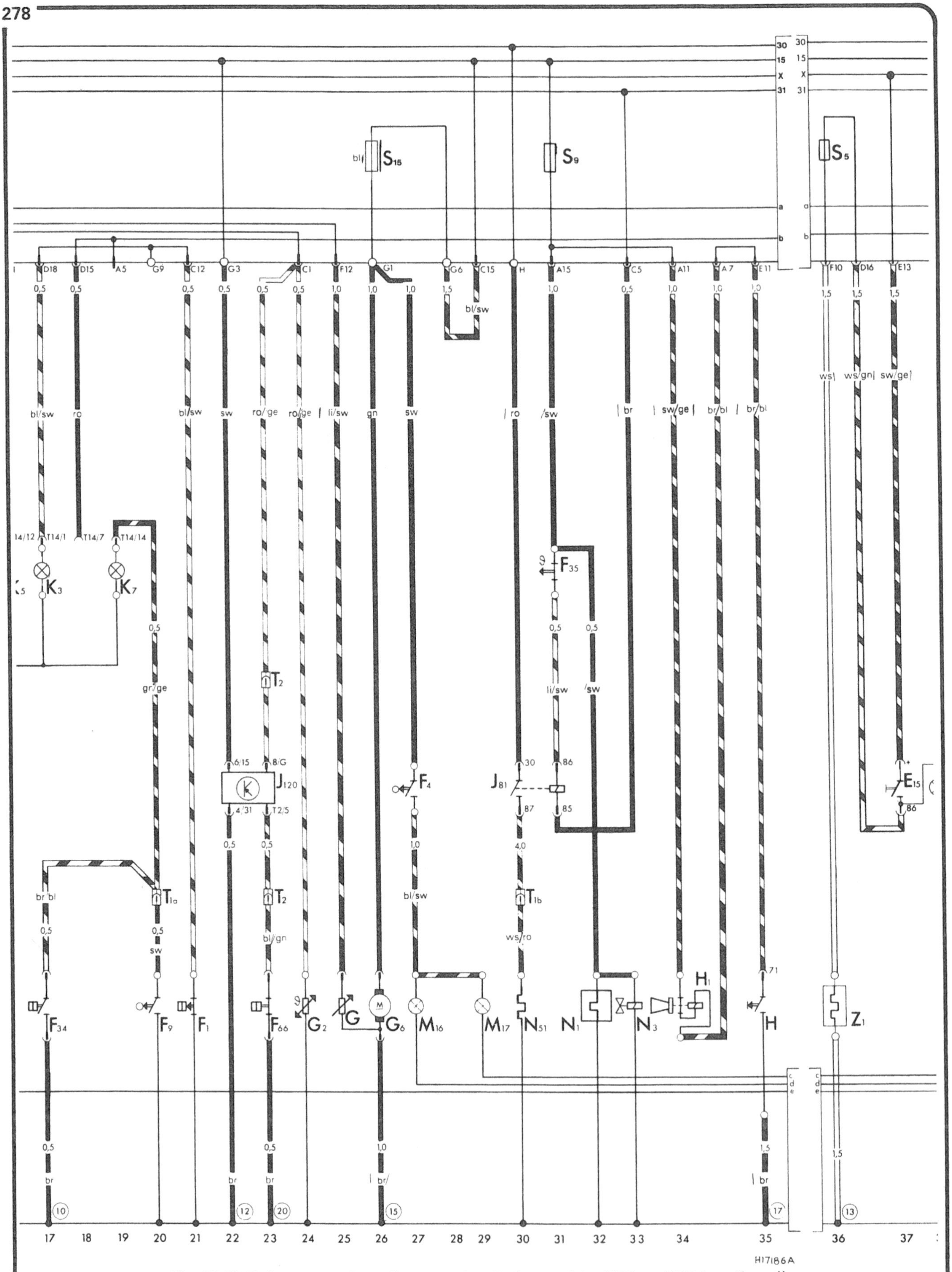

Fig. 13.78 Main current flow diagram – 6-cylinder models, 1983 to 1985 (continued)

H17186A

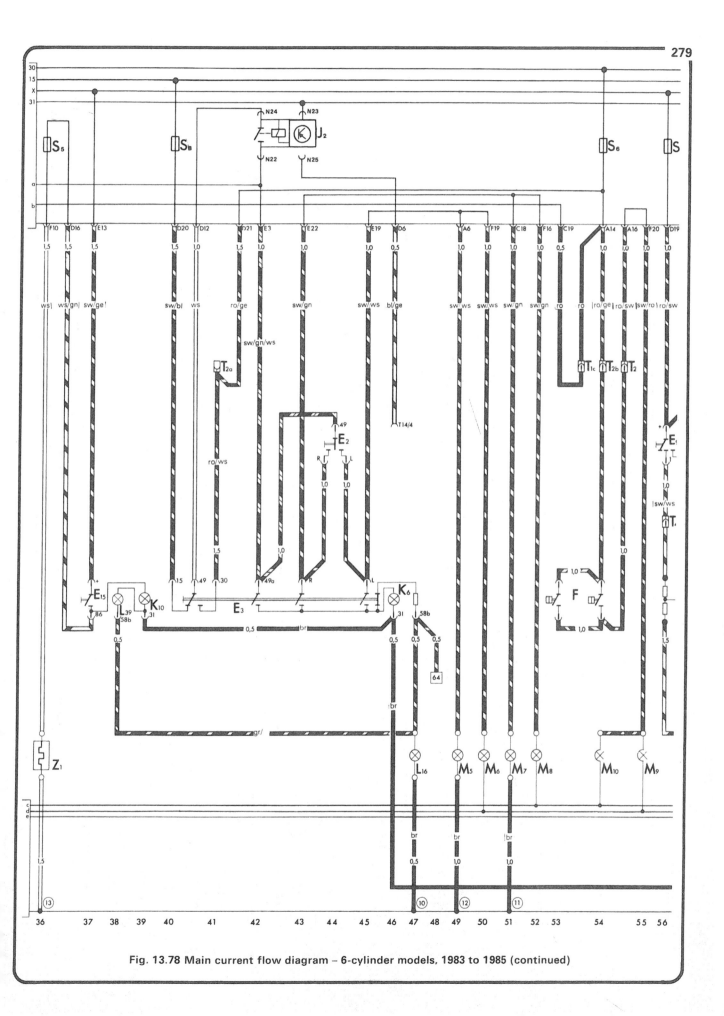

Fig. 13.78 Main current flow diagram – 6-cylinder models, 1983 to 1985 (continued)

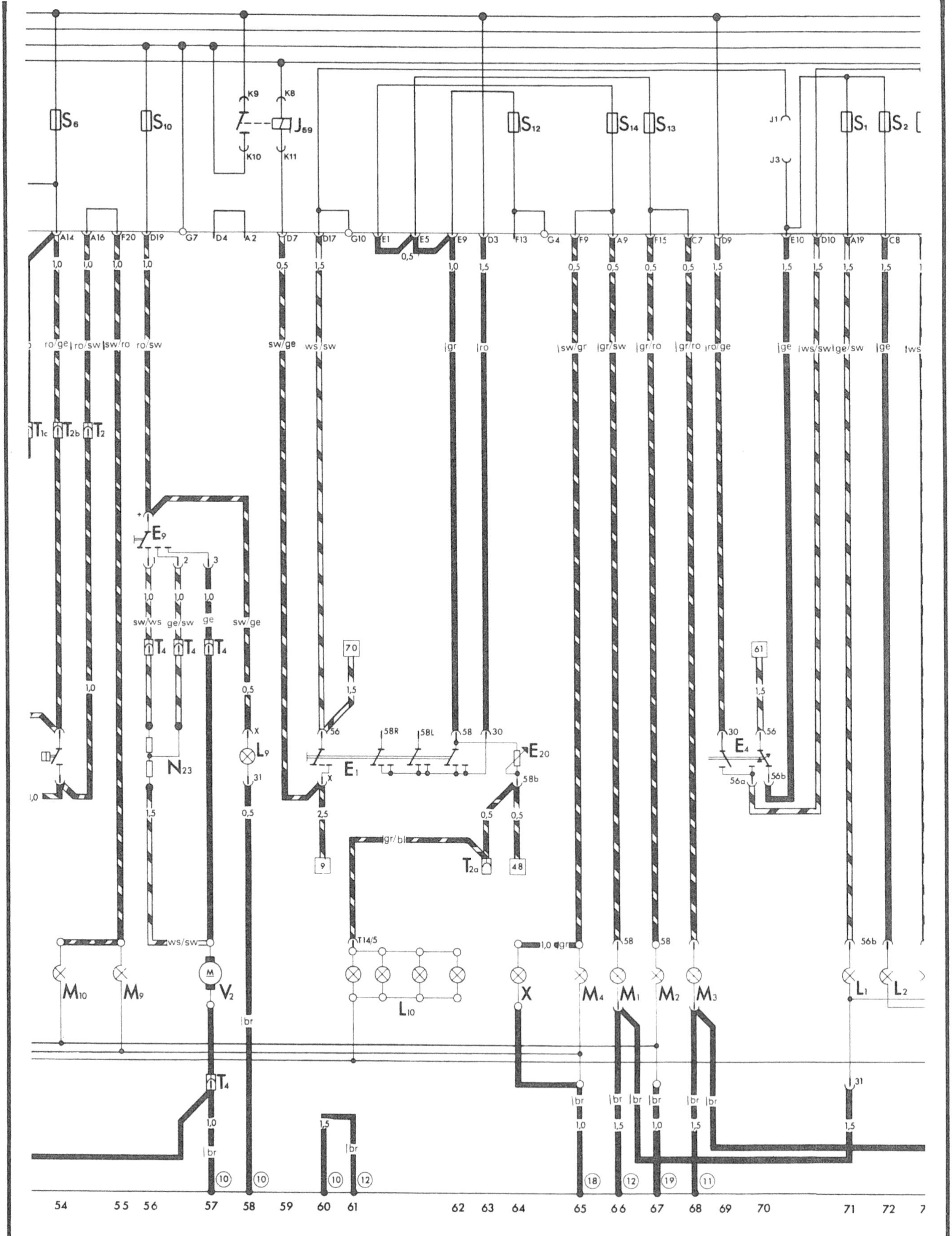

Fig. 13.78 Main current flow diagram – 6-cylinder models, 1983 to 1985 (continued)

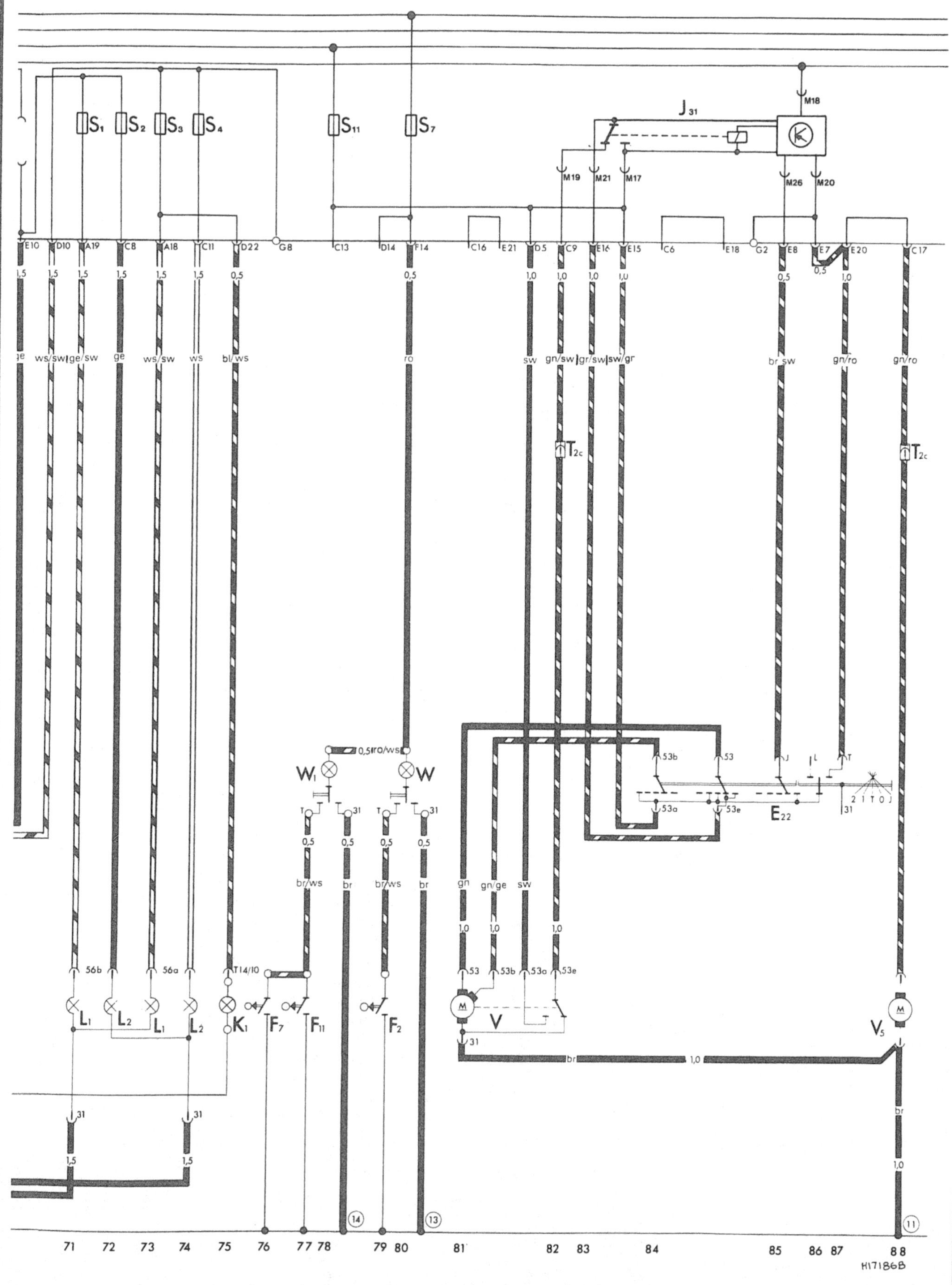

Fig. 13.78 Main current flow diagram – 6-cylinder models, 1983 to 1985 (continued)

H17186B

Key to Fig. 13.79

Designation		in current track
A	Battery	4
B	Starter	5 to 7
C	Alternator	2, 3
C1	Voltage regulator	2, 3
D	Ignition starter switch	10 to 14
E1	Lighting switch	76 to 80
E2	Turn signal switch	55, 56
E3	Emergency light switch	52 to 59
E4	Headlight dimmer/flasher switch	86, 87
E9	Fresh air blower switch	72, 73
E15	Heated rear window switch	46 to 48
E20	Instrument/instrument panel lighting control	80
E22	Intermittent wiper switch	100 to 102
F	Brake light switch	65, 66
F1	Oil pressure switch (1.8 bar)	42
F2	Front left door contact switch	95
F4	Reversing light switch	69
F7	Door contact switch, rear (sliding door)	92
F9	Handbrake warning system switch	40
F11	Rear right door contact switch	93
F22	Oil pressure switch (0.3 bar)	41
F26	Thermoswitch for automatic choke, front intake pipe	19
F34	Brake fluid level warning contact	39
F35	Thermoswitch for intake manifold preheating	16
F66	Coolant shortage indicator switch	23
F112	Thermoswitch for carburettor cooling blower	21
G	Fuel gauge sender	25
G1	Fuel gauge	28
G2	Coolant temperature gauge sender	24
G3	Coolant temperature gauge	29
G6	Electric fuel pump	26
H	Horn plate	44
H1	Horn	43
J2	Emergency light relay	53, 54
J6	Voltage stabiliser	28
J31	Intermittent wash/wipe relay	98 to 101
J59	Relief relay for X contact	74, 75
J81	Intake manifold preheating relay, on aux. relay adaptor	15,16
J114	Oil pressure monitor control unit	33, 34
J120	Switch unit for coolant shortage indicator on aux. relay adaptor	22, 23
J183	Relay for carburettor cooling blower (time relay)	20, 21
K1	Main beam warning lamp	37
K2	Generator warning lamp	31
K3	Oil pressure warning lamp	35
K5	Turn signal warning lamp	32
K6	Emergency light system warning lamp	58
K7	Dual circuit brake and handbrake warning lamp	39
K10	Heated rear window warning lamp	48
K28	Warning lamp for coolant temperature/coolant shortage	30
L1	Twin filament bulb, left	88, 90
L2	Twin filament bulb, right	89, 91
L9	Lighting switch light bulb	74
L10	Instrument panel insert light bulb	36 to 38
L16	Fresh air controls light bulb	70
L28	Cigarette lighter light bulb	50
L39	Heated rear window switch bulb	47
M1	Sidelight bulb, left	83
M2	Tail light bulb, right	84
M3	Sidelight bulb, right	85
M4	Tail light bulb, left	82
M5	Front left turn signal bulb	61
M6	Rear left turn signal bulb	62
M7	Front right turn signal bulb	63
M8	Rear right turn signal bulb	64

Designation		in current track
M9	Brake light bulb, left	67
M10	Brake light bulb, right	66
M16	Reversing light bulb, left	69
M17	Reversing light bulb, right	68
N	Ignition coil	8
N1	Automatic choke	18
N3	Bypass air cut-off valve	17
N6	Series resistance	8
N23	Series resistance for fresh air blower	72
N36	Series resistance for automatic choke	18
N51	Heater element for intake manifold preheating	15
O	Ignition distributor	9 to 13
P	Spark plug connector	9 to 13
Q	Spark plugs	9 to 13
R	Connection for radio	49, 51
S1 to		
S15	Fuses in fusebox	
S50	Fuse, terminal 58b, 10A, on aux. relay adaptor	80
S51	Fuse for carburettor cooling blower	20
T1	Connector single, in engine compartment	
T1a	Connector single, behind dash panel	
T1b	Connector single, behind relay plate	
T1c	Connector single, behind dash panel	
T1g	Connector single, near carburettor	
T1h	Connector single, near carburettor	
T1i	Connector single, behind dash panel	
T2	Connector 2 pin, behind relay plate	
T2a	Connector 2 pin, behind dash panel	
T2b	Connector 2 pin, behind dash panel	
T2c	Connector 2 pin, behind dash panel	
T4	Connector 4 pin, behind dash panel	
T4a	Connector 4 pin, behind dash panel	
T8/	Connector 8 pin, on dash panel insert	
T14/	Connector 14 pin, on dash panel insert★	
U1	Cigarette lighter	51
V	Windscreen wiper motor	97, 98
V2	Fresh air blower	73
V5	Windscreen washer pump	103
V62	Carburettor cooling blower, engine, left	20
W	Interior light, front	95, 96
W1	Interior light, rear/luggage compartment	93, 94
X	Number plate light	82
Y	Clock	36
Z1	Heated rear window	45

★The contacts in the Y14 connector are appropriate to the designations on the dash panel insert printed circuit and **not** the designations of the connector on the wiring loom.

1	Earthing strap, battery – body
3	Earthing strap, engine – body
15	Earth point on cylinder head
23	Earth wire above steering gear
30	Earth point next to relay plate
31	Earth point on dash panel insert
33	Earth point behind dash panel, right
65	Earth point on left hand longitudinal member, rear
68	Earth point on rear crossmember, left
69	Earth point on rear crossmember, right
71	Earth point on front roof rail
72	Earth point on roof rail
79	Earth point near interior light, rear

Refer to Figs. 10.32 and 10.33 for guidance

Not all items are fitted to all models

Colour code

sw	–	Black	gr	–	Grey
ws	–	White	ge	–	Yellow
be	–	Blue	pi	–	Pink
ro	–	Red	br	–	Brown
gn	–	Green	li	–	Lilac

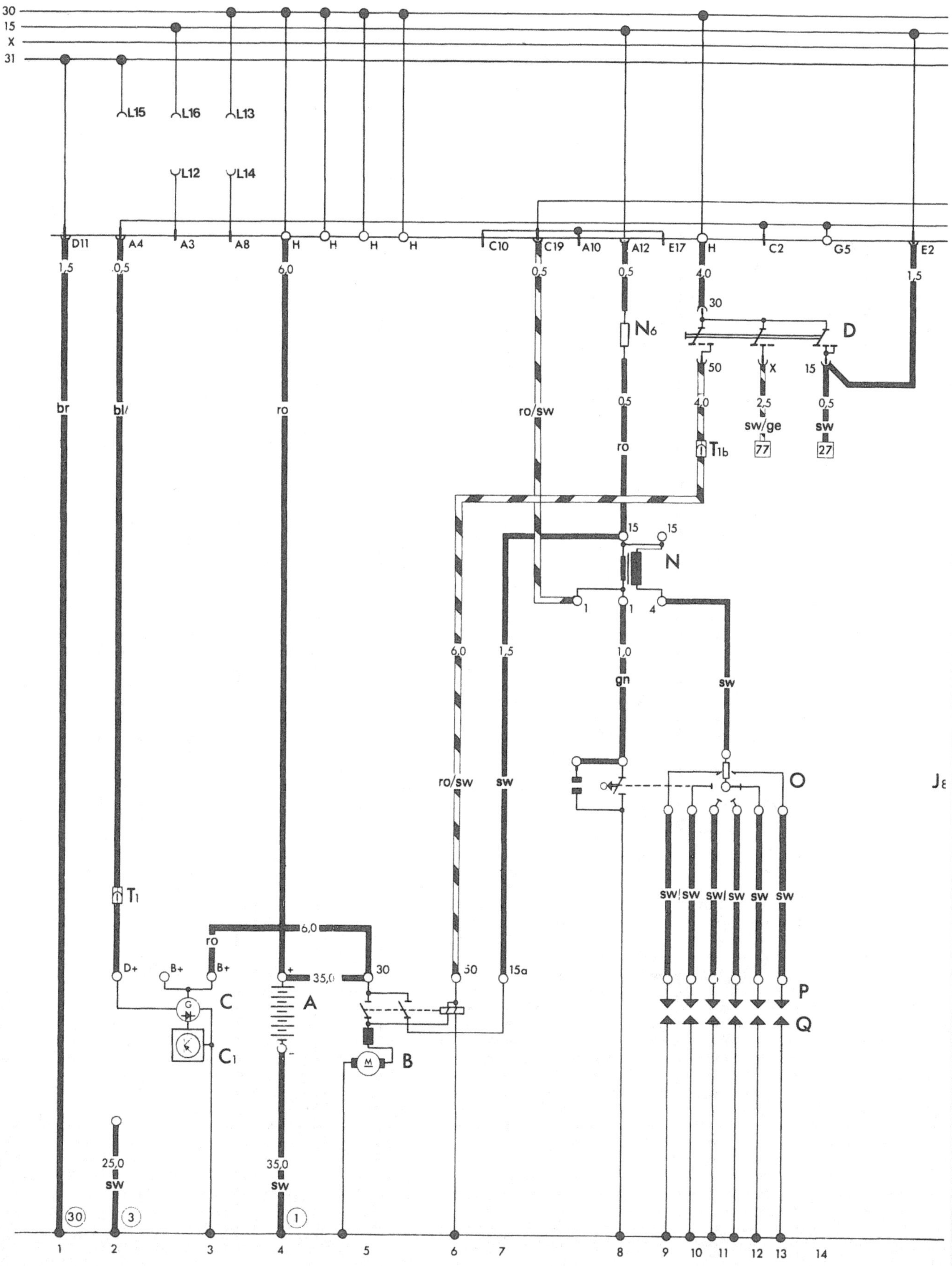

Fig. 13.79 Typical main current flow diagram – 6-cylinder models, 1986 and 1987

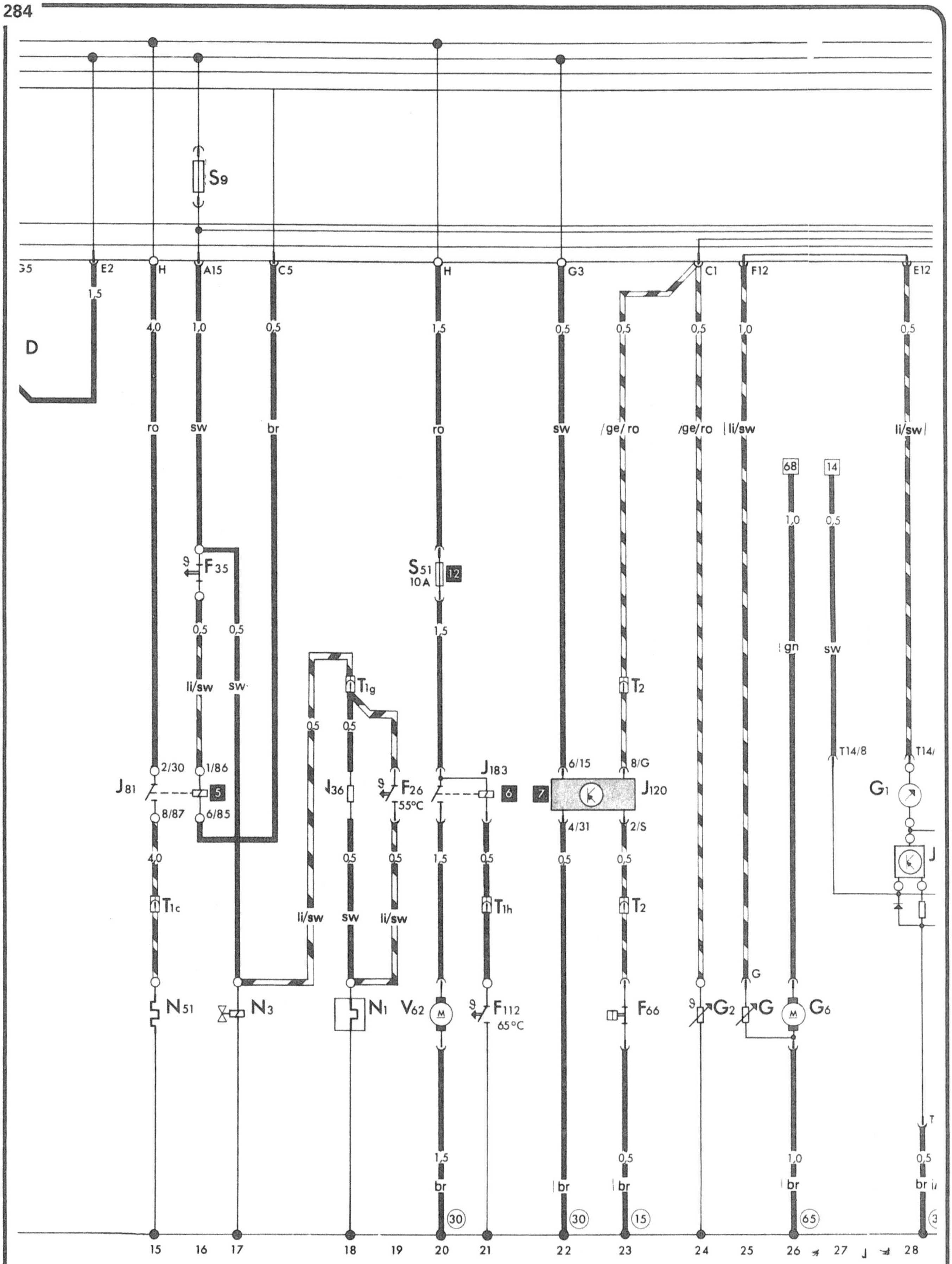

Fig. 13.79 Typical main current flow diagram – 6-cylinder models, 1986 and 1987 (continued)

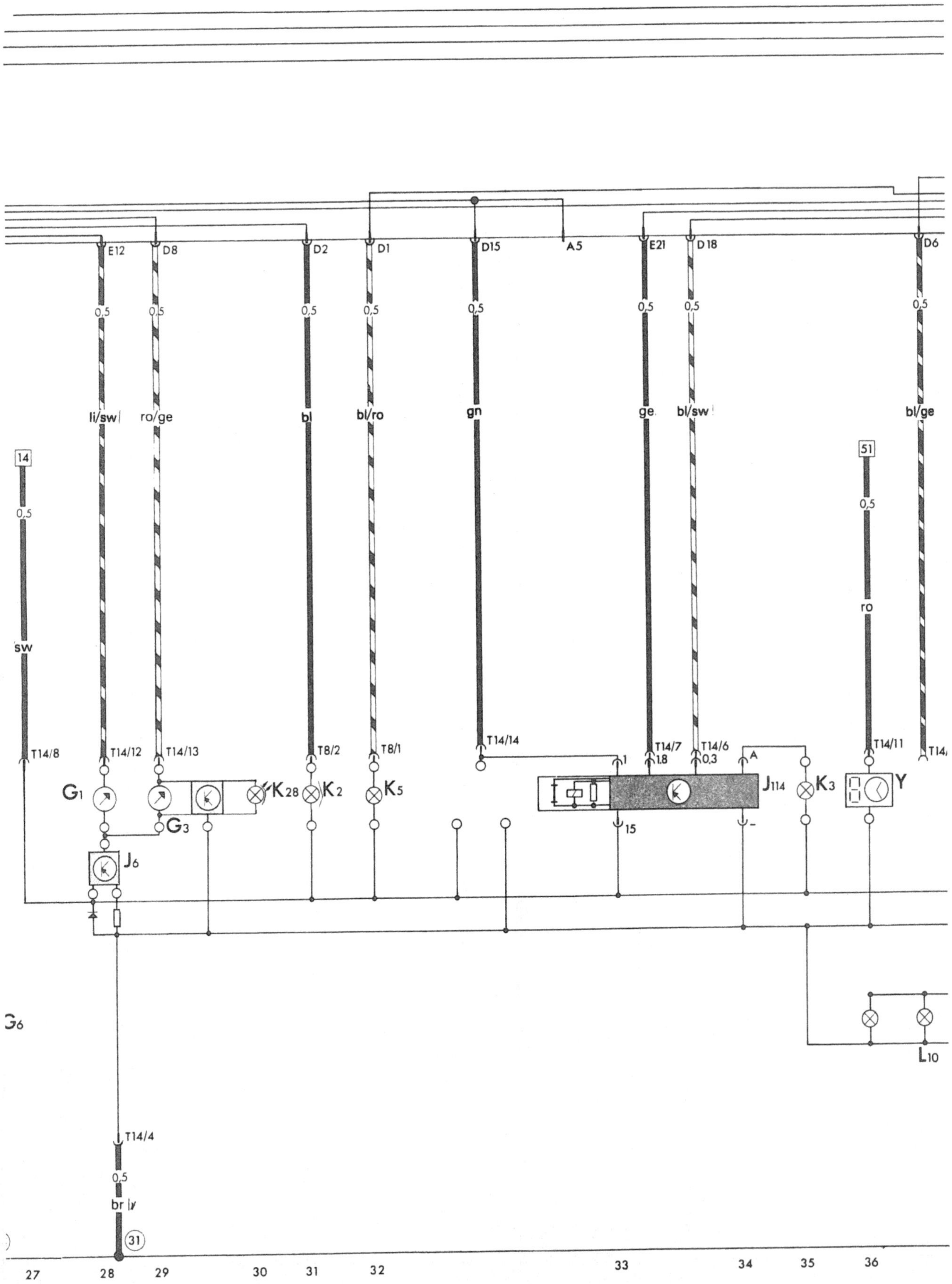

Fig. 13.79 Typical main current flow diagram – 6-cylinder models, 1986 and 1987 (continued)

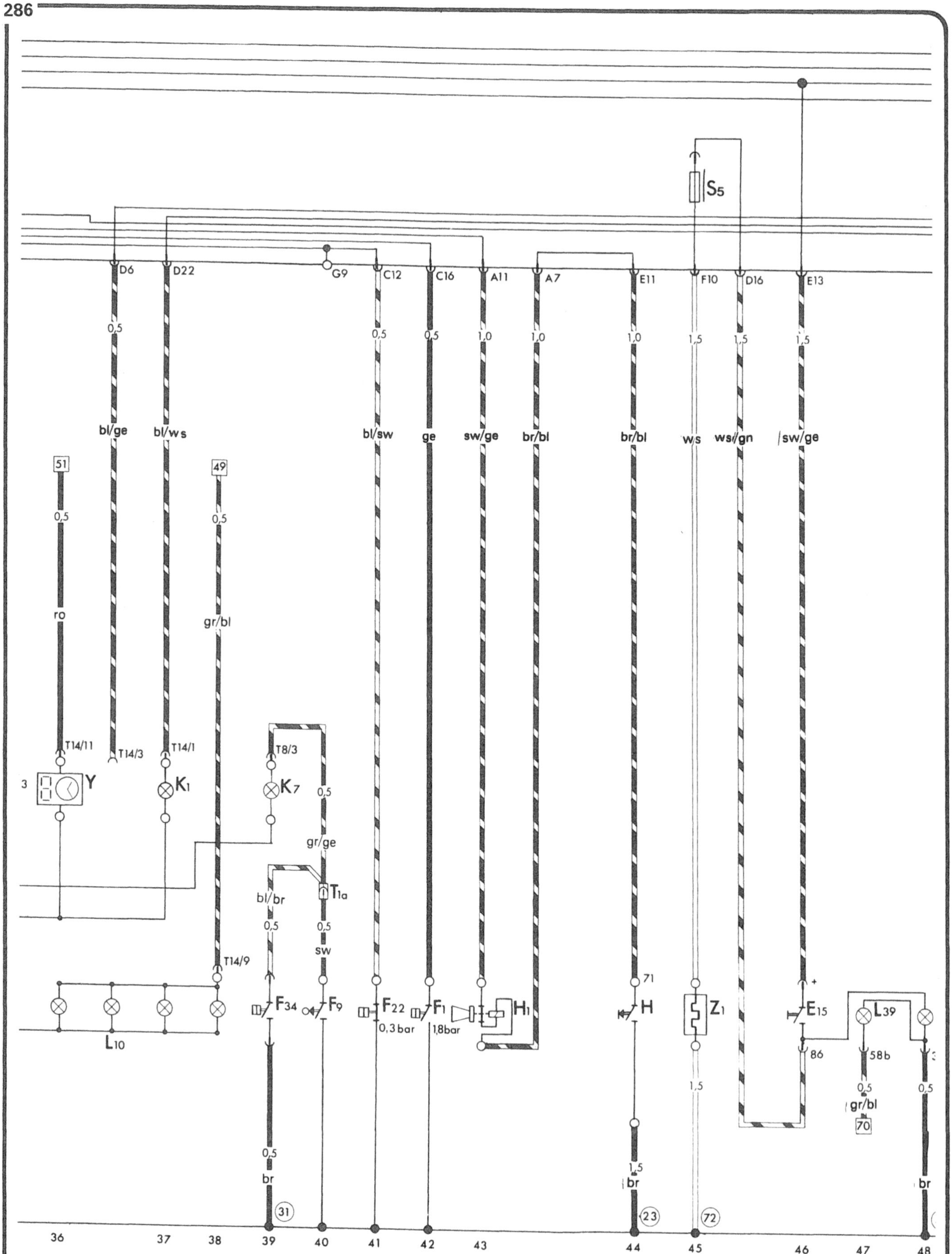

Fig. 13.79 Typical main current flow diagram – 6-cylinder models, 1986 and 1987 (continued)

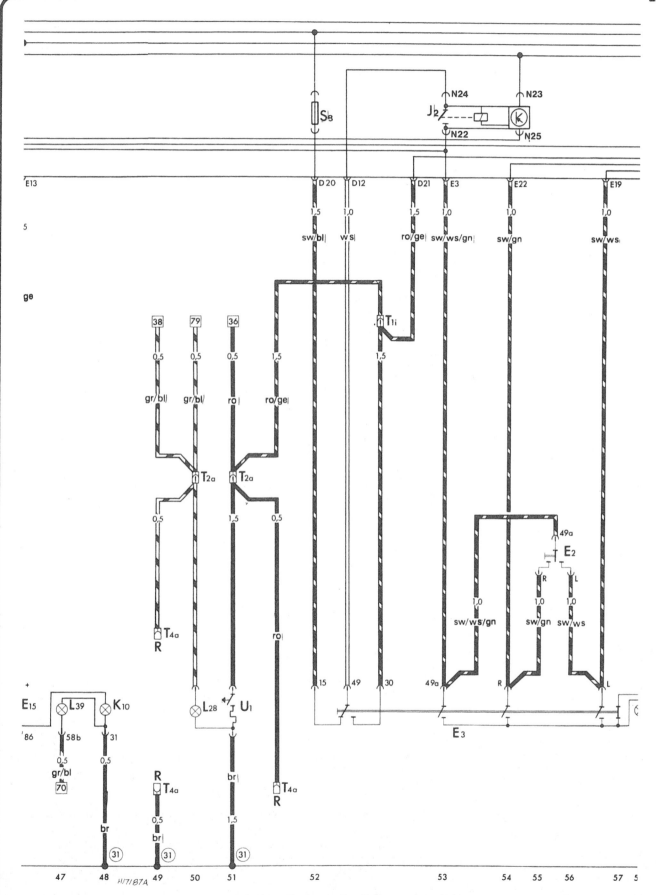

Fig. 13.79 Typical main current flow diagram – 6-cylinder models, 1986 and 1987 (continued)

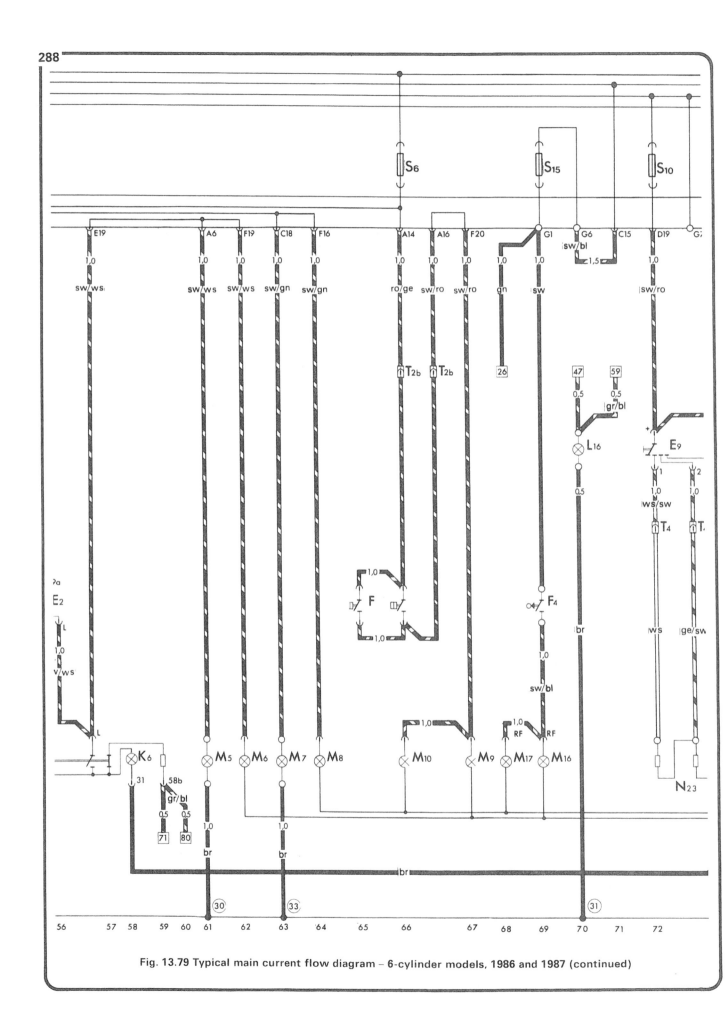

Fig. 13.79 Typical main current flow diagram – 6-cylinder models, 1986 and 1987 (continued)

Fig. 13.79 Typical main current flow diagram – 6-cylinder models, 1986 and 1987 (continued)

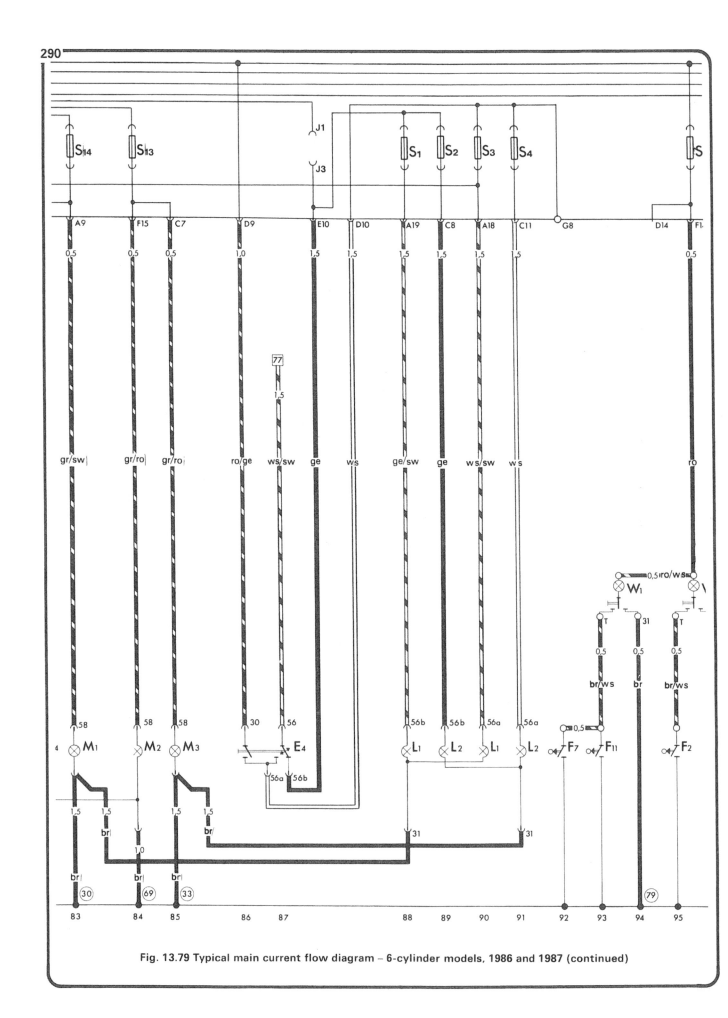

Fig. 13.79 Typical main current flow diagram – 6-cylinder models, 1986 and 1987 (continued)

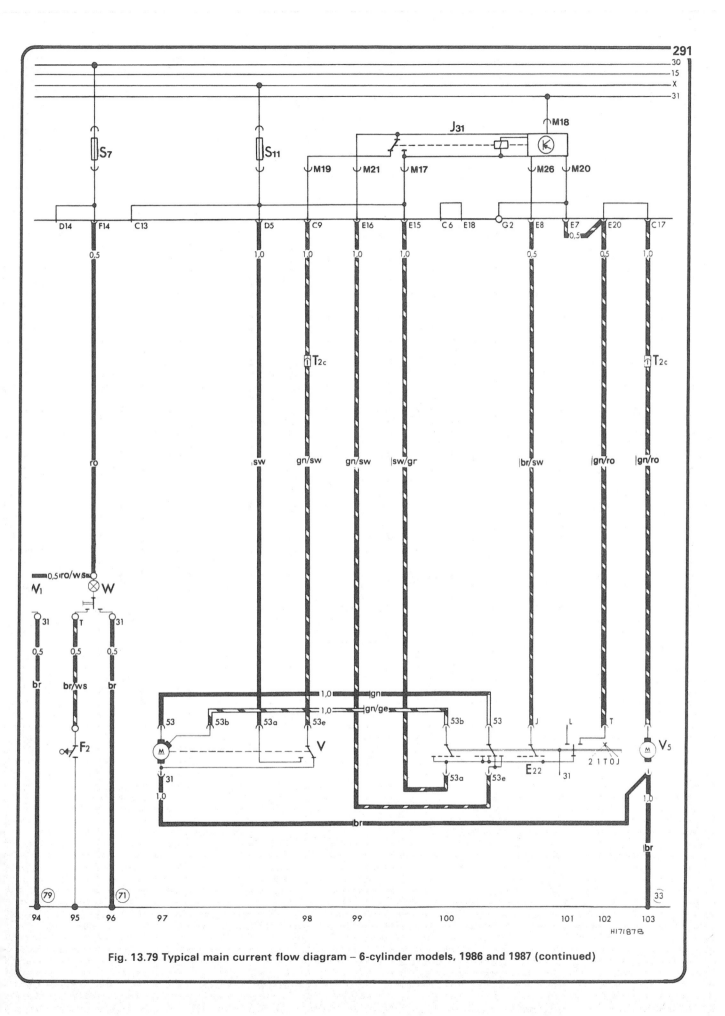

Fig. 13.79 Typical main current flow diagram – 6-cylinder models, 1986 and 1987 (continued)

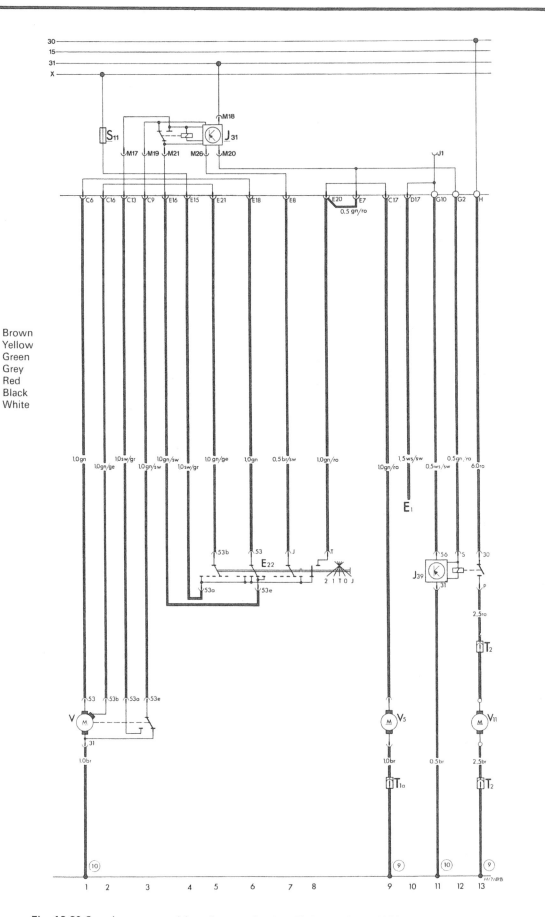

Colour code

br	–	Brown
ge	–	Yellow
gn	–	Green
gr	–	Grey
ro	–	Red
sw	–	Black
ws	–	White

Fig. 13.80 Supplementary wiring diagram for headlight washers, 1980 on

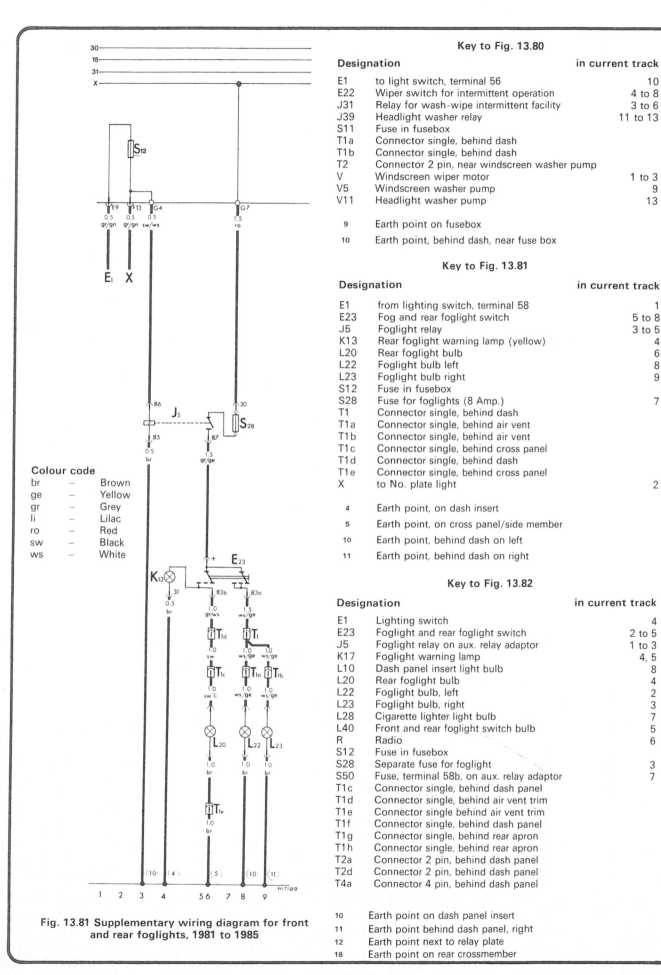

Key to Fig. 13.80

Designation		in current track
E1	to light switch, terminal 56	10
E22	Wiper switch for intermittent operation	4 to 8
J31	Relay for wash-wipe intermittent facility	3 to 6
J39	Headlight washer relay	11 to 13
S11	Fuse in fusebox	
T1a	Connector single, behind dash	
T1b	Connector single, behind dash	
T2	Connector 2 pin, near windscreen washer pump	
V	Windscreen wiper motor	1 to 3
V5	Windscreen washer pump	9
V11	Headlight washer pump	13
9	Earth point on fusebox	
10	Earth point, behind dash, near fuse box	

Key to Fig. 13.81

Designation		in current track
E1	from lighting switch, terminal 58	1
E23	Fog and rear foglight switch	5 to 8
J5	Foglight relay	3 to 5
K13	Rear foglight warning lamp (yellow)	4
L20	Rear foglight bulb	6
L22	Foglight bulb left	8
L23	Foglight bulb right	9
S12	Fuse in fusebox	
S28	Fuse for foglights (8 Amp.)	7
T1	Connector single, behind dash	
T1a	Connector single, behind air vent	
T1b	Connector single, behind air vent	
T1c	Connector single, behind cross panel	
T1d	Connector single, behind dash	
T1e	Connector single, behind cross panel	
X	to No. plate light	2
4	Earth point, on dash insert	
5	Earth point, on cross panel/side member	
10	Earth point, behind dash on left	
11	Earth point, behind dash on right	

Key to Fig. 13.82

Designation		in current track
E1	Lighting switch	4
E23	Foglight and rear foglight switch	2 to 5
J5	Foglight relay on aux. relay adaptor	1 to 3
K17	Foglight warning lamp	4, 5
L10	Dash panel insert light bulb	8
L20	Rear foglight bulb	4
L22	Foglight bulb, left	2
L23	Foglight bulb, right	3
L28	Cigarette lighter light bulb	7
L40	Front and rear foglight switch bulb	5
R	Radio	6
S12	Fuse in fusebox	
S28	Separate fuse for foglight	3
S50	Fuse, terminal 58b, on aux. relay adaptor	7
T1c	Connector single, behind dash panel	
T1d	Connector single, behind air vent trim	
T1e	Connector single behind air vent trim	
T1f	Connector single, behind dash panel	
T1g	Connector single, behind rear apron	
T1h	Connector single, behind rear apron	
T2a	Connector 2 pin, behind dash panel	
T2d	Connector 2 pin, behind dash panel	
T4a	Connector 4 pin, behind dash panel	
10	Earth point on dash panel insert	
11	Earth point behind dash panel, right	
12	Earth point next to relay plate	
18	Earth point on rear crossmember	

Colour code
br – Brown
ge – Yellow
gr – Grey
li – Lilac
ro – Red
sw – Black
ws – White

Fig. 13.81 Supplementary wiring diagram for front and rear foglights, 1981 to 1985

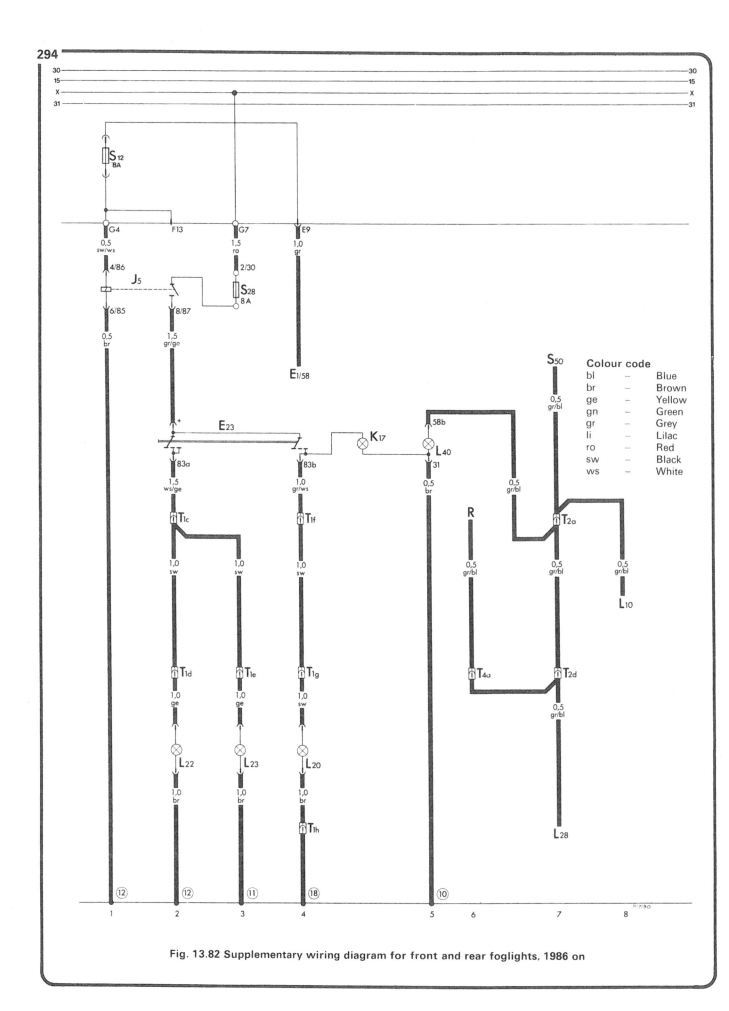

Fig. 13.82 Supplementary wiring diagram for front and rear foglights, 1986 on

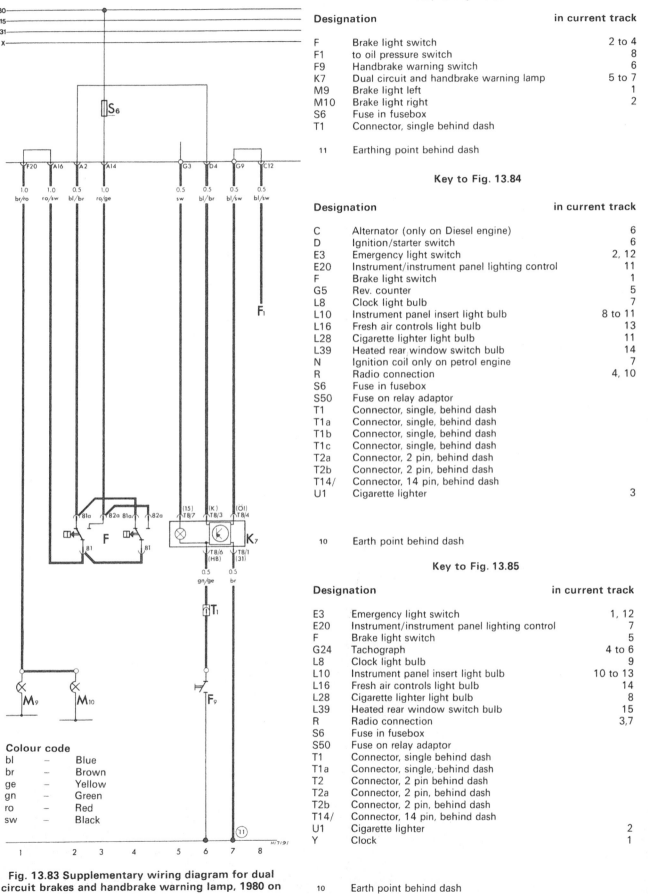

Fig. 13.83 Supplementary wiring diagram for dual circuit brakes and handbrake warning lamp, 1980 on

Colour code

bl	–	Blue
br	–	Brown
ge	–	Yellow
gn	–	Green
ro	–	Red
sw	–	Black

Key to Fig. 13.83

Designation		in current track
F	Brake light switch	2 to 4
F1	to oil pressure switch	8
F9	Handbrake warning switch	6
K7	Dual circuit and handbrake warning lamp	5 to 7
M9	Brake light left	1
M10	Brake light right	2
S6	Fuse in fusebox	
T1	Connector, single behind dash	
11	Earthing point behind dash	

Key to Fig. 13.84

Designation		in current track
C	Alternator (only on Diesel engine)	6
D	Ignition/starter switch	6
E3	Emergency light switch	2, 12
E20	Instrument/instrument panel lighting control	11
F	Brake light switch	1
G5	Rev. counter	5
L8	Clock light bulb	7
L10	Instrument panel insert light bulb	8 to 11
L16	Fresh air controls light bulb	13
L28	Cigarette lighter light bulb	11
L39	Heated rear window switch bulb	14
N	Ignition coil only on petrol engine	7
R	Radio connection	4, 10
S6	Fuse in fusebox	
S50	Fuse on relay adaptor	
T1	Connector, single, behind dash	
T1a	Connector, single, behind dash	
T1b	Connector, single, behind dash	
T1c	Connector, single, behind dash	
T2a	Connector, 2 pin, behind dash	
T2b	Connector, 2 pin, behind dash	
T14/	Connector, 14 pin, behind dash	
U1	Cigarette lighter	3
10	Earth point behind dash	

Key to Fig. 13.85

Designation		in current track
E3	Emergency light switch	1, 12
E20	Instrument/instrument panel lighting control	7
F	Brake light switch	5
G24	Tachograph	4 to 6
L8	Clock light bulb	9
L10	Instrument panel insert light bulb	10 to 13
L16	Fresh air controls light bulb	14
L28	Cigarette lighter light bulb	8
L39	Heated rear window switch bulb	15
R	Radio connection	3,7
S6	Fuse in fusebox	
S50	Fuse on relay adaptor	
T1	Connector, single behind dash	
T1a	Connector, single, behind dash	
T2	Connector, 2 pin behind dash	
T2a	Connector, 2 pin, behind dash	
T2b	Connector, 2 pin, behind dash	
T14/	Connector, 14 pin, behind dash	
U1	Cigarette lighter	2
Y	Clock	1
10	Earth point behind dash	

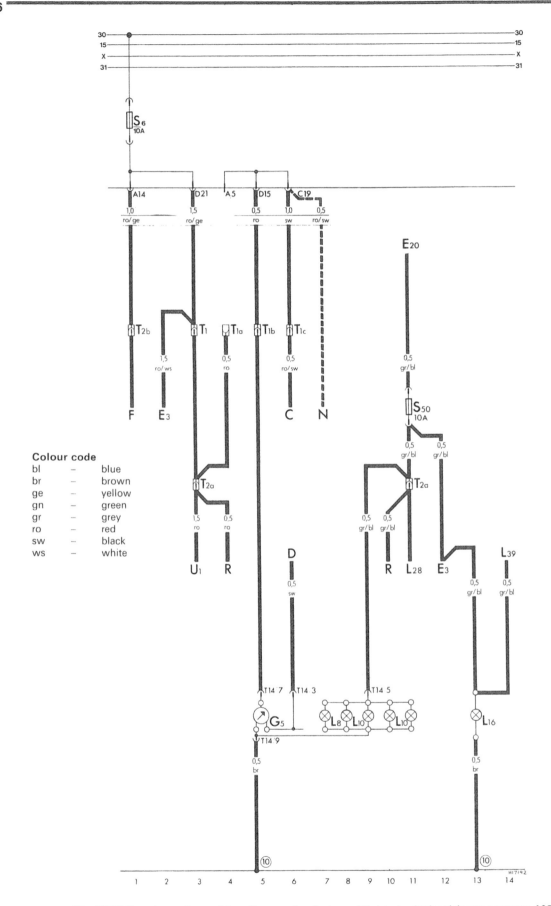

Fig. 13.84 Supplementary wiring diagram for fusing of lights in dash with rev counter, 1985 on

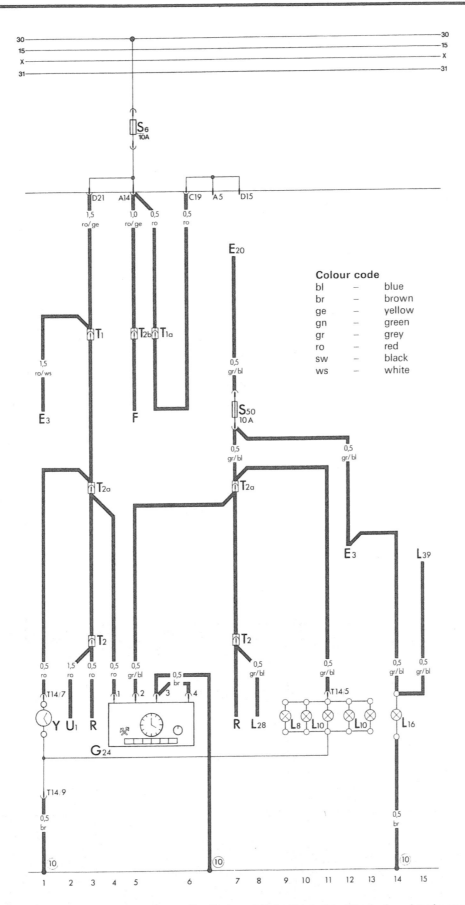

Fig. 13.85 Supplementary wiring diagram for fusing of lights in dash with clock and tachograph, 1985 on

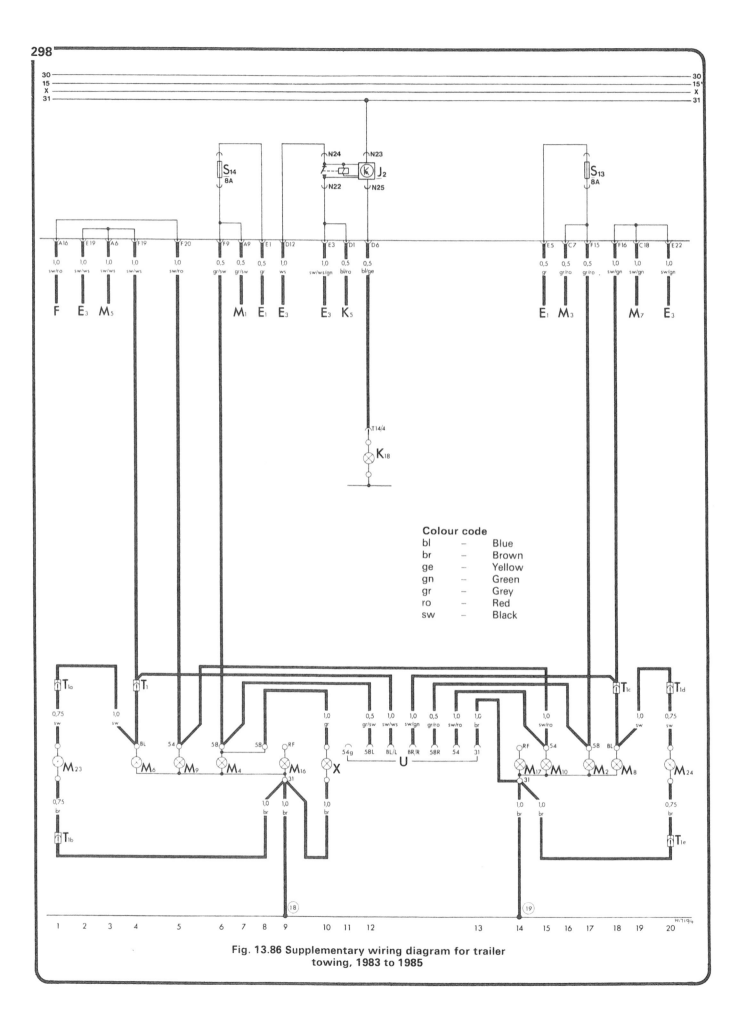

Fig. 13.86 Supplementary wiring diagram for trailer towing, 1983 to 1985

Key to Fig. 13.86

Designation		in current track
E1	Lighting switch	8,15
E3	Emergency light switch	2, 9, 10, 20
F	Brake light switch	1
J2	Emergency light relay	10 to 12
K5	Turn signal warning lamp	11
K18	Trailer operation warning lamp	12
M1	Sidelight bulb, left	7
M2	Tail light bulb, right	17
M3	Sidelight bulb, right	16
M4	Tail light bulb, left	6
M5	Front left turn signal bulb	3
M6	Rear left turn signal bulb	4
M7	Front right turn signal bulb	19
M8	Rear right turn signal bulb	18
M9	Brake light bulb, left	5
M10	Brake light bulb, right	15
M16	Reversing light bulb left	9
M17	Reversing light bulb right	14

Designation		in current track
M23	Bulb – flashing roof light left	1
M24	Bulb – flashing roof light right	20
S13	Fuses in fusebox	
S14	Fuses in fusebox	
T1	Connector, single, on rear cross bearer, left	
T1a	Connector, single, on rear roof frame, left	
T1b	Connector, single, on rear roof frame, left	
T1c	Connector, single, on rear cross bearer, right	
T1d	Connector, single, on rear roof frame, right	
T1e	Connector, single, on rear roof frame, right	
T14/	Connector, 14 pin, on dash panel insert	
U	Socket	11 to 13
X	Number plate light	10

18	Earthing point on rear left side member	
19	Earthing point on rear right side member	

14 Suspension and steering

General description (2.4 litre models)

1   The front suspension for the LT 28, 31, 35, 40 and 45 models remains essentially the same as described in Chapter 11, apart from the Specification changes given in this Supplement and the modifications given in this Section.

2   The LT 40, 45 and 50 front suspension is as shown in Fig. 13.87.
3   The main difference is in the mounting of the anti-roll bar which is on the front of the axle beam.
4   The rear suspension for the LT 28, 31 and 35 remains unaltered, again apart from the modifications given in this Section.
5   On LT 40, 45 and 50 models, the engine inclination angle was changed from 5° 30' to 27° 50' which meant changes to the rear suspension.

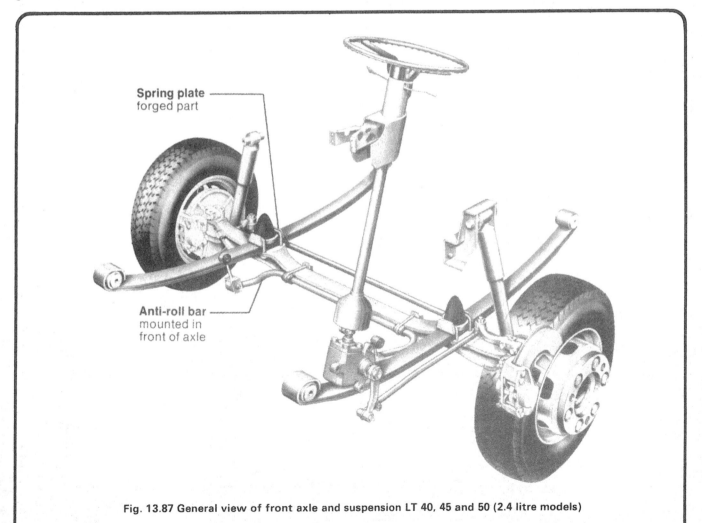

Spring plate
forged part

Anti-roll bar
mounted in
front of axle

Fig. 13.87 General view of front axle and suspension LT 40, 45 and 50 (2.4 litre models)

6   The propshaft and differential are out of the vehicle centre line, and the left-hand shock absorber is bolted to the differential housing. The rubber springs on the LT 40 and 45 are not fitted to the LT 50. No details of these modifications were available at the time of writing.

7   As with the suspension, the steering has remained basically unchanged, again except for the modifications given in this Section.

Front suspension king pin LT 40, 45 and 50

8   Beginning with production in June 1984, the spring pin which was used to lock the king pin in the axle beam is no longer used.

9   Axle beams produced after this date no longer have the drilling for the spring pin, nor does the king pin have the groove around its circumference (refer to Fig. 11.27).

10   When fitting a new style king pin to an old type axle beam, the spring pin hole should be filled with a sealing compound to prevent corrosion of the king pin.

Front hub components (LT 40, 45 and 50) – modification

11   From January 1981 the following modifications to the front hub and bearings came into effect.

12   The servicing operations in Chapter 11 are still applicable.

13   Larger taper bearings are used.

14   Bearing seats have been modified.

15   The tapered hole for the tie-rod in the steering knuckle has been modified to accept the new tie-rod (see later sub-section on tie-rod modification).

16   The oil seal, splash plate and hub cap have been matched to the new dimensions.

Front wheel bearing oil seals – modification

17   From approximately July 1980 a new oil seal has been fitted to the front hub bearings.

18   The new seal has a dust lip which faces away from the sealing lip and improves sealing..

19   The seal can be fitted to earlier models, but on LT 28, 31 and 35 models the new seal should not be fitted to vehicles which do not have the exchangeable ring (Fig. 13.88).

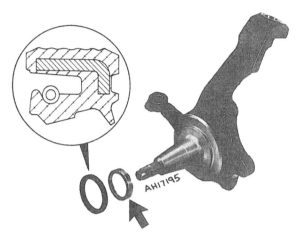

Fig. 13.88 Front wheel bearing oil seal with exchangeable ring (arrowed) (Sec 14)

20   On fitting these new seals, the space between the dust lip and the sealing lip should be packed with multi-purpose grease.

Idler arm bushes and seal (LT 28, 31 and 35) – modification

21   From February 1978 the seals in the bell crank lever shaft and bracket have been modified as shown in Fig. 13.89.

22   The bell crank lever and idler arm have also been modified to accommodate the new seals.

23   The fitting and removal instructions in Chapter 10, Section 24 remain unaltered, but components of different designs cannot be mixed.

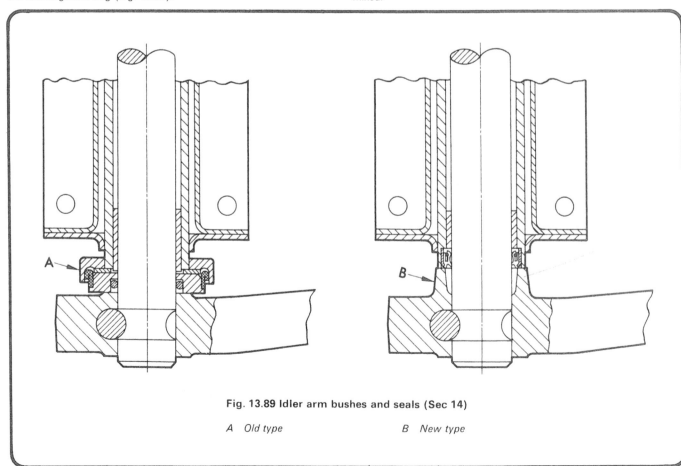

Fig. 13.89 Idler arm bushes and seals (Sec 14)

A   Old type   B   New type

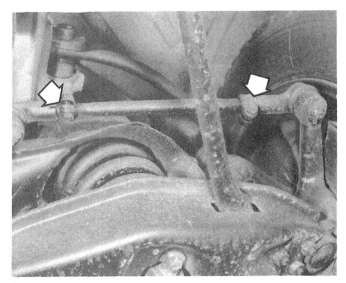

14.24 Tie-rod with locking clamps (arrowed)

Left-hand tie-rod (LT 28, 31 and 35) – modification
24 Models produced after September 1982 have a tie-rod with locking clamps in place of the lock nuts on earlier models (photo).

Tie-rods and drag links (LT 40, 45 and 50) – modification
25 Beginning in January 1981, modified tie-rods and drag links are fitted.
26 The locking nuts are replaced by clamps and the profile of both the drag link and tie-rod tube is different.
27 The balljoint pins are also of a different size and the threads are larger.
28 In connection with this, the connecting hole in the drop arm and steering knuckle are also larger.
29 New parts cannot be fitted to older vehicles.

Power-assisted steering – general description
30 The power-assisted steering gear is basically the same as the manual steering system, but has been adapted to receive power assistance from a hydraulic pump which is belt driven from the crankshaft pulley.
31 A hydraulic fluid reservoir is mounted behind the left-hand headlamp trim panel.
32 The components of the system are shown in Fig. 13.90. No repairs are possible to the individual components, and if they become defective, they should be renewed with a serviceable component, as described in the following paragraphs.
33 The following modifications have been made to the manual system to allow the power-assisted steering system to be fittted.
34 The holes in the chassis number through which the steering gear securing bolts pass have been enlarged.
35 The connecting disc between the steering box and steering column has been replaced by a universal joint.
36 The steering column has been shortened.
37 Should the power-assisted system fail at any time, manual steering is still available although this will necessarily be heavier. The power system should be repaired at the earliest opportunity.

Power-assisted steering fluid filter – removal and refitting
38 The filter in the hydraulic fluid reservoir should be renewed whenever the hydraulic fluid is changed, or any components renewed.
39 Remove the reservoir cap using a strap wrench; being careful not to lose the spring under the cover.
40 Lift out the washer, spring and filter cover, and then the filter.
41 Fit a new filter and reassemble the reservoir in the reverse order of removal.
42 Fill and bleed the system on completion.

Power-assisted steering system – filling and bleeding
43 Remove the trim panel from the left-hand headlamp.
44 With the engine running and the front wheels in the straight-ahead position, the fluid level should be at or near the 'MAX' mark on the side of the reservoir.
45 If the level is down to the 'MIN' mark, fill the system with the recommended fluid.
46 Switch off the engine, and raise the front wheels of the vehicle from the ground.
47 Turn the steering rapidly from lock to lock several times to allow air to escape.
48 Refill the reservoir to the 'MAX' mark. Start the engine, and watch the fluid level in the reservoir..
49 If it drops, refill immediately, keeping the level at 'MAX' until no more air bubbles are present as the steering is operated.
50 Stop the engine and remove the vehicle frrom jacks.
51 Refit the reservoir cap and headlamp trim panel.

Power-assisted steering system – leak test
52 Raise the vehicle onto jacks and insert a 30 mm (1.18 in) spacer bar between the steering knuckle and stop and the axle beam.
53 Start the engine and allow it to idle.
54 Hold the steering on full right lock; pinching the spacer bar. Do not hold the steering in this position for longer than 5 seconds at a time.
55 Inspect all parts of the system for leaks while it is being held under pressure.
56 On completion, remove the spacer and lower the vehicle from jacks.
57 More complex pressure and leak testing can be carried out by your VW dealer using specialist equipment.

Power-assisted steering pump – tensioning drivebelt
58 Slacken the pump pivot bolt, and then screw the adjuster bolt in or out so that, using moderate thumb pressure applied midway along the belt's longest length, it can be deflected by 10 mm (0.39 in).

Power-assisted steering pump – removal and refitting
59 The pump is bolted to a bracket which in turn is bolted to the chassis.
60 Remove the alternator drivebelt..
61 Remove the pump drivebelt by slackening the pivot bolt and loosening the belt adjuster.
62 Remove the inlet and outlet hydraulic hoses.
63 Remove the adjuster strut and pivot bolt and remove the pump.
64 Refit in the reverse order, using new copper washers on the hose connections, tighten all bolts to their specified torque, after adjusting the drivebelt.
65 On completion, fill and bleed the system, check for leaks and road test.

Power-assisted steering gear – removal and refitting
66 Raise the front of the vehicle onto jacks.
67 Remove the pinch-bolt on the universal joint between the steering gear and the steering column.
68 Detach the hydraulic fluid supply and return lines on the steering gear.
69 Remove the drag link from the drop arm.
70 On LT 28, 31 or 35 models, undo the pinch-bolt, and on LT 40, 45 and 50 models, bend down tthe locking tab and remove the nut before covering the drop arm off the steering pinion.
71 Remove the bolts securing the steering gear to the chassis member and remove the steering gear.
72 Refitting is a reversal of removal with the following points:
73 When installing the steering gear, the steering wheel must be pressed down to avoid damage to the column mounting.
74 Before fitting the universal joint between the steering gear and the steering column, set the steering wheel and the steering gear in the straight-ahead position. (The steering gear is in the centre position when it is turned back approximately two times from either step).
75 The lug on the dust cap must be aligned with the ridge on the casting on the steering gear.
76 On completion, fill and bleed the system and remove the vehicle from jacks.

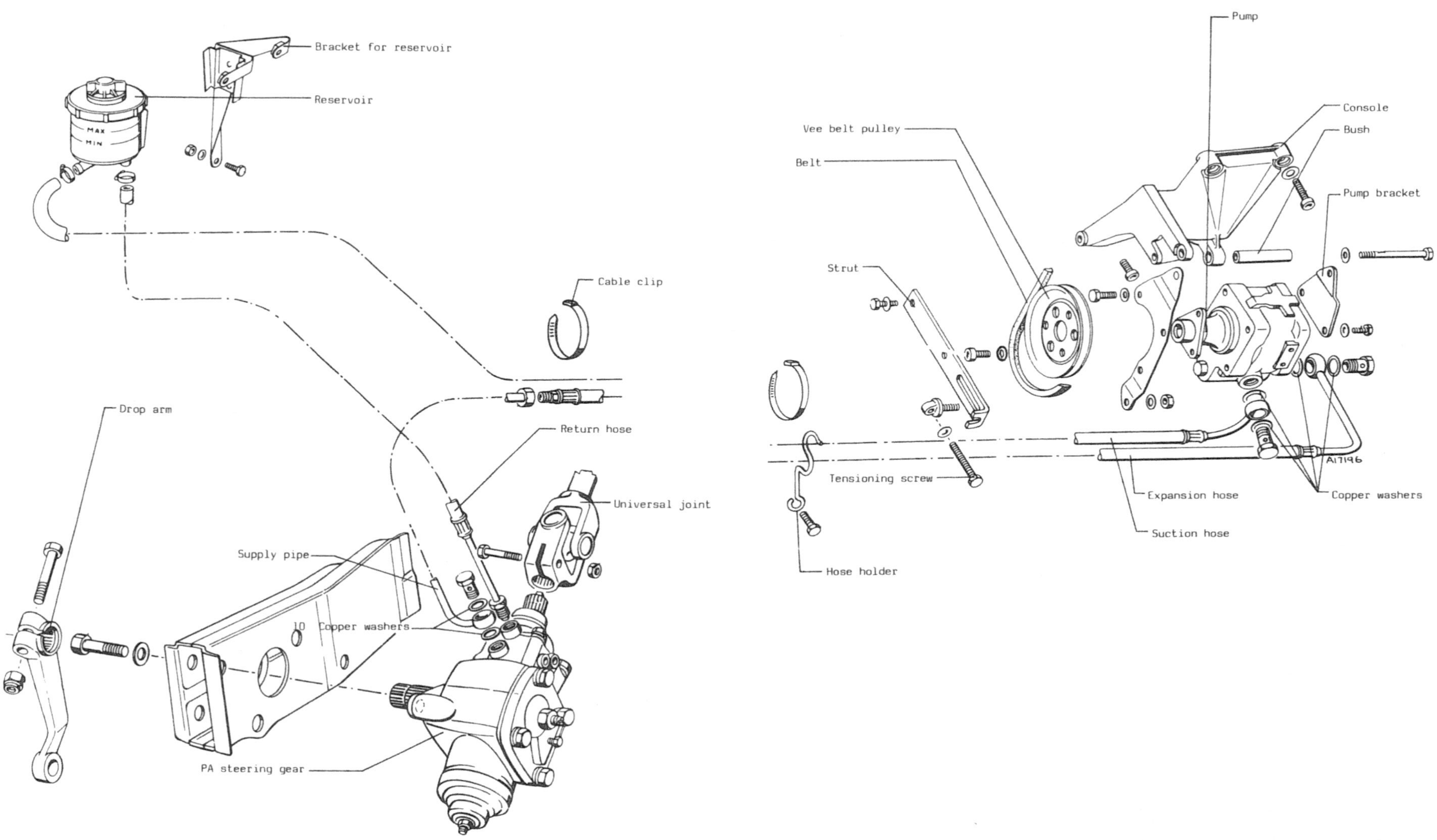
Bracket for reservoir
Reservoir
MAX
MIN
Cable clip
Drop arm
Return hose
Universal joint
Supply pipe
10
Copper washers
PA steering gear
Pump
Console
Bush
Vee belt pulley
Belt
Pump bracket
Strut
Tensioning screw
A17196
Expansion hose
Copper washers
Suction hose
Hose holder

Fig. 13.90 Exploded view of the power-assisted steering gear (Sec 14)

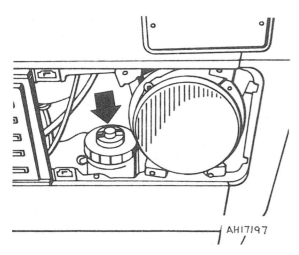

Fig. 13.91 Location of the power steering fluid reservoir –
arrowed (Sec 14)

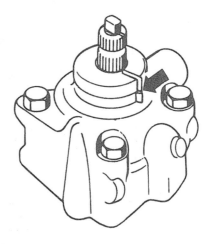

Fig. 13.93 Aligning the lug on the dust cap with the ridge
on the casting (Sec 14)

Wheel trims – 2.4 litre models
77 The plastic wheel trims are held on the wheel by four 'spring fingers' (photo).
78 To remove the trim, gently lever it off.
79 Refit by lining up the 'spring fingers' with the centre of the wheel, and then giving it a sharp tap home to engage the fingers in the wheel centre.

Wheels and tyres – general care and maintenance
Wheels and tyres should give no real problems in use provided that a close eye is kept on them with regard to excessive wear or damage. To this end, the following points should be noted.
Ensure that tyre pressures are checked regularly and maintained correctly. Checking should be carried out with the tyres cold and not immediately after the vehicle has been in use. If the pressures are checked with the tyres hot, an apparently high reading will be obtained owing to heat expansion. Under no circumstances should an attempt be made to reduce the pressures to the quoted cold reading in this instance, or effective underinflation will result.
Underinflation will cause overheating of the tyre owing to excessive flexing of the casing, and the tread will not sit correctly on the road surface. This will cause a consequent loss of adhesion and

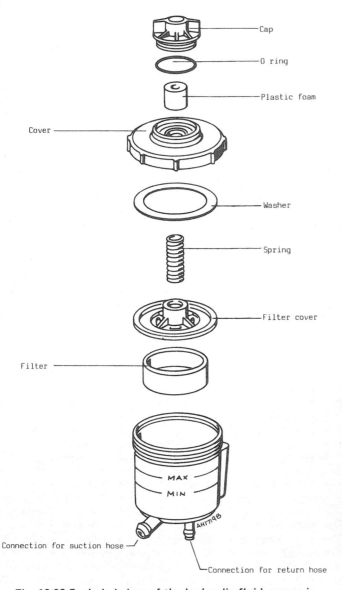

Cap
O ring
Plastic foam
Cover
Washer
Spring
Filter cover
Filter
MAX
MIN
Connection for suction hose
Connection for return hose

Fig. 13.92 Exploded view of the hydraulic fluid reservoir
(Sec 14)

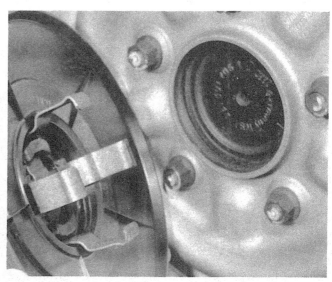

14.77 Wheel trim with 'spring fingers'

excessive wear, not to mention the danger of sudden tyre failure due to heat build-up.

Overinflation will cause rapid wear of the centre part of the tyre tread coupled with reduced adhesion, harsher ride, and the danger of shock damage occurring in the tyre casing.

Regularly check the tyres for damage in the form of cuts or bulges, especially in the sidewalls. Remove any nails or stones embedded in the tread before they penetrate the tyre to cause deflation. If removal of a nail *does* reveal that the tyre has been punctured, refit the nail so that its point of penetration is marked. Then immediately change the wheel and have the tyre repaired by a tyre dealer. Do *not* drive on a tyre in such a condition. In many cases a puncture can be simply repaired by the use of an inner tube of the correct size and type. If in any doubt as to the possible consequences of any damage found, consult your local tyre dealer for advice.

Periodically remove the wheels and clean any dirt or mud from the inside and outside surfaces. Examine the wheel rims for signs of rusting, corrosion or other damage. Light alloy wheels are easily damaged by 'kerbing' whilst parking, and similarly steel wheels may become dented or buckled. Renewal of the wheel is very often the only course of remedial action possible.

The balance of each wheel and tyre assembly should be maintained to avoid excessive wear, not only to the tyres but also to the steering and suspension components. Wheel imbalance is normally signified by vibration through the vehicle's bodyshell, although in many cases it is particularly noticeable through the steering wheel. Conversely, it should be noted that wear or damage in suspension or steering components may cause excessive tyre wear. Out-of-round or out-of-true tyres, damaged wheels and wheel bearing wear/maladjustment also fall into this category. Balancing will not usually cure vibration caused by such wear.

Wheel balancing may be carried out with the wheel either on or off the vehicle. If balanced on the vehicle, ensure that the wheel-to-hub relationship is marked in some way prior to subsequent wheel removal so that it may be refitted in its original position.

General tyre wear is influenced to a large degree by driving style – harsh braking and acceleration or fast cornering will all produce more rapid tyre wear. Interchanging of tyres may result in more even wear, but this should only be carried out where there is no mix of tyre types on the vehicle. However, it is worth bearing in mind that if this is completely effective, the added expense of replacing a complete set of tyres simultaneously is incurred, which may prove financially restrictive for many owners.

Front tyres may wear unevenly as a result of wheel misalignment. The front wheels should always be correctly aligned according to the settings specified by the vehicle manufacturer.

Legal restrictions apply to the mixing of tyre types on a vehicle. Basically this means that a vehicle must not have tyres of differing construction on the same axle. Although it is not recommended to mix tyre types between front axle and rear axle, the only legally permissible combination is crossply at the front and radial at the rear. When mixing radial ply tyres, textile braced radials must always go on the front axle, with steel braced radials at the rear. An obvious disadvantage of such mixing is the necessity to carry two spare tyres to avoid contravening the law in the event of a puncture.

In the UK, the Motor Vehicles Construction and Use Regulations apply to many aspects of tyre fitting and usage. It is suggested that a copy of these regulations is obtained from your local police if in doubt as to the current legal requirements with regard to tyre condition, minimum tread depth, etc.

15 Bodywork and fittings – 2.4 litre models

Seats – removal and refitting

1   The driver's seat is removed and fitted as described in Chapter 1, Section 4.

2   The passenger seat is integral with its support frame, which is bolted to the cab floor (photo).

3   Remove the four bolts and lift out the seat and support frame.

4   Refit in the reverse order.

Underbody splash panels

5   Protective panels are fitted to the underside of the engine and gearbox areas, to protect against road dirt.

6   These panels are bolted to the underframe (photos).

7   They are quite light and easy to handle, although because of their bulk, two people are needed when removing or fitting them.

Plastic components

8   With the use of more and more plastic body components by the vehicle manufacturers (eg bumpers, spoilers, and in some cases major body panels), rectification of damage to such items has become a matter of either entrusting repair work to a specialist in this field, or renewing complete components. Repair by the DIY owner is not really feasible owing to the cost of the equipment and materials required for effecting such repairs. The basic technique involves making a groove along the line of the crack in the plastic using a rotary burr in a power drill. The damaged part is then welded back together by using a hot air gun to heat up and fuse a plastic filler rod into the groove. Any excess plastic is then removed and the area rubbed down to a smooth finish. It is important that a filler rod of the correct plastic is used, as body components can be made of a variety of different types (eg polycarbonate, ABS, polypropylene).

9   If the owner is renewing a complete component himself, he will be left with the problem of finding a suitable paint for finishing which is compatible with the type of plastic used. At one time the use of a universal paint was not possible owing to the complex range of plastics encountered in body component applications. Standard paints, generally speaking, will not bond to plastic or rubber satisfactorily. However, it is now possible to obtain a plastic body parts finishing kit which consists of a pre-primer treatment, a primer and coloured top coat. Full instructions are normally supplied with a kit, but basically the method of use is to first apply the pre-primer to the component concerned and allow it to dry for up to 30 minutes. Then the primer is applied and left to dry for about an hour before finally applying the special coloured top coat. The result is a correctly coloured component where the paint will flex with the plastic or rubber, a property that standard paint does not normally possess.

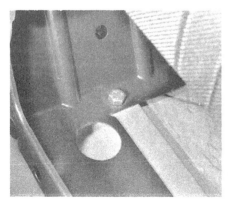

15.2 Passenger seat support frame bolted to the cab floor

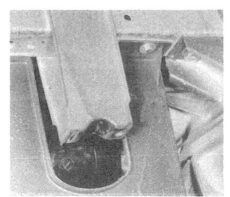

15.6A The splash panels ...

15.6B ... are bolted to the frame

General repair procedures

Whenever servicing, repair or overhaul work is carried out on the car or its components, it is necessary to observe the following procedures and instructions. This will assist in carrying out the operation efficiently and to a professional standard of workmanship.

Joint mating faces and gaskets

Where a gasket is used between the mating faces of two components, ensure that it is renewed on reassembly, and fit it dry unless otherwise stated in the repair procedure. Make sure that the mating faces are clean and dry with all traces of old gasket removed. When cleaning a joint face, use a tool which is not likely to score or damage the face, and remove any burrs or nicks with an oilstone or fine file.

Make sure that tapped holes are cleaned with a pipe cleaner, and keep them free of jointing compound if this is being used unless specifically instructed otherwise.

Ensure that all orifices, channels or pipes are clear and blow through them, preferably using compressed air.

Oil seals

Whenever an oil seal is removed from its working location, either individually or as part of an assembly, it should be renewed.

The very fine sealing lip of the seal is easily damaged and will not seal if the surface it contacts is not completely clean and free from scratches, nicks or grooves. If the original sealing surface of the component cannot be restored, the component should be renewed.

Protect the lips of the seal from any surface which may damage them in the course of fitting. Use tape or a conical sleeve where possible. Lubricate the seal lips with oil before fitting and, on dual lipped seals, fill the space between the lips with grease.

Unless otherwise stated, oil seals must be fitted with their sealing lips toward the lubricant to be sealed.

Use a tubular drift or block of wood of the appropriate size to install the seal and, if the seal housing is shouldered, drive the seal down to the shoulder. If the seal housing is unshouldered, the seal should be fitted with its face flush with the housing top face.

Screw threads and fastenings

Always ensure that a blind tapped hole is completely free from oil, grease, water or other fluid before installing the bolt or stud. Failure to do this could cause the housing to crack due to the hydraulic action of the bolt or stud as it is screwed in.

When tightening a castellated nut to accept a split pin, tighten the nut to the specified torque, where applicable, and then tighten further to the next split pin hole. Never slacken the nut to align a split pin hole unless stated in the repair procedure.

When checking or retightening a nut or bolt to a specified torque setting, slacken the nut or bolt by a quarter of a turn, and then retighten to the specified setting.

Locknuts, locktabs and washers

Any fastening which will rotate against a component or housing in the course of tightening should always have a washer between it and the relevant component or housing.

Spring or split washers should always be renewed when they are used to lock a critical component such as a big-end bearing retaining nut or bolt.

Locktabs which are folded over to retain a nut or bolt should always be renewed.

Self-locking nuts can be reused in non-critical areas, providing resistance can be felt when the locking portion passes over the bolt or stud thread.

Split pins must always be replaced with new ones of the correct size for the hole.

Special tools

Some repair procedures in this manual entail the use of special tools such as a press, two or three-legged pullers, spring compressors etc. Wherever possible, suitable readily available alternatives to the manufacturer's special tools are described, and are shown in use. In some instances, where no alternative is possible, it has been necessary to resort to the use of a manufacturer's tool and this has been done for reasons of safety as well as the efficient completion of the repair operation. Unless you are highly skilled and have a thorough understanding of the procedure described, never attempt to bypass the use of any special tool when the procedure described specifies its use. Not only is there a very great risk of personal injury, but expensive damage could be caused to the components involved.

Conversion factors

Length (distance)

	X			X		
Inches (in)	X	25.4	= Millimetres (mm)	X	0.0394	= Inches (in)
Feet (ft)	X	0.305	= Metres (m)	X	3.281	= Feet (ft)
Miles	X	1.609	= Kilometres (km)	X	0.621	= Miles

Volume (capacity)

	X			X		
Cubic inches (cu in; in^3)	X	16.387	= Cubic centimetres (cc; cm^3)	X	0.061	= Cubic inches (cu in; in^3)
Imperial pints (Imp pt)	X	0.568	= Litres (l)	X	1.76	= Imperial pints (Imp pt)
Imperial quarts (Imp qt)	X	1.137	= Litres (l)	X	0.88	= Imperial quarts (Imp qt)
Imperial quarts (Imp qt)	X	1.201	= US quarts (US qt)	X	0.833	= Imperial quarts (Imp qt)
US quarts (US qt)	X	0.946	= Litres (l)	X	1.057	= US quarts (US qt)
Imperial gallons (Imp gal)	X	4.546	= Litres (l)	X	0.22	= Imperial gallons (Imp gal)
Imperial gallons (Imp gal)	X	1.201	= US gallons (US gal)	X	0.833	= Imperial gallons (Imp gal)
US gallons (US gal)	X	3.785	= Litres (l)	X	0.264	= US gallons (US gal)

Mass (weight)

	X			X		
Ounces (oz)	X	28.35	= Grams (g)	X	0.035	= Ounces (oz)
Pounds (lb)	X	0.454	= Kilograms (kg)	X	2.205	= Pounds (lb)

Force

	X			X		
Ounces-force (ozf; oz)	X	0.278	= Newtons (N)	X	3.6	= Ounces-force (ozf; oz)
Pounds-force (lbf; lb)	X	4.448	= Newtons (N)	X	0.225	= Pounds-force (lbf; lb)
Newtons (N)	X	0.1	= Kilograms-force (kgf; kg)	X	9.81	= Newtons (N)

Pressure

	X			X		
Pounds-force per square inch (psi; lbf/in^2; lb/in^2)	X	0.070	= Kilograms-force per square centimetre (kgf/cm^2; kg/cm^2)	X	14.223	= Pounds-force per square inch (psi; lbf/in^2; lb/in^2)
Pounds-force per square inch (psi; lbf/in^2; lb/in^2)	X	0.068	= Atmospheres (atm)	X	14.696	= Pounds-force per square inch (psi; lbf/in^2; lb/in^2)
Pounds-force per square inch (psi; lbf/in^2; lb/in^2)	X	0.069	= Bars	X	14.5	= Pounds-force per square inch (psi; lbf/in^2; lb/in^2)
Pounds-force per square inch (psi; lbf/in^2; lb/in^2)	X	6.895	= Kilopascals (kPa)	X	0.145	= Pounds-force per square inch (psi; lbf/in^2; lb/in^2)
Kilopascals (kPa)	X	0.01	= Kilograms-force per square centimetre (kgf/cm^2; kg/cm^2)	X	98.1	= Kilopascals (kPa)

Torque (moment of force)

	X			X		
Pounds-force inches (lbf in; lb in)	X	1.152	= Kilograms-force centimetre (kgf cm; kg cm)	X	0.868	= Pounds-force inches (lbf in; lb in)
Pounds-force inches (lbf in; lb in)	X	0.113	= Newton metres (Nm)	X	8.85	= Pounds-force inches (lbf in; lb in)
Pounds-force inches (lbf in; lb in)	X	0.083	= Pounds-force feet (lbf ft; lb ft)	X	12	= Pounds-force inches (lbf in; lb in)
Pounds-force feet (lbf ft; lb ft)	X	0.138	= Kilograms-force metres (kgf m; kg m)	X	7.233	= Pounds-force feet (lbf ft; lb ft)
Pounds-force feet (lbf ft; lb ft)	X	1.356	= Newton metres (Nm)	X	0.738	= Pounds-force feet (lbf ft; lb ft)
Newton metres (Nm)	X	0.102	= Kilograms-force metres (kgf m; kg m)	X	9.804	= Newton metres (Nm)

Power

	X			X		
Horsepower (hp)	X	745.7	= Watts (W)	X	0.0013	= Horsepower (hp)

Velocity (speed)

	X			X		
Miles per hour (miles/hr; mph)	X	1.609	= Kilometres per hour (km/hr; kph)	X	0.621	= Miles per hour (miles/hr; mph)

Fuel consumption*

	X			X		
Miles per gallon, Imperial (mpg)	X	0.354	= Kilometres per litre (km/l)	X	2.825	= Miles per gallon, Imperial (mpg)
Miles per gallon, US (mpg)	X	0.425	= Kilometres per litre (km/l)	X	2.352	= Miles per gallon, US (mpg)

Temperature

Degrees Fahrenheit = (°C x 1.8) + 32   Degrees Celsius (Degrees Centigrade; °C) = (°F - 32) x 0.56

*It is common practice to convert from miles per gallon (mpg) to litres/100 kilometres (l/100km), where mpg (Imperial) x l/100 km = 282 and mpg (US) x l/100 km = 235

Index